# The New Communications Technologies

## Fifth Edition

# The New Communications Technologies: Applications, Policy, and Impact

## Fifth Edition

Michael M. A. Mirabito

Barbara L. Morgenstern

With a Foreword by Mitchell Kapor

AMSTERDAM • BOSTON • HEIDELBERG • LONDON
NEW YORK • OXFORD • PARIS • SAN DIEGO
SAN FRANCISCO • SINGAPORE • SYDNEY • TOKYO
Focal Press is an imprint of Elsevier Inc.

Focal Press is an imprint of Elsevier
200 Wheeler Road, Burlington, MA 01803, USA
Linacre House, Jordan Hill, Oxford OX2 8DP, UK

Copyright © 2004, Elsevier Inc. All rights reserved.

No part of this publication may be reproduced, stored in a retrieval system, or transmitted in any form or by any means, electronic, mechanical, photocopying, recording, or otherwise, without the prior written permission of the publisher.

Permissions may be sought directly from Elsevier's Science & Technology Rights Department in Oxford, UK: phone: (+44) 1865 843830, fax: (+44) 1865 853333, e-mail: permissions@elsevier.com.uk. You may also complete your request on-line via the Elsevier homepage (http://elsevier.com), by selecting "Customer Support" and then "Obtaining Permissions."

∞ Recognizing the importance of preserving what has been written, Elsevier prints its books on acid-free paper whenever possible.

**Library of Congress Cataloging-in-Publication Data**
Application submitted.

**British Library Cataloguing-in-Publication Data**
A catalogue record for this book is available from the British Library.

ISBN: 0-240-80586-0

For information on all Focal Press publications
visit our website at www.focalpress.com

04 05 06 07 08 10 9 8 7 6 5 4 3 2 1

Printed in the United States of America

To our families.

*To the men and women of the Space Shuttles* Challenger *and* Columbia *and all their fellow travelers.*
*May their dream live forever in the minds, hearts, and works*
*of the present and all future generations.*

# Contents

# Foreword

The accelerating pace of innovation in technology over the past several decades has brought about profound changes, perhaps nowhere more than in the information and communication media which pervade our lives. I am tempted to say we live in a truly wired world, except that the pace of development in wireless local area networks and advanced cellular transmission is rendering the actual presence of wires increasingly superfluous. Whether wired or wireless, we live in a connected world, whether we want to or not. Being disconnected and anonymous is less and less an option.

The audiovisual media of radio, television, movies, and music are increasingly being created, stored, produced, and distributed through digital means with profound impact on the choice and experience of consumers and the economics for content creators and producers. A new medium, the Internet, has become indispensable for electronic commerce, social interaction, and the delivery of news, information, and entertainment.

All this has profound implications in the civic sphere as well—for our privacy and freedom of expression as citizens. The First Amendment of the Bill of Rights guarantees freedom of the press and freedom of expression. As A.J. Leibling famously observed, "freedom of the press belongs to the person who owns one." In today's world, in which every personal computer can be a digital printing press (as well as a recording studio and duplication factory), it is not always simple to distinguish Constitutionally protected acts of expression from commercial infringement. Nor is it easy to decide how far the government is justified in increasing surveillance on its citizens in the name of national security.

It is therefore all the more important today that people be well informed about the social and legal issues arising out of new media technology. In this new edition, Professors Mirabito and Morgenstern continue to provide a comprehensive account of the technical bases of modern information and communication technology and its application. Equally important, the reader will benefit from their significantly increased focus on the critical social and legal issues arising out of new technology. *The New Communications Technologies* makes a valuable contribution to our understanding of the forces shaping our lives in the twenty-first century.

**Mitchell Kapor**
Cofounder Electronic Frontier Foundation
(www.eff.org)
Chair, Open Source Applications Foundation

# Preface

## WHAT'S NEW

Readers who are familiar with *The New Communications Technologies* will see a number of changes in the fifth edition, including the following:

- Chapters have been updated to explore new applications and technological trends.
- Topics have been updated where appropriate.
- The chapters contain new illustrations.
- The chapters have been reorganized to enhance the subject flow.
- Legal discussions have been expanded, with topics ranging from First Amendment issues to copyright and privacy. This is one of this edition's most important improvements. The legal discussions also cut across social and political issues. For example, in Chapter 1 we discuss how biometric technology is used to create more secure identification systems (e.g., to restrict access to a computer's data) and how these systems can have an impact on our individual privacy.

It is also important to note the World Trade Center tragedy plays a central, but sometimes hidden, role in certain scenarios. For example, this event has given birth to numerous government initiatives to combat terrorism. But while this may be the articulated goal, some individuals believe there are attendant civil liberty implications. It appears this is a growing challenge brought about by new technologies: *the same tools used to protect our freedom have the potential to curtail our freedom.* As communicators, it's important to be aware of these issues as we make determinations as to a technology's appropriate or inappropriate use and application.

### What's Unchanged

As in previous editions, the book still explores new technologies from a broad perspective. This includes the convergence factor—the relationships among different fields and how developments in one area can affect developments in another area.

Applications and underlying concepts also remain a focal point. The latter is particularly salient. Although new products may be introduced, fundamental principles may remain unaltered. Thus, by learning the concepts now, you may be able to work with new applications well into the future.

## THE POTENTIAL READERSHIP

The book is appropriate for communications technology courses in TV/radio, communication, journalism, and corporate communication departments. It can also serve as a primer for graduate courses in the same departments and as a supplementary text for legal, public relations/advertising, management, and instructional technology courses.

The book may also prove useful to communications professionals, including journalists, who want to gain a broad overview of the field. We also hope it will appeal to those who are interested in the communication revolution and its impact.

## ACKNOWLEDGMENTS

As with the first edition, we want to acknowledge those pioneers, those individ-

**Figure P.1**
*The new communications technologies have altered the way we can gain access to information. For example, you can now explore the world—such as Stromboli and its active volcano—via the Internet. (Courtesy of Juerg Alean and Stromboli.net.)*

uals, who helped launch the communication revolution itself.

Our students deserve thanks for their willingness to serve as sounding boards when new material was introduced and for providing us with valuable feedback. We also want to thank the readers whose comments about the various editions proved insightful.

We also appreciate the ongoing efforts and support of Ayn Miralano, our colleagues, and the numerous individuals at different companies who responded to photo requests. Thank you all.

Mitchell Kapor was also gracious enough to write the book's Foreword. We deeply appreciate his taking time out of a busy schedule for this project. As the founder of the Lotus Development Corporation and cofounder of the Electronic Frontier Foundation as well as the Chair of the Open Source Applications Foundation, Mr. Kapor's experience spans the technological, legal, and economic arenas. As one of the "sparks" of the communications and information revolution, he shares many important insights.

The editorial and production staffs at Focal Press also helped make this book a reality. The fifth edition is a milestone for us. *The New Communications Technologies* has been in publication since 1990, and we are particularly grateful to a number of individuals that we had the pleasure and privilege to work with over the past 13+ years. They include our first editor, Phil Sutherland, Marie Lee, as well as the editor of this edition, Amy Jollymore. Amy has been extraordinarily gracious and insightful in guiding the book's design. We would also like to acknowledge the valuable contributions of Kyle Sarofeen, this edition's production editor.

## DISCLAIMER

Some of the companies and products mentioned herein are trademarks or service marks. Such usage of these terms does not imply endorsements or affiliations. Adobe product screen shots reprinted with permission from Adobe Systems, Inc. The company's Premiere®, Photoshop®, PageMaker®, and Acrobat®, programs are covered in various chapters.

We also mention the New York Yankees at different points in the book when exploring a given topic. While this does not imply an endorsement or affiliation, we will admit we are fans of the most storied team in the annals of professional sports. This is, of course, a personal opinion . . . and just some food for thought.

# I

## FOUNDATION

# 1 Communication in the Modern Age

Much has been written about the communication revolution—some of it realistic, some of it not. Years ago, authors predicted that we would be using videophones, that satellites would create worldwide communications nets, and that we would be watching 3-D television and using personal computers (PCs). Although all of these predictions have not fully materialized, many have. Sophisticated satellites ring the globe and PCs have forever altered the way we work.

We are clearly living during a *communication revolution*. New and existing technologies and applications are shaping the communications industry and society. For example:

- The latest generation of PCs can produce sophisticated multimedia presentations.
- Optical disks offer increased storage and production capabilities.
- Telephone lines are channels for vast information pools.
- Satellites can serve as personal communications platforms.

Yet despite this revolution, some changes are actually evolutionary. As we'll discuss in this book, these changes include the upgrading or modernization of an industry's infrastructure. Consequently, the communication revolution can actually be viewed as a mix of revolution and evolution—the influx of new and the enhancement of older products. Together, they have brought about massive changes in the world around us.

Chapters 1–3 provide a foundation for the exploration of these topics and changes. These chapters introduce key concepts as well as explore digital and computer technologies and applications, the driving force behind many of the technological changes we see today.

## BASIC CONCEPTS

### The Communications System: An Extended Definition

A *communications system* is one such concept. It provides the means by which information, coded in signal form, can be exchanged. If we decide to telephone a friend, the communications system would include the telephone receivers, the telephone line, as well as other components. We use a variety of communications systems to exchange information. As covered throughout this book, they range from satellites to underwater fiber-optic lines.

In the context of this book, the term *communications system* also has a broad definition. (That is, it isn't limited to a description of information exchange systems.) It also includes the communications tools we use, their applications, and the implications that arise from the production, manipulation, and potential exchange of information.

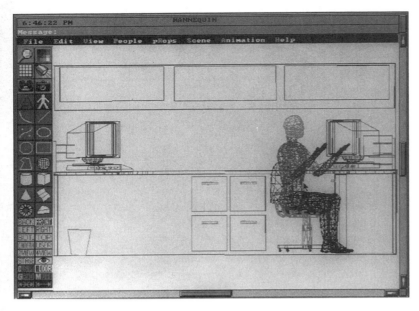

**Figure 1.1**

*PC software can be used in ergonomic design— the practice of developing equipment and systems around people. Such software can also create images suitable for other applications. (Courtesy of Biomechanics Corp., HumanCad Division; Mannequin.)*

### Information

*Information* can be defined as a collection of symbols that, when combined, communicates a message or intelligence. When you write a note to a friend about the New York Yankees, you combine letters and numbers to convey your thoughts or ideas, a message that has meaning to both of you. The combination of the letters W–o–r–l–d S–e–r–i–e–s is not just a collection of letters: It represents a concept.[1]

In our communications system, this information may be coded in a standardized form: in an electronic or electrical signal analogous to the coding of information by the printed letters and, ultimately, words in the note.

Next, the information can be relayed via a telephone line, satellite, or other communications channel. After it is received, it can be decoded or converted back into its original form.

This series of steps follows the traditional model of a communications system devised by Claude Shannon and Warren Weaver. The system consists of an information source, a transmitter, a channel to deliver the information, a receiver, and a destination. We'll discuss noise, another element, in the next chapter.

***Information as a Signal.*** When coded in signal form, information—a person's voice or a video camera's view of the Grand Canyon —is also compatible with communications equipment. For example, the camera's picture can be transmitted, stored, or altered by computer.

The communications system is also quite flexible in the representation of information. For example, in a fiber–optic system, light conveys information. In optical storage media (such as CDs and DVDs), a laser's light is used to retrieve information. We'll cover these topics in Chapters 5 and 9, respectively.

If you're interested in reading a more detailed explanation of what constitutes information, see the books by Shannon and Weaver and by Rogers listed at the end of this chapter. These books also cover the historical and technical elements of information theory and the mathematical theory of communication and information.

***Information: An Extended Definition.*** Information extends beyond the definition used in the general communications industry. Information is not just television programs, telephone conversations, and movies stored on digital versatile disks (DVDs). Information can also take the form of pictures manipulated by a computer that highlight details of the human body. Information can also take the form of a library of books stored on optical disks or stock market facts that can be accessed at home with a computer.

Information can also be viewed as a commodity. Traditionally, a television program has a financial value that varies according to its ratings. The information generated by nontraditional tools is also valuable. Companies use the telephone system to sell financial data to computer owners. Television networks purchase images created by remote sensing satellites to shed light on national and international news events.

Finally, information can be equated with power. If you know how to use a computer, you will be able to tap a greater wealth of information than a nonuser will, giving you a political or economic edge.

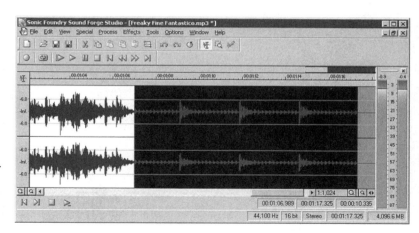

*Information Society.* Some technological developments, including the notion that information can be equated with power, have contributed to the creation of an *information society.* Even though this term has become cliché, it is nevertheless an accurate one. The information society is driven by information, be it the latest international news needed to keep abreast of a volatile world market or the creation of databases that can be accessed by computers. The information society has also created new job categories, such as web designers who use computers to create Internet sites and technical specialists who are hired to retrieve information from computer databases.

The information society has also influenced the U.S. economy. We are becoming a nation based on service rather than manufacturing. Tubing and steel plants have given way to the service industry. Hospitals, banks, and computer information companies fall into the latter category. For many businesses, using and processing information are integral elements of the services they offer. Still other companies produce the tools needed to sustain this information-based society. Meanwhile, new highways are constructed. But instead of carrying cars and trucks, these new "information highways" relay information to our work sites and homes.

## Implications of the Communication Revolution

The new information and communications technologies have also profoundly affected our social structure, and there is a growing interdependence among technology, information, and society. And new technologies have raised a series of ethical questions. As described later in Chapter 10, we can now use scanners (devices that input graphic information to a computer) to copy someone else's work.

In another scenario, some people are afraid that automation will accelerate the loss of jobs. This idea is typically linked to the belief that our society is becoming increasingly dehumanized as the market is flooded with computers and computer-controlled systems.

The communication revolution has also given birth to a global social class. In the past, distinctions between social groups were typically influenced by economic, political, and educational factors. These same forces are present, but our information society may make these distinctions even more pronounced. This is especially true if you do not have information- and computer-management skills.

Look at the world around you. Banks are shifting to computer-based teller systems, and communications lines are the keys to information pools. But unless you know

**Figure 1.2**
*Software and hardware developments provide us with new tools to manipulate audio and video information. (Courtesy of Sonic Foundry; Sound Forge Studio.)*

**FOUNDATION**

FOUNDATION

Diane L. Morgenstein in *Photonics Spectra*

**Photonics** — the technology of generating and harnessing light and other forms of radiant energy whose quantum unit is the photon. The range of applications of photonics extends from energy generation to communications and information processing.

**PHOTONICS** SPECTRA
VOICE OF THE PHOTONICS TECHNOLOGY

© Laurin Publishing Co. Inc., 1982

**Figure 1.3**
*This drawing highlights the interrelationships between photonics technology and its various fields. As stated in the caption, "photonics—the technology of generating and harnessing light and other forms of radiant energy whose quantum unit is the photon. The range of applications of photonics extends from energy generation to communications and information processing," some applications of which will be covered in the book. (Courtesy of Photonic Spectra; Diane L. Morgenstein in Photonics Spectra.)*

political, or educational resources to participate in the information age, may be similarly affected on a broader scale.

This situation presents us with a paradox: More people have access to information than at any other time in human history, yet entire slices of our society may not be able to partake of this information bonanza as we accelerate toward a world in which information is a lifeline.

Yet the communication revolution does have numerous positive implications. The same computer systems that some people fear can improve medical treatment through enhanced imaging techniques. Other computer systems help individuals with physical handicaps to better communicate with the world around them. It can also be argued that more people have greater access to information today than ever before, despite its unequal distribution.

The new technologies also have societal implications. While one of the book's focal points is technology's applications, the other is the communication revolution's fallouts, ranging from ethical to legal to political considerations. These considerations play key roles in defining our information society. The next section serves as an introduction to some of these topics, which are discussed at greater length in subsequent chapters.

## Other Factors

***Convergence of Technologies.*** The communication revolution accelerated the convergence of technologies and applications. As described in Chapter 11, powerful PCs, when combined with appropriate hardware and software, have created integrated desktop video systems. These tools allow us to create video productions. By combining this capability with desktop publishing and other applications, we are able to be more than just media consumers. We also become producers and editors.

how to tap these resources, you may join the ranks of the "information poor" and lack the skills necessary to compete for jobs that require computer proficiency. Entire nations, which may not have the economic,

This newfound capability also has aesthetic implications. Simply stated, *the technical capacity to create a project does not supplant underlying aesthetic principles.*

### The Democratization of Information.

The communication revolution has also promoted the free flow of information. You can use a desktop publishing system to print a newsletter, and, when combined with other equipment, to keep the world community apprised of fast-breaking political events.

This concept has also been extended to other media. As outlined in Chapter 17, an electronic democracy can be supported by holding electronic meetings between groups of people thousands of miles apart. A key element in this process is providing broad, affordable access to the system.

### Economic Implications.

The communication revolution has economic implications and can affect entire industries. In the United States, one issue is concerned with "spectrum allocations." We use the spectrum as a means to relay information. In our society, this capability also has an economic fallout.

### Intellectual Property.

The new communications technologies raise *intellectual property* questions as well, which are essentially the "rights of artists, authors, composers and designers of creative works."[2] These questions include the electronic copying of another's works as well as copyright and patent issues.

A related factor is the malleability of information. In our current system, information can be readily manipulated. While this capability can be used for creative purposes, it also highlights a problem: the same family of tools that can create a graphic or other product can be used to copy it illegally. Because computers and other such tools are ubiquitous, it's also impossible and,

in most cases, undesirable to monitor their use. For example, such monitoring could lead to censorship. Copyright laws address these issues, and violators can be prosecuted. The difficulty, however, may lie in enforcing the laws.

There's also a problem with recognizing intellectual property as actual property. To many people, intellectual property is intangible. You could spend a year creating a computer program that fits on a CD. On the surface, it may not look like much, just a single disk you can hold in your hand. But in reality, it may represent the "product of the creative intellect"—that is, intellectual property, another form or type of property.[3]

### First Amendment Issues.

The new communications technologies have also raised First Amendment questions that are covered in different chapters. The following are a number of such questions:

- Should an electronic information service be treated as a publisher or a distributor?
- What legal measures can help ensure that First Amendment rights are supported and nurtured?
- How do you reconcile First Amendment rights with potential censorship?
- Has global censorship become more prevalent?

### Privacy.

As our communications systems become more sophisticated, so too does the potential to invade our privacy, as covered in the next section of this chapter. Other facets of this topic are subsequently explored in later chapters.

### Other Issues.

Our information society will also be influenced by other factors. Will teleconferencing and telecommuting reduce human communication to a machine-dominated format? Do computers dehumanize the creative process? Or do they extend our creative capabilities by helping us to transform an idea or vision into an actual product?

## SYSTEMS APPROACH: ELECTROMAGNETIC PULSE AND BIOMETRICS

To wrap up this chapter, it's important to discuss one technique the book uses to explore the communication revolution: a *systems approach*.[4] Although you can examine individual applications and implications, you should also explore related areas to gauge interrelationships and their impact on the overall communications system. Without this broad perspective, you may be looking at an incomplete picture. Two examples illustrate this point.

### Electromagnetic Pulse

Our society's vulnerability to world events highlights the importance of a systems approach. It also points out the interdependence between technology and our information society. In brief, our information and communications infrastructure could be crippled and silenced by a single high-altitude nuclear explosion.[5] This paralysis would be caused by a powerful *electromagnetic pulse* (EMP), a by-product of a nuclear detonation.

Scientists have observed this phenomenon during weapons tests. They have subsequently used simulations to gauge the vulnerability of electronic components, equipment, and systems that could be affected by such a pulse, with the goal of developing protection mechanisms.

A step in this process was the Federal Emergency Management Agency's (FEMA's) program to protect radio stations in the Emergency Broadcast System (EBS) and other communications centers. Theoretically, if a nuclear burst ever paralyzed our communication capabilities, the protected EBS stations would be activated to supplement the damaged channels. A potential problem, however, is the receiving device. To pick up a broadcast, you may have to rely on a portable, battery-operated radio, which is one of the few modern electronic devices that may not be "readily damaged" by the EMP.[6] But many modern radios are actually part of a stereo system, which may itself be damaged. If not, there is a good chance the power plant that supplies the electricity to run the stereo would not function.

Thus, as a society that depends on and is driven by information, the very tools we use to create, manipulate, and deliver this information would be disrupted, if not rendered inoperable, by a nuclear explosion. The issue is multifaceted, and a broad systems perspective enables us to examine it from different angles. We can also cover the key implications both inside and outside the communications field.

In keeping with this approach, it's also ironic to note that vacuum tubes are relatively impervious to EMP effects. But in our contemporary communications system, such tubes have been universally replaced by solid state chips, one of an EMP's favorite snacks. Thus, as our communications system was modernized, it was also made more vulnerable to world events and, potentially, to random acts of terrorism.

### Biometric Systems

*Biometrics, by their nature, are generally inconsistent with anonymity. Yet the starting point for privacy is the ability of citizens to go about their business freely and unobserved.*

—Malcolm Crompton, Australia's third Federal Privacy Commissioner[7]

The importance of adopting a systems approach is also illustrated by *biometrics*. Biometrics can be defined as the "automatic identification of an individual based on [his/her] physiological . . . traits."[8] Examples of biometric systems include fingerprint identification and voice recognition systems. Some biometric systems compare and subsequently match the unique characteristics of a person's eyes to stored information.

One goal of biometric systems is to verify a person's identity without using passwords. In another application, law enforcement agencies have employed biometric technology in an attempt to identify criminals on city streets, in airports, and in other locales. In the wake of the World Trade Center attacks, the widespread adoption of such security and identification measures has been offered as a partial solution to this problem.

We have actually seen biometric systems in use for years in the movies. Often, a character places his or her hand on a sensor next to a locked door. The sensor then scans the hand for specific characteristics and opens the door upon identification.

However, biometric systems didn't enter mainstream society until the late 1990s and early 2000s. A common example is a small unit that scan fingerprints and compare the data against stored information. Targus, a company that specializes in computer accessories, recently introduced such a unit for under $125 for portable and desktop computer systems. With this device, you can limit other users' access to your computer, much like you would use a password to restrict access.

The field of biometric technology calls into focus the collision between technological developments and their societal implications. As such, it should be explored from a broad perspective to accurately gauge its current and potential impact. Thus from a systems perspective, biometric technology

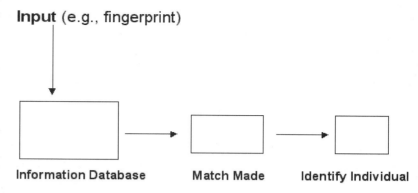

carries with it a broad range of technical, legal, and ethical implications.

**Figure 1.4**
*Biometric System Operation. This diagram highlights, in a simplified form, a biometric system's operational steps.*

***Technological Implications.*** Electronic biometric systems employ various technologies and tools. For example, facial recognition systems involve "the use of a highly automated computerized process to measure angles and distances between geometric points on the face—eye corners, the nostrils, the ends of the mouth—to identify an individual."[9] A typical system is composed of a camera that is linked to a computer system running specialized software. A photo taken with the camera is processed so that it is compatible with the computer. Next, the computer's software controls the identification process to determine whether the photo matches the individual's photo stored in the computer's database.

Some organizations and critics have questioned the technological accuracy of these systems. The American Civil Liberties Union (ACLU), for one, contends that such systems may be inaccurate. In one trial use in Tampa, Florida, a facial recognition system was set up to help police at a specific locale. However, during this trial run, the system produced a number of false matches (that is, individuals were not properly identified).[10]

In essence, while this type of system may work well in a controlled environment, it may not work as well in the real world. Lighting conditions may not be optimal or the angle in which a given individual is "shot" may make proper identification difficult.

Other critics have questioned what would happen if the information in the database is compromised or inaccurate.[11] If such stored information is compromised, it would be analogous to an individual stealing your password to gain access to your computer. Once a password is compromised, the door to your private information is unlocked. Likewise, biometric systems are dependent on the quality of the stored data. Information may be inaccurately keyed, or, from a technological standpoint, it may have defects (such as a poor-quality photo of a suspected criminal stored in a database).

***Legal Implications.*** Such technological developments produce legal implications, such as privacy concerns. While a biometric system may help identify criminals, what is the cost in regard to privacy? Is the technology too invasive? Or does it have a legitimate law enforcement role if used appropriately? Such systems pose important Fourth Amendment implications.[12]

***Ethical and Political Implications.*** Biometric technology raises an ethical, and, by extension, a political question. Since biometric systems can be used for identification purposes, the data they generate can track an individual's movements from place to place over time. According to the Electronic Frontier Foundation (EFF), this has serious implications since a "society in which everyone's actions are tracked is not, in principle, free. It may be a livable society, but would not be our society [current U.S. society]."[13]

***Broader Implications.*** From a system's perspective, biometric systems should also be viewed as components of broader information-based initiatives that raise additional questions. In a position paper about *Total Information Awareness* (TIA), a government-sponsored program that would purportedly compile volumes of personal, medical, and, potentially, biometric information about U.S. citizens, the ACLU states:

Virtual dragnet programs like TIA . . . are based on the premise that the best way to protect America against terrorism is for the government to collect as much information as it can about everyone.[14]

According to the ACLU, these information pools would have a detrimental effect on privacy and could be used for monitoring our buying habits and for other commercial enterprises. The ACLU says this would "represent a radical departure from the centuries-old Anglo-American tradition that the police conduct surveillance only where there is evidence of involvement in wrongdoing."[15] By weighing this factor with other surveillance capabilities, some critics believe Orwell's notion of "Big Brother" may become a reality.

It is also important to note this issue is somewhat fluid. For example, Senator Russell Feingold introduced legislation (the Data-Mining Moratorium Act of 2003) to "immediately suspend data-mining in the Department of Defense and the Department of Homeland Security until Congress has conducted a thorough review of Total Information Awareness and the practice of data-mining."[16] Data mining, in this context, refers to a comprehensive computer search for a particular commodity. In this case, that commodity is information itself.

Much like the ACLU and EFF's viewpoint the concern was over privacy and other matters. The Act did, however, recognize technology's role in combating potential terrorist activities. The key question was how broadly this tool would be used without the appropriate constraints and, ultimately, Congressional oversight and review.

***Supporters and an Update.***    While the TIA program has critics, others believe it and related programs do not put our personal privacy at risk. Such supporters contend that privacy measures were integrated in the TIA when it was created. According to one report, the TIA will keep "the protection of civil liberties at its forefront while providing a valuable tool for investigating suspected terrorists and improving communication."[17]

The TIA program was also renamed the *Terrorism Information Awareness* program. Supporters believe this name more accurately reflects the program's original intent: to participate in antiterrorism activities rather than collecting information about ordinary U.S. citizens.[18] Thus, from a proponent's perspective, such new technologies and their applications actually protect and enhance our "open society" as well as our freedom.

However, despite assurances from TIA supporters, the U.S. House and Senate voted to impose operational limitations and to eliminate funding for the TIA program, respectively. While a joint conference session will make the final determination, privacy issues and the government's slow response to such concerns helped trigger this reaction.[19]

***Summary.***    Biometric technology has numerous applications and implications. Much like the EMP, we must explore the topic from a broad systems perspective to fully understand its applications and societal impacts. If we only examine one facet of the field, such as its technological developments, we ignore other key issues.

*A systems approach can also help generate an information pool so we can ask pertinent questions about technologically driven issues.* In the case of the TIA and similar initiatives, these questions include:

- Where do you draw the line between protecting an individual's rights and ensuring that a law enforcement agency can conduct its work effectively?
- How do you strike a balance between a program's accountability and the need for flexibility in a rapidly changing world order?
- How do you prevent the misuse of collected data?
- Are there safeguards to correct errors (such as innaccurate data)?
- Can the same technology base that enhances our lives provide a government with unprecedented power and tools to delve into our private lives?

These and other similar questions are difficult to answer. Nevertheless, they are questions that are relevant to any free society.

## CONCLUSION

This chapter has served as an introduction to the new technologies universe and some of the implication raised by the communication revolution. The next chapters focus on more specific topics and will follow a preset pattern. When relevant, we will discuss the technical underpinnings of a given technology, which will help us explore and understand its applications. This knowledge may also prove valuable when deciding whether a technology could be used for a specific application and, possibly, for developing applications that are as yet undiscovered. We will then examine the implications raised by the technology and its applications.

*Finally, it is important to repeat a point raised in the Preface:* While new products may be introduced and new implications may sur-

face, fundamental principles may remain the same. The same scenario can play out in the social, political, and legal fields. For example, the TIA may not be an issue five years from now. Nevertheless, the underlying forces that drove the program's design as well as proponents' and opponents' arguments may remain. Thus, if you understand these concepts now, you may be able to work with their implications well into the future, regardless of the program's name.

**Figure 1.5**
*A computer-generated image of Mount St. Helens. You can use computers to create realistic and surrealistic views of the earth and other worlds. (Software courtesy of Virtual Reality Laboratories; Vista Pro.)*

## REFERENCES/NOTES

1. Frederick Williams, *The New Communications*. (Belmont, CA: Wadsworth Publishing Co., 1984), 9.

2. Westlaw, Intellectual Property Database, downloaded.

3. "Patents: Protecting Intellectual Property," *OE Reports* 95 (November 1991), 1.

4. For more information, see Ervin Laszlo, *The Systems View of the World*. (New York: George Braziller, Inc., 1972).

5. A single nuclear explosion could theoretically blanket the United States and parts of Canada and Mexico. It could disrupt the countries' communications infrastructure even if the weapon did not cause any direct damage.

6. Samuel Glasstone and Philip J. Dolan, *The Effects of Nuclear Weapons*. (Washington, DC: U.S. Government Printing Office 1977), 521.

7. Malcolm Crompton, Australia's Third Federal Privacy Commissioner, "Biometrics and Privacy: The End of the World as We Know It or the White Knight of Privacy?," downloaded from http://www.biometricsinstitute.org/bi/cro mptonspeech1.htm.

8. Anil K. Jain et al., "Biometrics: Promising Frontiers for Emerging Identification Market," *Computer* 33 (February 2000), downloaded.

9. Roberto Iraola, "Dedication to the Small Town Attorney: New Detection Technologies and the Fourth Amendment," *South Dakota Law Review*, 47 D.D.L. Rev. 8, downloaded from LEXIS.

10. American Civil Liberties Union, "Drawing a Blank: Tampa Police Records Reveal Poor Performance of Face-Recognition Technology," Press Release, January 3, 2002, downloaded from www.aclu.org/Privacy/Privacy.cfm?ID=10210&c=39&Type = s.

11. Mandy Andress, "Biometrics at Work?" *InfoWorld* 22 (May 28, 2001), 75.

12. Please see Roberto Iraola, "Dedication to the Small Town Attorney: New Detection Technologies," for an in-depth discussion about Fourth Amendment Implications for a variety of identifications systems, including those that tap biometric technology. The Fourth Amendment itself is more fully defined in this book's e-mail chapter, Chapter 18.

13. See Electronic Frontier Foundation, "Biometrics Who's Watching You?," downloaded from www.eff.org/Privacy/Surveillance/biometrics.html. Basically, how much surveillance is too much surveillance?

14. American Civil Liberties Union, "Q&A on the Pentagon's 'Total Information Awareness' Program," downloaded from www/aclu.org/Privacy/Privacylist.cfm?c=130.

15. Ibid.

16. Congressional Record, "Statements in Introduced Bills and Joint Resolutions," January 16, 2003 (Senate), pp. A1071–S1085, downloaded from www.eff.org/Privacy/TIA/feingold-s188.php.

17. Michael Scardaville, "No Orwellian Scheme Behind DARPA's Total Information Awareness System," WebMemo #175, November 20, 2002, downloaded from www.heritage.org/Research/Homeland/Defense/wm175.cfm.

18. "Executive Summary: EFF Review of May 20 Report on Total Information Awareness," downloaded from www.eff.org/Privacy/TIA/20030523_tia_report_review.php.

19. Dan Verton, "Senate Votes to Kill Antiterror Data Mining Program," *Computerworld* (July 18, 2003), downloaded from www.computerworld.com/securitytopics/security/privacy/story/0,10801,83205,00.html?nas=AM-83205.

## SUGGESTED READINGS

Balaban, Dan. "Should Smart Cards Carry Their Own Biometric Sensors?" *Card Technology* (November 2001), 24–28; Paul Festa. "All Eyes on Face Recognition." (March 26, 2003), downloaded from http://zdnet.com.com/2100-1105-99111.html; J.R. Wilson "Airport Security Designs Revolve Around Biometrics." *Military & Aerospace Electronics* 13 (September 2002), 15–23. These publications examine biometric systems and their applications. A smart card is a credit-card–sized information system that may include personal and financial data as well as a built-in fingerprint biometric sensor. You insert the card in a reader, touch the sensor with your finger, and are subsequently identified via a stored print.

Carter, A.H., and members of the Electrical Protection Department. *EMP Engineering and Design Principles*. Whippany, NJ: Bell Telephone Laboratories, Inc., Technical Publication Department, 1984; John R. Pierce, Chairman, Committee on Electromagnetic Pulse Environment; Energy Engineering Board; Committee on Engineering and Technical Systems; National Research Council. *Evaluation of Methodologies for Estimating Vulnerability to Electromagnetic Pulse Effects*. Washington, DC: National Academy Press, 1984. These publications examine different elements of the EMP issue. The first provides a good overview of the creation and technical implications of an EMP as well as different equipment and protection schemes. The second covers a wide range of subjects, including the role of statistics in trying to predict potential equipment and system failures.

Inglis, Andrew F. *Behind the Tube*. Boston: Focal Press, 1990; National Telecommunications and Information Administration. *Telecommunications in the Age of Information*. NTIA Special Publication 91–26, October 1991; John V. Pavlik. *New Media Technology*. Boston: Allyn and Bacon. 1996; Lenore Tracey. "A Brief History of the Communications Industry." *Telecommunications* 31 (June 1997), 25–36. These publications provide excellent coverage of the communications industry and various issues.

Mazor, Barry, "Imaging for Biometrics Security: The Impact of the Privacy Issue." *Advanced Imaging* 17 (August 2002), 10–11. A roundtable discussion of industry experts about the privacy issues raised by biometric systems. See the ACLU and EFF web sites (www.aclu.org and www.eff.org) for additional information about biometric systems and their legal implications. Also see www. heritage.org for what may be, at times, opposing viewpoints.

Rogers, Everett M. *Communications Technology*. New York: The Free Press, 1986. An excellent reference and resource, this book examines the history of communications science, the social impacts of communications technologies, and research methods for studying the technologies.

Shannon, Claude, and Warren Weaver. *The Mathematical Theory of Communication*. Urbana, IL: University of Illinois Press, 1949.

**The** book about information theory. Also see Ramachandran Bharath. "Information Theory," *Byte* 12 (December 1987), 291–298, for a discussion of the information theory; and Edward Tufte. *Envisioning Information*. Cheshire, CT: Graphics Press, for an indepth examination of the visual representation of information.

Telecommunications Act of 1996, 104th Congress, 2nd session, January 3, 1996.

Vonder Haar, Steven. "Censorship Wave Spreading Globally," *Inter@ctive Week* 3 (February 12, 1996): 6.

## GLOSSARY

*Biometrics:* The automatic identification of an individual based on physiological traits or characteristics, including fingerprints.

*Communications System:* The means by which information, coded in signal form, can be exchanged. In the context of this book, the communications system also encompasses the communications tools we use, their applications, and the various implications that arise from the production, manipulation, and potential exchange of information.

*Electromagnetic Pulse (EMP):* A by-product of a nuclear explosion; a brief but intense burst of electromagnetic energy that can disrupt and destroy integrated circuits and related components.

*Information:* A collection of symbols that, when combined, communicates a message or intelligence.

*Information Society:* A society driven by the production, manipulation, and exchange of information. Information can be viewed as a social, economic, and political force.

*Intellectual Property:* The rights of artists, authors, and designers of creative works; the products of the creative intellect.

FOUNDATION

# 2 Technical Foundations of Modern Communication

This chapter introduces the technical elements that are the foundation of our modern communications system. It is also a continuation of Chapter 1, but concentrates more on technical information and digital communication.

The chapter concludes with a discussion of the importance of technical standards. Standards can promote the growth of a communications technology and industry. If standards are not adopted, the industry's growth could be hampered.

## BASIC CONCEPTS

### The Transducer

A *transducer* is a device that converts one form of energy into another form of energy. When you talk into a microphone, it converts your voice—sound or acoustical energy—into electrical energy, or in more familiar terms, an electrical signal. A speaker, also a transducer, can convert this signal back into your voice.

The "spoken words," the sound waves, can also be considered a natural form of information, information that our senses can "perceive."[1]

A transducer can convert this category of natural information, among others, into an electrical representation. The transducer acts as a link between our communications system and the natural world.

Certain transducers can also be considered extensions of our physical senses. They can convert what we say, hear, or see (for example, a camera) into signals that can be processed, stored, and transmitted.[2]

### The Characteristics of a Signal

If a microphone's operation were visible, we would see what appears to be a series of waves traveling through the connecting line. The waves, the electrical representation of your voice, have distinct characteristics. Two that are pertinent to our discussion are amplitude and frequency.

The amplitude is a wave's height, and in our example corresponds to the signal's strength or the volume of your voice. The frequency, the pitch of the voice, can be defined as the number of waves that pass a point in 1 second. If a single wave passes the point, the signal is said to have a frequency of 1 cycle per second (cps). If a thousand waves pass the same point, the signal's frequency is 1000 cps.

The frequency, the cps, is usually expressed in hertz (Hz), after Heinrich Hertz, one of the pioneers whose work made it possible for us to use the electromagnetic spectrum, a keystone of our communications system. Hence, a frequency of 10 cps is written as

10 Hz. Higher frequencies are expressed in kilohertz (kHz) for every thousand cycles per second, megahertz (MHz) for every million cycles per second, and Gigahertz (GHz) for every billion cycles per second.

### Modulation, Bandwidth, and Noise

Modulation is associated with, but not limited to, the communications systems that are the most familiar to us, including AM and FM radio stations. It can be defined as the process by which information, such as a DJ's program, is superimposed or impressed on a carrier wave for transmission.[3] A physical characteristic of the carrier is altered to convey, or to act as a vehicle to carry, the information. Prior to this time, it did not convey any intelligence.

After the signal is received, the original information can be stripped, in a sense, from the carrier. We can hear the DJ over the radio.

For the purpose of our discussion, a communications channel's bandwidth, its capacity, dictates the range of frequencies and, to all intents, the categories and volume of information the channel can accommodate in a given time period.[4]

A relationship exists between a signal's frequency and its information carrying capability. As the frequency increases, so too does the capacity to carry information. The signal must then be relayed on a channel wide enough to accommodate the greater volume of information. A television broadcast signal, for example, has a higher bandwidth requirement than either a radio or telephone signal.[5] Consequently, under normal operating conditions, a standard telephone line cannot carry a television signal.

During the information exchange, noise may also be introduced, potentially affecting the transmission's quality. If the noise is too severe, it may distort the relayed signal or render it unintelligible, and the information may not be successfully exchanged.

For example, if the snow on a television screen—the noise distorting the relayed signal—is very severe, it may become impossible to view the picture.

Noise can be internal, introduced by the communications equipment itself, or external, originating from outside sources. The noise may be machine generated or natural.

Lightning is one such natural source. Lightning is fairly common and may be manifested as static that disrupts a radio transmission. The sun is another source, creating interference through solar disturbances and other phenomena. Machine-generated noise, on the other hand, can derive from the electric motors in vacuum cleaners or large appliances.

When discussing noise relative to a communications system, the term *signal-to-noise ratio* is frequently encountered. It is a power ratio, that of the power or strength of the signal versus the noise. For information to be successfully relayed, the noise must not exceed a certain level. If it does, the noise will disrupt the exchange to a degree dependent on its strength. It will have an impact on a communications channel's quality and transmission capabilities.

### The Electromagnetic Spectrum

The *electromagnetic spectrum* is the entire collection of frequencies of electromagnetic radiation, ranging from radio waves to X rays to cosmic waves. Infrared and visible light, radio waves, and microwaves are all well-known elements and forms of electromagnetic energy that compose the spectrum.

We tap into the spectrum with our communications devices and use the electromagnetic energy as a communications tool—a means to relay information. For example, radio stations employ the radio-frequency range of the spectrum to broadcast their programming.

Spectrum space, allocated to television stations and other services, can also be viewed as a commodity in our information society. It can generate income, and like oil, gas, and other valuable natural resources, is scarce. In this light, the spectrum has an intrinsic worth and monetary value.

The issue of spectrum scarcity, which has also been used as a basis for broadcast regulation, is a reflection of our communications system. We cannot use all the spectrum for communications purposes, and the available portions are divided nationally and internationally.[6]

New and emerging technologies are exacerbating this situation. The demand is so great that when a small portion of the spectrum opened up for mobile communications services in 1991, the Federal Communications Commission (FCC) received approximately 100,000 applications for this space in 3 weeks.[7]

Because the spectrum is such a valuable resource, government agencies called for new spectrum management policies. These included imposing fees for spectrum usage and auctioning allocations.

Proponents initially claimed that auctioning would generate additional revenues, and mechanisms would ensure that the public interest mandate, an important regulatory ideal, would still be met.[8] Opponents countered that auctioning would conflict with the notion that the spectrum was a public resource. It was not private property that could be bought and sold.

Regardless of the viewpoint, Congress and the FCC embraced auctioning. In the late 1990s and the early 2000s, the situation also heated up. Traditional television allocations became a prime target as high definition television (HDTV) and other digital television services were slowly implemented. As described in a later chapter, broadcasters received new spectrum space to host such services. After a specified time

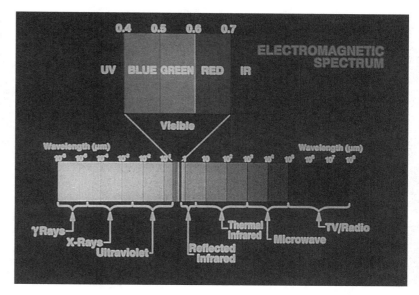

**Figure 2.1**
*Electromagnetic Spectrum. (Courtesy of the Earth Observation Satellite Company, Lanham, MD.)*

period, broadcasters would relinquish the old allocation, valuable electronic real estate that could be auctioned to other users.

It should also be noted that the idea of selling spectrum space is not a new one. In 1964, for instance, Ayn Rand wrote about the broadcast industry that "the airwaves should be turned over to private ownership. The only way to do it now is to sell radio and television frequencies to the highest bidders (by an objectively defined, open, impartial process)."[9] Rand, an author, philosopher, and advocate of capitalism, indicated that private rather than public ownership of the airwaves would have protected the electronic media from government regulation and would have ensured an open, competitive, and free marketplace.

In another twist, some individuals called for switching services: Television should be delivered by cable, thus freeing the spectrum for wireless communication. This proposal, also called the *Negroponte switch*, after Nicholas Negroponte, the director of the Massachusetts Institute of Technology's Media Lab, was offered as a potential spectrum scarcity and management solution.[10]

This type of switch, though, could not

take place overnight. Over-the-air television is also "free." Will we still receive free programming if it is relegated to a cable-based delivery system?

## DIGITAL TECHNOLOGY

Digital technology has fueled the development of new communications lines, information manipulation techniques, and equipment. Preexisting communications channels and devices have also been affected. It is one of the communication revolution's driving forces.

### Analog and Digital Signals: An Introduction

Many communications devices, such as telephones and microphones, are analog devices that create and process analog signals. As stated by Simon Haykin in his book *Communication Systems*, "Analog signals arise when a physical waveform such as an acoustic or light wave is converted into an electrical signal."[11]

For a microphone, this signal, an electrical representation of your voice, is said to be continuous in amplitude and time. The amplitude, for example, can assume an enormous range of variations within the communications system's operational bounds. The signal is an "analogue" (analog); that is, it is representative of the original sound waves. As the sound waves change, so too do the signal characteristics in a corresponding fashion.

A digital signal, in contrast, is "a noncontinuous stream of on/off pulses. A digital signal represents intelligence by a code consisting of the sequence of discrete on or off states . . ."[12] A digital system uses a sequence of numbers to represent information and, unlike an analog signal, a digital signal is noncontinuous.

Digital and analog signals, and ultimately equipment and systems, are generally not mutually compatible. This mandates the use of analog-to-digital and digital-to-analog conversion processes. They help us to use a mixed bag of analog and digital equipment in the overall communications system. This capability is crucial. Digital information has certain advantages, and the conversion processes allow us to tap these advantages even when using a microphone or other analog devices.[13]

Different elements of our communications structure are also somewhat based on an analog standard. The conversions allow us to integrate digital technology in this system.

Finally, different categories of information can be represented and consequently transmitted over the appropriate channels, in both analog and digital forms. An analog audio signal, for instance, can be converted into a digital representation and subsequently relayed.

### Analog-to-Digital Conversion

In an *analog-to-digital conversion*, the analog signal is converted into a digital signal. Binary language is the heart of digital communication. It uses two numbers, 1 and 0, arranged in different codes, to exchange information.

The 1s and 0s are also called *bits*, from the words *binary digits*, and they represent the smallest pieces of information in a digital system. They are also the basic building blocks for a widely used digital information system, pulse code modulation (PCM). A simplified description follows.

Pulse code modulation is a coding method by which an analog signal can be converted into a digital representation, a digital signal.[14] PCM information consists of two states, either the presence or the

FOUNDATION

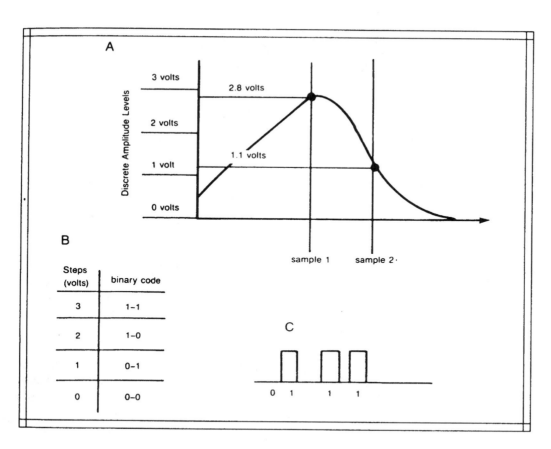

**Figure 2.2**
*A PCM operation. The diagram has been simplified for illustration purposes. The two sampling points have been indicated by samples 1 and 2 (A). Sample 1 is 2.8 volts and is assigned to the nearest step, 3 volts. Sample 2, 1.1 volts, is assigned to the nearest step, 1 volt. Based on the chart (B), the samples are coded as 1–1 and 0–1, respectively. The result is the PCM signal (C).*

absence of a pulse, which can also be expressed as "on" or 1 and "off" or 0.

When the analog signal is actually digitized, it is sampled at specific time intervals. Rather than converting the entire analog signal into a digital format, it is sampled or segmented, and only specific parts of the signal are examined and converted. Enough samples are taken, however, to obtain a sufficiently accurate representation of the original signal.

The samples are then compared to, for illustrative purposes, a preset scale composed of a finite number of steps. The steps represent different values or amplitudes the original analog signal could assume.[15] A sample is assigned, in a sense, to the step that matches or is closest to its amplitude. Each

step, in turn, corresponds to a unique word composed of binary digits (for example, 11 or 01).

A sample is then coded and represented by the appropriate word. The word can be relayed as on and off pulses, and when the information reaches the end of the line, the receiver detects it. Ultimately, a value corresponding to the original analog signal at the sampling point has been transmitted since the word represents a known quantity. (*Note:* For more specific details, see the books and articles at the end of this chapter.)[16]

Morse code functions in a similar fashion. Information, a message, is coded as a series of dots and dashes. Following its relay, an operator can reconstruct the original message since the code represents a known value.

***Sampling and Frequency.*** The sampling rate, which is the number of times the analog signal is sampled per second, is central to the reproduction process. The primary goal of sampling is to reflect accurately the original signal through a finite number of individual samples.

A survey conducted to predict the outcome of a presidential election can serve as an analogy. A sample of individual voters represents the voting population just like the samples of the analog signal represent the original signal.

The sampling rate is based on a signal's highest frequency in a given communications system. If a signal is sampled at a rate that is at least twice its highest frequency, the analog signal could be accurately represented. This is called the *Nyquist rate*.

For a standard telephone line, the communications channel carries only frequencies below 4 kHz, and the sampling rate is 8000 samples per second. Enough samples are generated to reproduce the analog signal and a person's voice. Note that higher sampling rates can also be employed.

***Steps.*** In many communications systems, when the analog signal is digitized, it is generally coded as either 7- or 8-bit words. A relationship exists between the number of bits in a word and the number of steps: As bits are added, the number of steps that can represent the analog signal increases in kind.

With an 8-bit-per-word code, essentially 256 levels of the signal's strength can be represented. Derived from the binary system, an 8-bit code is the equivalent of two to the eighth power ($2 \times 2 \times 2 \times 2 \times 2 \times 2 \times 2 \times 2$). A different combination of eight 1's and/or 0's represents each step.

An increase in the number of steps can lead to a more precise representation of the original signal. Certain communications systems and data also require a large number of steps, and consequently levels, for an accurate representation. Nevertheless, they're limited in comparison with the range an analog signal could assume.

***Analog-to-Digital and Digital-to-Analog Converters.*** For the purpose of our discussion, the analog signal is converted into a digital format by an analog-to-digital converter (ADC). Once the coded information is relayed, it can be reconverted into the original analog signal by a digital-to-analog converter (DAC) to make the signal compatible, once again, with analog equipment and systems. ADCs and DACs act as bridges between the analog and digital worlds.

***The Transmission.*** After the analog signal is sampled and coded, the final transmission can be composed of millions of bits. In one format, the telephone industry has employed an 8 bit-per-word code when the analog signal is digitized, and a telephone con-versation can be transmitted at the rate of 64,000 bits per second (64 kilobits/second). Other digitized analog signals similarly generate high bit rates.

**Figure 2.3**

*An ADC and DAC operation. An analog signal is converted into a digital domain via the ADC, and back into analog, with the DAC.*

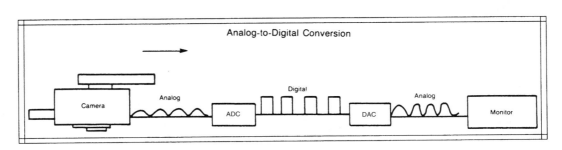

This high volume of information poses a problem for some communications systems. They may not have the capacity to transmit the information, and special lines are used. One such line is the T1 carrier. It can transmit 24 digitally coded telephone channels at a rate of 1.544 million bits per second (1.544 megabits/second). This data stream is also composed of bits that ensure the data's integrity and satisfy other technical and operational parameters.

The T1 line is one of the workhorse channels of the communications industry, and transmissions take place both inside and outside of the telephone system. It is also a flexible standard and can integrate voice and data so one communications channel can carry different types of information. It can also accommodate video.[17]

## ADVANTAGES OF A DIGITAL COMMUNICATIONS SYSTEM

Digital communications equipment and systems have numerous advantages. These features have and will continue to provide the impetus for their continued use and growth.

### Computer Compatibility

Once digitized, a signal can be processed by a computer. The capability to manipulate digitally coded information, such as a video camera's pictures, is central to the video production industry and other industries.

For a camera, the video signal, an electrical representation of the light and dark variations, the brightness levels of the scene the camera is shooting, is converted into a digital form. The digital information represents picture elements (pixels); a number of small points or dots that actually make up the picture.[18] In a black-and-white system, a pixel assumes a specific level or shade of

**Figure 2.4**
*The visual effect of reducing the number of gray levels in an image.*

the gray scale. The gray scale refers to a series or range of gray shades and to the colors black and white, all of which compose and reproduce the scene. A pixel is represented by a binary word that is equivalent to a level of this scale.[19]

The actual number of gray levels is determined by the number of bits assigned to a word. If too few bits are used, only a limited number of gray shades will be supported, and the image may not be accurately reproduced.

Once the picture has been digitized, it can be computer manipulated. A special effect can be created, or in a science discipline, picture qualities and defects can be enhanced and corrected—an unfocused picture can be sharpened.

Information in digital form also lends itself to mass storage, and the information can be duplicated, as is the case with video copies, without generation loss. Both topics are covered in later chapters.

### Multiplexing

Digital and analog signals can be multiplexed; that is, multiple signals can be relayed on a single communications line. The signals share the line and, thus, fewer lines have to be constructed and maintained.

Two advantages offered by multiplexing are its cost and labor-saving properties. In frequency-division multiplexing (FDM), a

FOUNDATION

**Figure 2.5**
*The* Mariner 4 *space probe. As described in the text,* Mariner 4 *helped pioneer the exploration of Mars. (Courtesy of the National Space Science Data Center— NSSDC.)*

communications line is divided into separate and smaller channels, each with its own unique frequency. For a telephone system, the various signals are assigned to these separate channels, processed, and relayed.

In time-division multiplexing (TDM), time, not frequency assignments, separates the different signals. A TDM scheme can make for a very effective relay when conducted in the digital domain. An example is the T1 line as employed by the telephone industry.

The T1 system is digital in form, but the typical telephone receiver and parts of the local telephone system work with analog information. This requires the analog signal to be digitized. Briefly, groups of signals, the different conversations, are divided into smaller pieces for transmission. Each signal is relayed a piece at a time, in specific time slots.[20] This digital multiplexing system is fast, efficient, and clean, with regard to the signal's quality. It can also be controlled and monitored by computers with all the attending advantages—computer accuracy and speed.

### Integrity of the Data When Transmitted

A relay may necessitate the use of a repeater(s), which with an analog signal, strengthens or amplifies the signal as it travels along the transmission path. But the signal may be ". . . degraded in this process because of the ever-present noise . . ."[21] Noise can also accumulate, and the signal's quality can progressively deteriorate.

Digital systems are not similarly affected. Instead of boosting a signal, the pulses are regenerated. New pulses are created and relayed at each repeater site. This process makes a digital transmission much hardier. A digital transmission is also less susceptible to noise and interference in general, further contributing to its superior transmission capabilities.

This factor led to the adoption of a digital relay by the National Aeronautics and Space Administration (NASA) during the 1964 *Mariner 4* mission to Mars. *Mariner 4* provided us with the first close-up views of the planet, and the digital relay helped preserve the information's integrity as it traveled through millions of miles of space.[22]

### Flexibility of Digital Communications Systems

Digital systems are flexible communications channels that can carry information ranging from computer data to digitized audio and video. In an all-digital environment, analog signals, such as those produced by telephones, would be digitized, whereas computer data, already in digital form, would be accommodated without this type of processing. A digital configuration would also eliminate the use of a conventional modem, even though a special adapter may be required to connect a computer, for example, to the line.

The *modem*, an acronym of the words modulator and demodulator, is used to relay computer data over a standard telephone line. It converts a computer's information into a form that complies with the technical characteristics of the line. At the other end, the information is reconverted by a second modem.

In the early 1980s, typical consumer and business data relays were conducted at 300- to 1200-bits-per-second speeds. Contemporary relays greatly exceed this figure and equate to more efficient and cost-effective operations.[23] Nevertheless, modem-based relays are limited, and users want to gain access to more information in shorter time periods. The changing nature of the information itself is accelerating this process. Earlier relays were generally text-based. You sat in front of a monitor, typed a command, and depending on the relay's speed, watched the requested information slowly fill the screen.

This text-dominated stream has evolved. It is now composed of text, audio and video clips, graphics, animations, and other data types. More pointedly, this new information requires higher capacity and faster communications channels—*broadband* communications systems or highways—that support the high-speed delivery of information and entertainment services to consumers and businesses.

One goal is to break through bottlenecks caused by

- slower communications channels,
- the ever increasing data load, and
- data intensive media types (e.g., video).

A complementary goal has been to develop cost-effective communications systems that are flexible, can integrate different information on a single line, and are easier to set-up and use.

An example of an earlier platform or network, which taps the telephone infrastructure and satisfies certain of these criteria, is the Integrated Services Digital Network (ISDN). The ISDN "can be thought of as a huge information pipe, capable of providing all forms of communications and information (voice, data, image, signaling). . . . It is an information utility accessible from a wall outlet . . . with a variety of devices able to be simply plugged and unplugged.[24]

As envisioned, pictures, computer graphics, and data would be exchanged with the same regularity and ease with which a telephone is used for conversations. The communication process would also be more transparent. Just like a car or telephone is typically used without thinking about the underlying technologies, a range of information would be easily relayed without calling in a computer or communications expert.

While the idea is good, ISDN technology was not universally embraced. The problems ranged from standards and pricing issues to availability.

But the rapid growth of the Internet, covered in a later chapter, helped fuel the development of more consumer/general business-oriented systems, which generally embody these goals. In one example, cable companies developed a cable-based delivery system, where a computer taps the Internet through a cable link and a cable modem. Telephone companies, for their part, introduced Digital Subscriber Line (DSL) systems by using "existing telephone lines to deliver Net [Internet] data at speeds up to 50 times faster than a 28.8 Kbps modem."[25] Different DSL flavors exist, but the upshot is that we can now relay and receive information at faster rates over our telephone lines.[26] The satellite industry is also a player in this field. With the proper configuration, you can gain access to the Internet through a satellite relay, even though the telephone line may be used for the information request in one configuration.

While a given application may still not be fully transparent, in regard to its implementation, DSL and cable-based systems are the high-speed Internet/information delivery vehicles of choice for general consumer and business use. In many instances, they can also be "plug and play" processes—you attach or plug the device in, and you may be able to use the application with a minimum of fuss.[27]

It is also important to note two other developments:

1. ISDN applications are still widely used in the business world to support various applications.[28] However, the original dream of creating a flexible information pipeline is actually more apropos to the Internet and its complementary communications systems.
2. We still use conventional modems. They are valuable tools, particularly if you don't have access to a high-speed link or if you are travelling and have to contact your home or office. Geographical restrictions (distance) also play a role in this situation, and it may not be possible to establish a link between a customer and a telephone company's central office for DSL or other telephone-based services. In the case of a cable operation, a high-speed link may not be supported, or if you live in a rural area, you may not have access to cable. High-speed links are also more expensive than conventional modem-based operations.

Finally, the modernization of a nation's telecommunication infrastructure is crucial. As stated, information is a commodity. The country that can create an efficient national and international information platform will have a decided advantage when competing in the world market.

### Cost Effectiveness

As digital equipment is mass produced and manufacturing costs are reduced, digital systems become increasingly cost efficient to build and maintain. They're also generally more stable and require less maintenance than comparable analog configurations.

This superior stability and durability is partly due to the integrated circuit (IC). An IC is a semiconductor, a solid-state device that is one of the driving forces behind our information and communications systems. In essence, the IC is one of the much talked about chips that helped launch the communication revolution.

The creation of the IC chip had an enormous impact on technology. For example, instead of wiring thousands of individual components to build a piece of equipment, a single chip is used. If the equipment malfunctions, only one chip may have to be replaced, and it is easier to isolate component failures.

But ICs are not without their own set of problems. Initial development costs may be high, and until the chips are mass produced, the early units can be expensive. The chips also tend to be vulnerable to static charges and power surges. If exposed to such an environment, a chip's operational capabilities may be temporarily or permanently disrupted.

Political issues may also have an impact. In one example, the price of computer memory chips soared during the 1980s. The Reagan administration tried to protect American chip manufacturers from Japanese competitors, and a short-term outcome of this political maneuvering was a jump in chip prices. At one point, a single, common memory chip was more expensive than an ounce of silver.

The 1990s witnessed a similar situation. Small, portable computers fitted with liquid crystal display (LCD) panels became popular. A group of American manufacturers subsequently charged the Japanese with dumping inexpensive panels on the market, and a tariff was levied on such imports to help support American manufacturers.

Although the tariff could have helped manufacturers in the long term, there was a consequence. Imported computers already fitted with the displays were exempt from the tariff, and American computer companies threatened to move their manufacturing facilities offshore to avoid the additional fee.[29] Thus, government intervention led to unforeseen circumstances.

## DISADVANTAGES OF DIGITAL COMMUNICATIONS

With any technology, there are disadvantages as well as advantages. Digital technology is no exception.

### Quantization Error

The digitization process may introduce a quantization error if not enough levels are used to represent the analog signal. If, for instance, a video system is governed by a 2-bit word code, only four colors would be reproduced. The original analog signal and scene would not be accurately represented. To correct the problem, the number of levels can be raised. But since this may increase the relay and/or storage requirements, a compromise is usually made between these factors and the accuracy of the digitization process.

### Dominance of the Analog World and Standards

We live, to a certain extent, in an analog world. Many forms of information, besides the devices and systems that produce and relay information, are analog. These include telephones, televisions, and radios. This necessitates the use of ADCs and DACs.

### Public Investment

The public investment issue must also be weighed. An overnight switch to an all-digital standard would necessitate the immediate replacement of our telephones, television sets, and radios or the use of special converters. The same principle applies to communications organizations. The industry's retooling would cost billions of dollars and would disrupt both industry and society.

Thus, the change to a digital standard has generally been evolutionary rather than revolutionary. As described in different chapters, many of the new technologies and their products will be integrated in the current communications structure, which may ease the impact of their introduction.

A case in point is the recording industry and the compact disk (CD) market. If an individual owned a turntable and large record collection, an overnight switch to a CD format would have made the original system obsolete. The old records, the LPs, were not compatible with a CD player, and the owner may not have been able to purchase additional records.

Accordingly, CD players and disks were integrated in the industry. Companies began manufacturing CD players while maintaining their conventional equipment lines. The record industry likewise supported the new medium but continued to produce vinyl LPs.

As more CD players were sold, the CD market expanded. Record companies increased their CD production and curtailed their vinyl lines. This trend accelerated, and the digital format, along with audiocassettes, dominated the industry. But because the process was somewhat gradual, the industry was not disrupted. Owners did not suffer immediate losses, and equipment manufacturers continued their support of LP systems, albeit at a reduced level (a similar scenario is playing out in the consumer videotape and DVD markets).

Other factors do play a role in the acceptance of a new technology. This may include whether a product is cost effective when compared with the product it is replacing. For the CD, a player's cost fell over a relatively short time.

Another potential factor is whether a product has certain advantages over preexisting ones. For the average user, the CD satisfied this criterion, which contributed to its popularity.

Finally, while the gradual integration of a new technology and its products may have benefits, it may not always be the best technological solution. In one case, the issue of backward compatibility chained the color television standard to the past, to the detriment of a picture's perceived quality. Similar concerns were raised that this scenario could have been played out in the development of a new television standard, as discussed in Chapter 13.

On the one hand, a bold, technological step forward, free of past constraints, could revolutionize elements of the communications industry. The downside, though, could be industry disruption and the public investment factor.

## STANDARDS

This final section is devoted to the idea of standards, another central concept that will resurface in future chapters. Simply stated, standards are a series of technical parameters that govern communications equipment and systems. The standards dictate how information is generated, stored, and exchanged. The Society of Motion Picture and Television Engineers (SMPTE) and other national and international organizations have been engaged in this task.

A standard's influence can vary. Some are mandatory and legally enforced. Other standards are voluntarily supported or may be *de facto* in form. Various technical and economic forces may generate industry-wide support for the latter, thus helping to avoid the chaos that would arise if multiple standards were adopted.

The idea of allowing an industry to adopt its own standards, especially in the telecommunications industry, had also become more prevalent. In the view of the National Telecommunications and Information Administration (NTIA) in a 1991 document,

[t]he task of standards-setting is best left to the private sector. . . . We recognize that there may be rare cases where FCC or NTIA action to expedite the standards process could be justified. . . . Government intervention could include mediation among conflicting interests or a mandate to industry to develop standards by a time certain [sic], leaving the actual development of standards to the private sector. . . . Any intervention, however, should be limited to cases where there is a specific and clearly identified market failure, and where the consequences of that failure outweigh the risk of regulatory failure (i.e., forcing the resolution of a standard too early in the development of a technology).[30]

The private sector, industry, should have the freedom to develop appropriate standards in an open market. But according to some individuals and agencies, if standards do not emerge—to the detriment of a technology and an industry—then government intervention may be appropriate. The adoption of this regulatory stance could be important for new products that have to compete in a marketplace with established competitors.

The standards issue is vital for several reasons. First, if standards didn't exist, it would be almost impossible to develop an electronic communications system and to foster the idea of equipment compatibility. Without standards, the telephone and television you buy may not be compatible with the local telephone system and television station.

Second, standards can promote the growth of a communications system. If a standard is adopted, both manufacturers and consumers can benefit. A manufacturer would be ensured that its television set would work equally well in New York City and Alaska. The consumer would be willing to purchase the set for the same reason.

There can also be multiple standards in the same industry and, depending on the circumstances, they may not adversely affect its growth. The parallel development of the Beta and VHS videotape formats serves as an example. While incompatible, they created their own market niches, despite VHS's emergence as the *de facto* consumer standard.

The presence of two strong standards in a market may also accelerate an industry's growth. The competition may spur a manufacturer to introduce improved equipment to gain a larger market share.

Although multiple standards could be beneficial, this is not always true. This factor could have a devastating effect on an organization that is introducing a new application in a competitive field. In one case, competing standards may splinter a small, initial market.

Third, an accepted standard helps guarantee that a piece of equipment you buy today will generally not be replaced by a new and incompatible device tomorrow. While developments do take place and equipment may become obsolete, this process may be gradual. This protects industry and the consumer.

It is also appropriate to point out that obsolescence does not necessarily mean incom-patibility. An older video camera may not have all the latest technological "bells and whistles," but it may still be compatible with your system.

There's a point, though, at which a standard may be replaced by another, as was the case with the $\frac{1}{2}$-inch black-and-white videotape recorder (VTR). In the 1970s, this portable VTR and companion camera were popular with video artists and schools. The

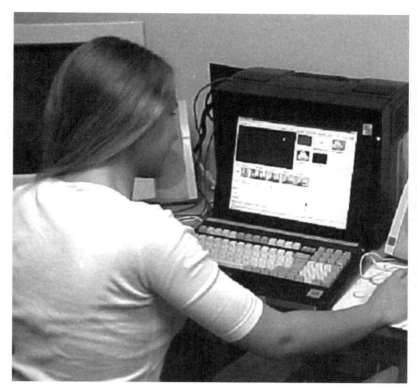

**Figure 2.6**
*A fallout of the digital revolution is the introduction of cost-effective digital tools. These include portable video editing systems that can support a high-quality output.*

system was easy to use and affordable. But it became obsolete. The equipment and governing standards were superseded.

Fourth, standards have an impact in the international video markets. For example, the industry has been dominated by incompatible standards: NTSC, SECAM, and PAL. A program produced under one system must undergo a conversion process to make it compatible with another system.[31] This arrangement has posed some problems for the international market. It creates a roadblock in the exchange of programs and adds to a program's overall cost.

Thus, even though the adoption of a single standard might have fueled a country's domestic video industry, the variety of standards became a hindrance. This problem has assumed a greater degree of importance as the communications industry becomes increasingly international in scope, in keeping with the automotive, computer, and other major industries.

FOUNDATION

## REFERENCES/NOTES

1. Frederick Williams, *The New Communications* (Belmont, CA: Wadsworth Publishing Co., 1984), 133.

2. Ken Marsh, *The Way the New Technology Works* (New York: Simon and Schuster, 1982), 26.

3. Herbert S. Dordick, *Understanding Modern Telecommunications* (New York: McGraw-Hill Book Publishing Co., 1986), 32.

4. Technically speaking, the bandwidth is the difference between the highest and lowest frequencies (that is, the range of frequencies) a communications channel can accommodate. The term *bandwidth* is likewise applied to the signals that are relayed over these channels. Note also that a time factor plays a role. In essence, a communications channel can accommodate or relay a specific volume of information in a given time period based on the channel's capacity and the noise present on the line.

5. For example, it has a greater information content.

6. For example, in the United States, governmental agencies/services may also receive allocations.

7. "Al Sikes's Grand Agenda," *New York Times* (June 2, 1991), Section 3, 6. *Note:* All over-the-air communications systems do not require FCC allocations. Commercial over-the-air lasers discussed in a later chapter fall in this category.

8. Basically, the public would still be served by various types of programming, such as news and public information shows.

9. Ayn Rand, "The Property Status of Airwaves," *The Objectivist Newsletter* 3 (April 1964), 13; from the reprint of *The Objectivist Newsletter*, Vols. 1–4 (New York: The Objectivist, Inc., 1971).

10. Gary M. Kaye, "Fiber Could Be Winner in the Battle for the Spectrum," *Photonics Spectra* 25 (November 1991), 79.

11. Simon Haykin, *Communication Systems* (New York: John Wiley & Sons, 1983), 6.

12. Tom Smith, *Telecabulary* 2 (Geneva, IL: abc TeleTraining, Inc., 1987), 28.

13. The analog information is converted into digital information, as described in the next section.

14. Bernhard E. Keiser and Eugene Strange, *Digital Telephony and Network Integration* (New York: Van Nostrand Reinhold Company Inc., 1985), 19.

15. William Flanagan, "Digital Voice and Multiplexing," *Communications News* (March 1984), 38E.

16. Thanks to Jim Loomis, director of Telecommunications Facilities, Ithaca College, for his suggestions for this section.

In a more detailed look at this process, three phases are completed. In the sampling phase, the analog signal (continuous in time and amplitude) is sampled. This creates a pulse amplitude modulation (PAM) signal. The amplitudes of the individual and discrete PAM pulses correspond to the variable amplitude of the original analog signal at the sampling points. Thus, the amplitudes of the PAM pulses are continuously variable, like the original signal. In the next phase, the quantization stage, the PAM pulses are assigned to the nearest steps or levels in reflection of their amplitudes. This phase converts the wide range of amplitudes to a finite and limited number of amplitudes or values. The final phase is the coding process. The samples are coded in binary form. Consequently, the original analog signal, which is continuous in time and amplitude, is made non-continuous by the sampling and quantization phases.

In this type of system, the information is also composed of pulses with identical amplitudes. This is a reflection of PCM information where the amplitudes do not convey the information.

17. The T1 line is only one of a digital family of lines.

18. Arch C. Luther, *Digital Video in the PC Environment* (New York: McGraw-Hill Book Publishing Co., 1989), 51. The picture is organized in memory as a series of pixels. *Note:* For specific technical details, see Frederick J. Kolb, Jr., et al., "Annotated Glossary of Essential Terms," *SMPTE Journal* 100 (February 1991), 122.

19. The level, in turn, is delineated by the brightness value of the corresponding section

of the original scene now represented by the pixel.

20. In this particular configuration, digitized information from 24 channels is transmitted one after the other on the line. The information is organized in a grouping (frame). Following the transmission, the 24 channels are separated. For details, see Tom Smith, *Anatomy of Telecommunications* (Geneva, IL: abc TeleTraining, Inc., 1987), 107.

21. Henry Stark and Franz Tuteur, *Modern Electrical Communications* (Englewood Cliffs, NJ: Prentice-Hall, Inc., 1979), 162.

22. Michael Mirabito, *The Exploration of Outer Space with Cameras* (Jefferson, NC: McFarland and Co., 1983), 30. *Note:* Digital communications systems similarly played an important role in later missions. Digital tools were also used to enhance and manipulate the pictures produced by a spacecraft's camera system after the data were received on Earth.

23. The terms *bits per second* and *baud* are associated with a modem's relay speed. There are differences between the two, however. For a detailed discussion, see Brett Glass, "Buyer's Advisory," *InfoWorld* 13 (October 21, 1991), 147.

24. Don Wiley, "The Wonders of ISDN Begin to Turn into Some Real-World Benefits as Users Come On Line," *Communications News* (January 1987), 29. Please see Bill Baldwin, "Integrating ISDN Lines for Financial Users," *Telecommunications* 25 (June 1991), 34, for more specific information about ISDN systems, including its data rates: "Basic Rate Interface (2B + D): two 64 kilobit/second B channels for information (for example, voice) and one 16 kilobit/second D channel for signaling and control information. Primary Rate Interface (23B + D): 23 64 kilobit/second B channels and one 64 kilobit/second D channel. For Europe, the Primary Rate Interface is 30B + D."

25. Michelle V. Rafter, "Users, Start Your Modems," *The Industry Standard* 2 (October 11, 1999), 16.

26. This includes, in a trial operation of Bell Atlantic's Asymmetric Digital Subscriber Line (ADSL) service, the delivery of video information. Please see Salvatore Salamone, "Higher Data Speeds Coming in Plain Phone Lines," *Byte* 21 (January 1996), 37, for details.

27. In reality, this may not always be the case, and you still may have to configure your computer to properly function.

28. This includes, as described in a later chapter, videoconferencing.

29. "Dropping the Color LCD Tariff Will Save Jobs," *PC Week* 9 (February 10, 1992), 76. *Note:* American chip manufacturers experienced a recovery during the early 1990s (for example, a consortium of companies formed to promote research and development).

30. NTIA, *Telecommunications in the Age of Information*, NTIA Special Publication 91–26, October 1991, xvi.

31. The standards are national in origin and have been used by different countries. For analog systems they are United States, NTSC; England, PAL; and France, SECAM.

## SUGGESTED READINGS

Berst, Jessie. "The Broadband Contenders." Downloaded from ZDNET; Linden deCarmo. "Packet Protector." *Emedia Magazine* (November 2001), 36–42; Michelle V. Rafter. "Users, Start Your Modems." *Telecom* 2 (October 11, 1999), 116–121. Coverage of DSL systems, and for the deCarmo article, an implication for an Internet-based application (video streaming as described in the Internet chapter, Chapter 17).

Bigelow, Stephen J. *Understanding Telephone Electronics*. Carmel, IN: SAMS, 1991. A comprehensive overview of the telephone, from basic concepts to digital and network operations.

Clement, Fran. "Digital Made Simple." *Instructional Innovator* (March 1982), 18–20; Bill Gibson. "Sampling Rates." (from The AudioPro Home Recording Course), downloadedfromwww.digitalproducer.com/2002/01_jan/features/01_14/artistprosampling.ht.

Good primers about digital information, signals, and the digitization process.

Cosper, Amy C. "One-Way Data Traffic Over DBS Satellites." *Satellite Communications* 21 (May 1997), 32–34. Satellite data relays (including the Internet).

Edwards, Morris. "What Telecomm Managers Need to Know About Europe '92." *Communications News* (June 1991), 64–68. An earlier look at European communications systems.

Haykin, Simon. *Communication Systems*. New York: John Wiley & Sons, 1983, 408–428. Technical descriptions of PCM and multiplexing operations, among other communications topics.

McLachlan, Wayne. "Analog Video 101 and 102 for All." *SMPTE Journal* 110 (March 2001), 151–157. An excellent overview of analog video (broadcast) and technical issues on the road to a digital standard.

"Radio Wave Propagation." *Hands-On Electronics* (November/December 1985), 32–38, 104. A primer on how a radio wave can travel from its source to the receiver.

Rogers, Tom. *Understanding PCM*. Geneva, IL: abc TeleTraining, Inc., 1982. One of a series of abc TeleTraining publications. This particular publication provides an excellent overview and explanation of PCM.

Streeter, Richard. "Is Standardization Obsolete?" *Broadcasting* (February 9, 1987), 30. A brief article about the importance of standards.

Tropiano, Lenny, and Dinah McNutt. "How to Implement ISDN." *Byte* 20 (April 1995), 67–74. ISDN growth and setting-up a connection.

Winch, Robert G. *Telecommunication Transmission Systems*. New York: McGraw-Hill, 1998. An excellent examination of telecommunications systems and their operation.

## GLOSSARY

*Analog Signal*: A continuously variable and varying signal. Many of the communications devices and systems we are most familiar with, such as telephones and conventional radio stations, produce and process analog signals.

*Analog-to-Digital Converter (ADC):* An ADC converts analog information into a digital form. ADCs work with DACs to bridge the gap between analog and digital equipment and systems.

*Binary Digit (bit):* A bit is the smallest piece of information in a digital system and has a value of either 0 or 1. Bits are also combined in our communications systems to create codes to represent specific information values.

*Channel*: A communications line. The path or route by which information is relayed.

*Communications Channel Bandwidth*: A communications channel's bandwidth, its capacity, dictates the range of frequencies and, to all intents and purposes, the categories and volume of information the channel can accommodate in a given time period.

*Digital Signal:* A digital signal is noncontinuous and assumes a finite number of discrete values. Digital information is represented by bits, 1s and 0s.

*Digital Subscriber Line:* A high-speed and high-capacity communications link used by consumers and businesses.

*Digital-to-Analog Converter (DAC):* A DAC converts digital information into an analog form.

*Electromagnetic Spectrum*: The entire collection of frequencies of electromagnetic radiation.

*Frequency*: The number of waves that pass a point in a second. The frequency of a signal is expressed in cycles per second or, more commonly, in hertz.

*Integrated Services Digital Network (ISDN):* A digital communications platform that could seamlessly handle different types of information (for example, computer data and voice).

*Modem*: The device used to relay computer information over a voice-grade telephone line. A modem is used at each end of the relay.

*Modulation*: The process by which information is impressed on a carrier signal for relay purposes.

*Multiplexing*: The process whereby multiple signals are accommodated on a single communications channel.

*Picture Element (Pixel):* A pixel is a segment of a scan line.

*Pulse Code Modulation (PCM):* A digital coding system.

*Standards*: The technical parameters that govern the operation of a piece of equipment or an entire industry. A standard may be mandated by law, voluntarily supported, or *de facto* in nature.

*T1 Line:* One of the workhorse and important digital communications channels. Information is relayed at a rate of 1.544 megabits/second.

*Transducer*: A device that changes one form of energy into another form of energy. Transducers are the core of our communications system and include microphones and video cameras.

# 3 Computer Technology Primer

The computer played a pivotal role in launching the communication revolution. This is particularly true of the microcomputer, also known as the personal computer. The PC has made it possible for us to complete jobs ranging from designing spreadsheets to creating graphics for a news show.

PCs have also influenced a new generation of audio and video equipment. As described in later chapters, PCs, other computer systems, and microprocessors have touched almost every phase of the production process.[1] Consequently, this chapter serves as an introduction to computer technology and its complementary communications applications. It also provides a foundation for exploring the computer's roles at work, at home, and at play. But before we begin, three points about terms and other matters, as used by the book, must be made:

1. The generic term *personal computer* (PC) is used for Apple's Macintosh series (Macs) as well as IBM PCs and clones (e.g., Compaq, Dell, etc.). The Amiga also falls under this generic term.[2]
2. The terms *data* and *information* are used as has been defined by Alan Freedman in his valuable reference work *The Computer Glossary*. Information is "the summarization of data. Technically, data are raw facts and figures that are processed into information. . . . But since information can also be raw data for the next job or person, the two terms cannot be precisely defined. Both terms are used synonymously and interchangeably."[3] This book follows suit.
3. When appropriate, specific computer products may be used to describe applications. They simply serve as springboards to explore these topics.

## HARDWARE

PCs are equipped with a number of components. Some of the important ones, forms of which may be used by other computer systems, are discussed in the following subsections.

### Memory

A computer's memory is its internal storage system and workspace. The most important

Figure 3.1
*Contemporary software has simplified numerous operations. In this case, you can tap a PC-based Help System to quickly find information about a program's features and operation. Simply move the cursor to a term and click a mouse button—the linked information (the description)—appears. (Software courtesy of Adobe Systems, Inc.; Photoshop.)*

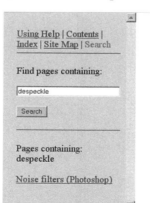

Using Help | Contents | Index | Site Map | Search

Find pages containing:

despeckle

Search

Pages containing: despeckle

Noise filters (Photoshop)

Previous | Next Applying Filters for Special Effects

## Noise filters (Photoshop)

The Noise filters add or remove *noise*, or pixels with randomly distributed color levels. This helps to blend a selection into the surrounding pixels. Noise filters can create unusual textures or remove problem areas, such as dust and scratches, from an image.

**Add Noise**

Applies random pixels to an image, simulating the effect of shooting pictures on high-speed film. The Add Noise filter can also be used to reduce banding in feathered selections or graduated fills or to give a more realistic look to heavily retouched areas. Options memory is random access memory (RAM). A program and the data generated during a work session are stored in RAM. These may include a word processing program and a letter you are writing to a friend. The RAM workspace is also considered a temporary storage area. Once the computer is turned off, the program and letter are cleared from memory and cannot be recalled unless they were previously saved on a permanent storage system.

Contemporary PCs are generally furnished with 512 or more megabytes (MBs) of RAM. A computer must also be equipped with a specific quantity of RAM to run a given program, and many operations are enhanced if additional memory is available.[4]

### Central Processing Unit

The central processing unit (CPU) is the computer's brain. It also dictates the type of software the computer can run and affects the computer's overall data processing speed. Additional performance factors include the computer's storage system, bus speed, and the use of graphics accelerators and other auxiliary systems. The former can speed-up the graphics creation process.

### Expansion Boards

Some computers are designed with internal slots that can be fitted with expansion boards or cards. The cards support basic functions, such as generating a monitor's display, or enhance the computer's operation. They may be required or they may supplement and complement built-in capabilities.

A video display or graphics card for an IBM PC, for instance, dictates various technical parameters. These include the resolution of graphics and alphanumeric information as well as the number of supported colors. Alphanumeric information encom-

passes alphabetic and numeric characters (letters and numbers). Other cards can serve as internal modems, and as covered later, video editing components.

Portable computers, described in the next section, similarly benefit from expansion products. In most cases, they are fitted with *PCMCIA* slots. Small credit card-sized cards, called PC cards, are inserted in the slots and can support data communications and other operations.

### Data Storage

Different systems can be used to store your data once the computer is turned off. Common examples include floppy, hard, and removable disk drives, all of which are governed by magnetic principles.[5]

Floppy disks were once the primary software distribution vehicle. Programs stored on one or more disks would subsequently be installed on the PC. However, CD-ROMs have largely superceded floppy disks in this application. CD-ROMs and other optical media are discussed in Chapter 9.

Hard drives are the most common and popular data storage system. Contemporary drives can store gigabytes of data and support fast storage and retrieval operations. As covered in the multimedia and video production chapters, specialized systems are used in video editing and other, data-intensive tasks.

Inexpensive, removable magnetic storage systems were widely adopted in the latter half of the 1990s. Although they had existed for years, newer and cost-effective models, which could even potentially handle demanding video applications, were introduced.

In operation, a removable system functions much like a floppy. Once a cartridge is filled with data, it is replaced. But unlike a floppy, removable cartridges are true mass storage devices with fast, data access times.

## Monitors

A monitor is the computer's display component, and different standards exist in the PC world. These include the VGA (640 × 480 resolution) and Super VGA (800 × 600 + resolution) standards.

A key development in the PC market is the advancement of display technologies. The average monitor's size and corresponding viewing area is increasing. So too is the resolution, that is, an on-screen image's apparent sharpness. Contemporary graphics cards and monitors are also easier on the eyes. They produce a higher resolution display with less flicker to help reduce eyestrain.

A graphics card may also speed up or accelerate certain software operations. It is also equipped with its own memory, which has an impact of the number of colors it may support and other performance characteristics.[6]

At this time, 24-bit systems are an industry standard. The 24-bit figure refers to "a display standard in which the red, green, and blue dots that compose a pixel each carry 8 bits of information, allowing each pixel to represent one of 16.7 million colors."[7] Basically, instead of working with a limited palette or range of colors, 24-bit systems can gain access to millions of colors. (The actual number of on-screen colors, however, is lower than 16.7 million.) Visually, a picture is more vivid. When used with appropriate software, a landscape with clouds or other image may appear to be lifelike or photorealistic.

A controversy also surrounded the use of traditional CRT monitors, which produce electromagnetic fields. Concerns were raised that emissions could play a role in miscarriages, birth defects, and cancer. But the data have been conflicting and inconclusive. The frequencies that concern us are very low frequency (VLF) and extremely low frequency (ELF) emissions. Much of the research has focused on the magnetic field, and studies have been conducted to determine its effect on the human body.

Until this issue is finally resolved, it may be wise to take some elementary precautions. You can buy a nonemission display, or as became the trend in the mid-1990s, a monitor that conforms to strict emission standards.

An example of the former is an LCD panel, formerly relegated to notebook computers. In the early 2000s, LCD monitors for desktop PC systems flooded the market. Their size increased as prices fell.

While CRT monitors still have some advantages as of this writing (e.g., lower cost), LCD units have become a popular option. They weigh less and have a smaller footprint and larger viewing area than comparably sized conventional monitors.[8]

## THE MICROCOMPUTER

Prior to the PC era, the computer industry was dominated by mainframe and minicomputer systems. Both may still be used by organizations with multiple users and extensive processing needs. But the PC is now a fixture in almost every market.

The PC's popularity can be partly traced to the introduction of the original, widely available, IBM microcomputer in 1981. People were already familiar with IBM products, and the corporation's entry in this field legitimized the PC in the eyes of the business community.

Other factors that contributed to the industry's growth over the years include

- the production of more advanced models;
- powerful software that filled specific needs;[9]
- the PC's low cost in comparison with other systems, which helped extend computer processing to a broader user base; and

FOUNDATION

• falling PC prices as PCs became more sophisticated.[10]

The computer industry is one of the few manufacturing areas where this type of price versus performance ratio prevails—to our benefit.

## Portable Personal Computers

Besides desktop PCs, manufacturers produce portable, battery-operated notebook computers. Contemporary models are equipped with high quality color displays, internal modems, network connections, enhanced sound capabilities, and a CD-ROM or other optical drive. Another possible feature is a desktop docking station, which provides additional expansion capabilities.

In keeping with the downward or miniaturization trend in the PC world, portable computers are becoming smaller and lighter as their processing power increases. These systems have spawned even smaller computers, including handheld units of varying capabilities.

Although this miniaturization trend will continue, there are, at least for the near future, size constraints. These may have more to do with the human-PC interface than with the capability to design ever smaller computers. In general, as keyboard and screen sizes decrease, their effectiveness, at least for the operator, may decrease in kind. A partial solution may lie in adopting pen computing, speech recognition, and other, more advanced interfaces.

## Human-Computer Interface

As stated, one consideration in using a computer is the human-computer interface: how we communicate with and control the computer. On one level, interfaces are the hardware tools we use. They range from keyboards to graphics tablets to virtual reality systems, in which we can actually become part of a computer-generated world.

Software is the other level, the driving force behind human–computer interfaces. A computer's operating system provides the basic link; other software families are discussed in later chapters.[11]

The most common interface is the keyboard. Alphanumeric and function keys are used for control and software operations. You can type a command, press two keys to magnify an on-screen graphic, and type a letter with a word processing program.[12]

A second category supplements and complements the keyboard. A mouse, for example, is a small device that is interfaced with the computer and sits on a desk. As you move the mouse, a cursor on the monitor's screen moves correspondingly. In one application, you use the mouse to select a command listed on a pull-down menu. You highlight the specific command, click a mouse button, and the command is carried out. A mouse also functions as a drawing tool for graphics software.

A more effective device for this task is the graphics tablet, a drawing pad typically interfaced with an electronic pen. You can literally sketch a picture in freehand, among other options. Some configurations provide very fine degrees of control when combined with specific software.

The touchpad, another interface, works much like a mouse. As implied by the name, a touchpad is a small, sensitive pad (physical area). You can use your finger with the touchpad to emulate mouse functions.

Another interface is the touch screen. Unlike a mouse or graphics tablet, you interact directly with the monitor. The touch screen enhances this process, eliminating the use of a manual tool. In a standard application, a menu with a list of commands appears on the screen. You select the command by touching the screen at the appropriate location. In a sample application, consider an interactive computer program for a museum. When you touch the museum's rooms, which are displayed on a monitor, the exhibits at each location are listed.

The touch screen can be an intuitive interface because you do not have to learn how to use it. Simply touch what you want, and the hardware and software complete the task.

Another human-computer interface uses a pen metaphor, pen-based computing. In one configuration, you hold a computer much as you would hold a clipboard. But instead of writing on paper, you write on a flat-panel display with a pen. This interface can simplify routine jobs. The express mail/package industry, for one, uses electronic forms and pen-based systems. As deliveries are made, people working in the field can check off boxes on forms and fill in other information. The data are then stored and can be transferred to a central computer.

Future pen-based systems may also draw on the idea of an electronic or eBook. It can be rapidly updated, annotated with your own notes, and integrated with graphics.[13] Like a touch screen, this type of interface is intuitive since we are using a familiar communication form, basically an electronic version of paper.

Additionally, portable computers, including a new generation of tablet PCs, can recognize handwriting and special markings used for editing.[14] These devices marry computer technology with the convenience of using a pen and paper, but now, they're in an electronic form.

A system's processing capabilities range from that of a traditional notebook computer to small, handheld units that are typically used for storing addresses, short messages, and similar data. The latter, called Personal Digital Assistants (PDAs), became popular in the early 2000s. While they lack the processing capabilities of larger computers, they can be fitted with small color screens, keyboards, and even still video camera devices.

PDAs can also communicate with a desktop computer for data exchanges, and when properly configured, can be used for e-mail and connecting to high-speed communications networks without the use of a cable (wireless networking). In some cases, PDA functions, and even certain PDAs, have been combined with cellular telephones (cell phones).[15] As such, the PDA became a comprehensive information and communications tool—you could use it to process information and for voice/data communication.

Another element of the human–computer interface touches on ergonomic design, the philosophy of developing equipment and systems around people, that is, making equipment conform to an operator's needs, and not the other way around. Desks, chairs, monitors, and keyboards are typical computer equipment influenced by ergonomic design.

Sound designs can help prevent some of the problems associated with working with computers. With carpal tunnel syndrome, your wrist can be damaged through repetitive keyboard motions. It can be alleviated, or possibly even avoided, by ergonomically sound keyboard and desk designs. Adopting good work habits can also help. Maintain a good sitting posture, and take a break every hour.

The latter can also reduce eye fatigue, as can proper light placement and intensity. A high-resolution display/board combination, which may also produce less flicker as a result of a high screen refresh rate, can likewise help.

**Figure 3.2**
*Human-computer
interfaces allow us to
explore new techniques,
in this case, the potential
to create an animated
figure based on an
individual's movements.
(Courtesy
of Polhemus, Inc.)*

## COMPUTER SOFTWARE

For organizational purposes, software can be divided into two categories: general release programs, the focus of this section, and programs geared for specific communications applications. The latter include hypertext and multimedia authoring software.

### Operating Systems and Graphical User Interfaces

The operating system (OS) is the most important piece of computer software. It controls the computer as well as specific data and file management functions.

MS-DOS, for example, emerged as a standard in the PC market through the widespread integration of IBM PCs.[16] MS-DOS supported a text-based interface. Keywords were typed to carry out various functions.

A graphical user interface (GUI), in contrast, uses a visual metaphor. Popularized by Apple and its Macintosh line, Apple helped establish what we now associate with a GUI:

- A pointing device, typically a mouse;
- On-screen menus;
- Windows that graphically display what the computer is doing;
- Icons that represent files, directories, etc.; and
- Dialog boxes, buttons, and other graphical widgets that let you tell the computer what to do and how to do it.[17]

A windowing operation generates multiple windows, one or more framed workspaces on the monitor's screen. Windows can be moved, resized, or removed. Windows can enhance file copying and other procedures, and you can switch between programs with the click of a button. A program is run or displayed in its own window.

An icon is a small picture, a pictorial representation of, for example, a disk drive. Instead of typing a command to display the drive's files, you click on the icon. Instead of typing a command to delete a file, you move the appropriate icon to another icon, typically a trash can.

Apple's popularization of the GUI carried over to other PC platforms. Different products have sported GUIs, including Microsoft Corporation's Windows family.[18] In fact, Apple was involved in a multiyear lawsuit with Microsoft over an intellectual property issue. Apple contended that Microsoft borrowed the look and feel of its GUI when Microsoft developed Windows.

### Database and Spreadsheet Programs

A database program can organize a consultant's client list, the titles of a radio station's carts, and other information. Song titles, for instance, can be filed under a singer's name or even under the types of music a station plays.

Once stored, the information can be organized and manipulated. You can also define and explore the associations between information categories. For the radio station, these associations can include the sales staff, their clients, and appropriate sales figures.

A database may also accommodate graphics. Visual and textual information can be merged. In one application, a specification sheet that highlights a house's features may be integrated with a picture of the site. This configuration would enable a real estate company to maintain a written and pictorial database of houses on the market.

A spreadsheet program, in contrast, is primarily a financial tool. Data are entered via a table format, in columns and rows, and are tabulated and manipulated through a series of built-in mathematical and financial functions.

With a spreadsheet, you can rapidly finish jobs that would normally require hours to complete, as may be the case when it is used as a forecasting tool. As various figures on the spreadsheet are changed to reflect higher advertising rates, for example, all the pertinent figures are automatically recalculated.

A spreadsheet may also support a graphing capability. The data are portrayed as a line graph, pie chart, or other form. Viewing data in this fashion may make it easier to discover "hidden" relationships. The same graph may also create a more powerful presentation. You can now "see" the numbers instead of just rows and columns on a page.

## Word Processing Programs

A word processing program is used to write letters, news stories, and many other types of documents. Some programs are also equipped with a mail merge option, which is used to merge a mailing or address list with a standard form letter.

A program has other functions that enhance the writing process. You can typically move, copy, and delete blocks of text, produce tables, and incorporate graphics, spreadsheet files, and other data on a page.

Most word processing programs can optionally save data in an international standardized code called the American Standard

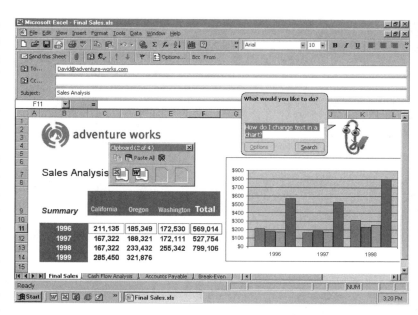

Code for Information Interchange (ASCII, pronounced "AH-skee"). Once saved, the information can be retrieved by other computers and programs. But underlined and italicized text, as well as other text/document formatting information, is lost. A program may, however, be able to import and use another program's native files to retain the formatting.

## Web Browsers

If you are connecting to the Internet, you explore and interact with this electronic information and communications universe with a web browser. As covered in the Internet chapter, Chapter 17, browsers use a visual metaphor, analogous to a GUI, to simplify Internet navigational and operational tasks. Prior to this time, a series of text commands were typed to carry out different functions.

## Programming Languages

Application programs are created through programming languages. A programming language is considered a control language. It

**Figure 3.3**
*Microsoft Excel, a spreadsheet program with enhanced graphing capabilities. (Courtesy of Microsoft Corporation; Excel.)*

provides the computer with a set of instructions to perform a series of operations.

Common languages include Visual Basic, Pascal, COBOL, C++, Java, and Ada. Each language has its own characteristics and is typically geared toward select applications.

Java and COBOL have been widely used with the Internet and business worlds, respectively. Visual Basic is the first programming language many people learn, and more esoteric languages, such as Lisp and Prolog, have been the domain of the artificial intelligence community.[19]

Ada, a language sponsored by the U.S. Department of Defense, is well suited for complex systems. Ada is named after Augusta Ada Byron, considered to be the world's first programmer.[20]

## Graphics Programs

Graphics programs are used to create a company's logo, a rendition of the space shuttle, and other drawings. A graphic can also highlight next year's car model and can chart the U.S. population growth.

There are different types of graphics programs and formats (e.g., TIFF and GIF), some of which are introduced in this section. Most programs can also import more than one format, and it is possible to share files between IBM and Mac platforms.[21]

Another trend is the convergence between program categories. A single program may now support multiple functions. Instead of using two or more of the programs listed in this section, a single program may suffice.

*Paint Programs.* A paint program addresses or manipulates the individual pixels on a screen. The pixels can be assigned specific colors and can be controlled to produce numerous effects. Applications include manipulating video-based images and computer art, in which an artist paints

with an electronic rather than a traditional medium.

The paint program creates bitmapped graphics. Bitmapping is, in part, the computer's capability to manipulate the individual pixels that make up a graphic. It also describes the method by which the graphics information is stored. As succinctly stated by Gene Apperson and Rick Doherty, "The bitmapped image is represented by a collection of pixel values stored in some orderly fashion. A fixed number of bits represents each pixel value. The display hardware interprets the bits to determine which color or gray levels to produce on the screen."[22] Thus, the values, which essentially are the pixels' colors, are coded, stored, and eventually retrieved and interpreted by the computer to create the graphic.

A paint program offers a selection of brush shapes and sizes for freehand drawing. Additional tools can be used to create geometric shapes and for other functions.

You can also control the palette of colors. The number of usable colors can vary from system to system, and you can alter a color to fit your project. One common method is to select a color and to change its red, green, and blue values by moving R-G-B slider bars.

*Image Editing Programs.* An image editing program's primary function is that of an image editor or electronic darkroom. This software family can be viewed as the word processor of images. Just as you can edit and move text, a picture can be rotated, rescaled, and manipulated prior to its printing on paper and/or videotape.

For example, you can use special filters to either sharpen or blur an image (or select parts). In one operation, you can throw a distracting background out of focus. Programs also feature color editing and correction tools, paint modules, and possibly a hook to a scanner.

As is later described, a scanner can digitize photographs and other still images, which are subsequently fed to a computer. With this software and hardware combination, a picture can be scanned and then retouched or altered.

***Drawing Programs.*** A drawing program does not address individual pixels. It treats a graphic as a series of individual geometric shapes or objects that can be manipulated and moved to other screen locations. Graphics are vector based, and picture information is expressed and stored mathematically, not as bitmaps.[23]

Drawing programs are typically used for creating ads and illustrations. They also have powerful text handling tools: Words can have a three-dimensional appearance and can follow designated nonlinear paths.

Like paint software, drawing programs have their own advantages and application areas in which they excel. For example, when a graphic produced by a drawing program is enlarged or reduced, the lines will remain sharp and well defined.[24]

***Computer–Aided Design Programs.*** A computer-aided design (CAD) program is similar to a drawing program in that it manipulates geometric shapes. But it has enhancements that make it a powerful architectural and industrial design tool.

There are two-dimensional (2-D) and three-dimensional (3-D) CAD programs. The latter can be used to create a 3-D view of a building. The image can then be rotated to show the building from different perspectives.

Initially, 3-D views were generally limited to wireframe drawings, images composed only of lines, with no solid appearance. Computer advances broke through these restrictions, and we can add physical surfaces and solid attributes. In 3-D, the building in our example would look more like a "real"

building and would help the designers to better evaluate its physical characteristics.

A program may also have an animation feature. It may likewise be possible to tie into another program to reveal the area(s) of a designed part that may be subject to stress.

Finally, a car part or other information could be fed to a series of computer-controlled tools. At this point, an engineer's vision, a drawing on a computer monitor's screen, is transformed into the actual physical part by the tool. This function is an element of computer-aided manufacturing (CAM), an area closely allied to CAD systems. In many instances, CAD and CAM are linked in a CAD/CAM system.

***Animation and 3-Dimensional Programs.*** An animation can be described as a series of images that, when viewed in sequence, convey motion. We are all familiar with Saturday morning cartoons, such as Bugs Bunny and the X-Men. With computers, we can now tap animation techniques to produce our own projects.

Contemporary programs have also simplified this process. Some allow you to create predefined paths that a figure will follow.

**Figure 3.4**

*An image editing program can support special effects. The two pictures on the right have been manipulated by the software. (Software courtesy of Image-In, Inc.; Image-In Color.)*

You select the number of frames, determine the starting point, and set up the path or direction(s) of the movement. The program will then draw the individual frames, out of the specified total, as the object progresses down the path.

A program may also support other functions. You may be able to speed up or slow down the animation at specific points for smoother transitions and enhanced realism. Or you may be able to define two distinct shapes and have the software draw the intermediate frames so one changes into the other (morphing).

Animations can also be created in 3-D, adding realistic depth. A picture can also be wrapped around a 3-D object, and wood and other textures can be applied.

Programs may accommodate 24-bit images, and you typically have control over light sources and camera parameters. Light sources are individual lights that illuminate a scene. The number of lights can vary, as can their color, intensity, and type.[25] The camera parameters affect what we see on the monitor. A view could be changed from wide angle through telephoto and, in an animation, the camera itself could move.

Lifelike animations and still images can be created through ray tracing, a rendering technique that "literally traces the paths of thousands of individual rays of light through a three-dimensional scene via computation."[26] Ray tracing can produce very realistic images with accurate shadows and reflections.

The only penalty associated with ray tracing is time. It may take hours to produce or render the final scene, depending on the PC's capability. But you can initially create the scene in a faster mode. When finished, the scene can be reviewed, changes can be made, and the final sequence can then be produced and saved.

A program may also save an animation as a series of individual frames (files). In one operation, they can then be imported by a nonlinear editing system, discussed in a later chapter, where they are processed and used as an animation.

**Presentation Programs.** Presentation programs can generate computer-based charts, graphs, and electronic slides. Templates are available to format the data or you can create your own designs. Interactive links may also be supported. When you activate an on-screen button, an action is triggered. This action may include playing a digitized audio and video clip.

If you have a series of on-screen graphics, you can control the length of time each image is displayed as well as transitional effects.[27] You can also incorporate audio and video clips and can typically export the presentation for replay on the Internet—the project is converted by the software for this environment. In this case, the same presentation can be used across multiple distribution venues.

**Visualization Programs.** Scientific visualization is the "ability to simulate or model 3-D images of natural phenomena on high

**Figure 3.5**
*Presentation programs can be used to produce PC-based slide shows. A program may also support an Internet export feature—you can also distribute/play the presentation via the Internet. (Courtesy of Microsoft Corporation; PowerPoint.)*

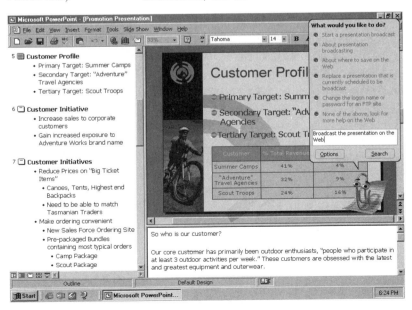

performance graphics computers."[28] Data are graphically represented, and this visual representation can provide insights into how our world and the universe function.

Instead of looking at pages of numbers, a scientist can view the data graphically. For example, while designing a new space plane, airflow and thermal characteristics could be observed.[29] Essentially, the visual image makes it easier to interpret the data because you actually see, in a sense, the data brought to life. The invisible is made visible.

*Implications.*    To wrap up this discussion about graphics programs, it is appropriate to talk about creativity and computer systems. Certain advantages accrue when using computer-based technologies. In the graphics area, a PC with the appropriate software can help you transform an idea or vision into a reality, an actual product. It is the marriage between the conceptual and the concrete.

The same system can help you reach this goal even if you are not a graphic artist. You may have an idea for a poster but may not be adept at creating the 3-D letters your project demands. A computer with the appropriate software could help. In this case, the PC is functioning as a tool. You provide the guiding thought, and the PC helps to implement your idea.

But it is also important, regardless of your artistic ability, to at least have a basic grasp of underlying aesthetic principles relative to the job at hand. Otherwise, an ad's message, or other product, could be lost in a maze of words and graphics. In another example, you can use a PC-based system to create a video production. But unless you understand lighting, camera, and audio techniques, the final product may lack integrity.

A PC may be helpful, but it *does not* circumvent knowledge and aesthetics. A PC could extend your creative vision, but *you* still have to supply the imagination and skills.

Finally, this use of computers has raised questions. Does a computer dehumanize or actually enhance the creative process? Can traditional and computerized methods coexist?

## PRINTERS AND LOCAL AREA NETWORKS

A computer system may have components other than those described thus far. These include the printer and a network that links two or more computers in a communications system.

### Printers

A printer produces a paper or hardcopy of information. Three types of printers have dominated the general business and consumer markets, and there's a fourth category of important, yet less widely adopted, machines.

The first major category, the laser printer, can produce near-typeset quality documents and graphics. A small laser is the heart of a printer's engine, the actual printing device. Other elements include a photoconductor drum, toner particles, which make up the image much like the toner particles of a copy machine, and the paper.

The laser printer helped trigger a publishing revolution. When combined with a PC and desktop publishing program, individuals and organizations had access to enhanced publishing tools. Even though the final copy did not equal a commercial publication, it was superior to the typical PC printer's output.

The second major category, the ink-jet printer, uses a reservoir of ink for printing. Its output can almost match a laser printer in certain areas, and ink-jet printers are cost effective and can support color.

The dot-matrix printer, the third category, was a favorite choice during the earlier years of personal computing. It produces alphanumerics and graphics through a matrix of closely spaced dots. They are still used, even though they have generally been supplanted by laser and ink-jet models.

The fourth category of printers is the plotter. A plotter uses a series of pens to create large-scale architectural and technical prints, as well as other line drawings. The output can range from plans for a sailboat to a new piece of machinery.

Color printers have also become popular. Most software programs can handle color output, and cost-effective inkjet and laser printers have been marketed to meet a growing demand. Printers are further discussed in the desktop publishing chapter, Chapter 10.

### Local Area Networks

In brief, a local area network (LAN) is an information and communications system that can tie together a group of offices or other defined physical area (e.g., a building or group of buildings). It can link PCs so they can be used to share equipment and exchange information. Printers, data storage drives, and other devices are included on the network, and program and data files can be tapped by the network's users.

A LAN's design and implementation is analogous, in certain respects, to a multiuser system, in which a central computer is typically connected to terminals. A terminal simply serves as a keyboard and a monitor, an interface device in this environment, since the central computer performs the processing tasks.[30]

A LAN, like a multiuser configuration, makes it possible to share system resources. The LAN has an advantage, though. In the typical multiuser environment, the whole system may come to a crashing halt

if the central computer "goes down." In a LAN, each PC has its own processing capability and may be able to operate independently if the network is rendered inoperable by equipment failure. This advantage, when weighed with cost-effective PCs/network components and other factors, have made the LAN the networking tool of choice. A wide area network (WAN) extends these capabilities over greater geographical distances and can tie LANs together through different communications channels.[31]

**Operation.** A LAN's topology, its physical layout, actually dictates how the system is tied together. The topology has usually been based on a star, ring, or a bus design, and the information flow typically takes place over twisted-pair, coaxial, or fiber-optic cable.

Twisted-pair cable has the lowest information capacity, while coaxial cable is a shielded and superior relay line.[32] Fiber-optic cable is a newer contender, and one of its major strengths is a large channel capacity. As described in a later chapter, wireless LANs have also been created.

As the data are relayed through a network to the various PCs, the nodes, the data must be routed to the correct destinations. In this respect, a LAN can be considered a superhighway overseen by a traffic cop. Because multiple PCs are interconnected, the data relays and requests must be organized to facilitate the information flow.

Finally, an important element in this configuration is the server. The server is a computer that helps manage and control this flow. It also stores program and data files.

In a typical operation, you connect to the network by logging on, which may include using a password. At this point, you can gain access to the server's programs. Depending on the setup, you use either the server or your PC to store the data files. When finished, you disconnect from the network by logging off.[33]

## CONCLUSION

Computer technology will exert an even greater influence on the communications field—and inevitably society—as more sophisticated computers are purchased by an expanding user base. Prices will continue to fall as technology continues to advance. This combination of factors will eventually contribute to the remaking of the computer world, especially at the PC level.

The refinement of technology has also progressed at a dizzying pace. The processing, graphics capabilities, and software sophistication of the new generation of computers represent a substantial design leap over earlier models. Eventually, the characteristics that separate the different computer classifications may blur—it may be difficult to distinguish one computer family from another. At this time, the PC matches the capabilities of computers, including the workstations pioneered by Apollo Computers and Sun Microsystems, that were once the domain of engineering and design firms.

However, as our computing capabilities are enhanced, so too are elements that could adversely affect these capabilities. Two examples are *viruses* and the overall *industrial infrastructure*. For the former, the number of computer viruses has been rising. A computer virus is a program that attaches itself, in a sense, to other programs. A virus may simply display a harmless message or, worse, cause valuable data to be lost.

Viruses can be spread by various means, including infected disks inadvertently shipped by commercial manufacturers. A

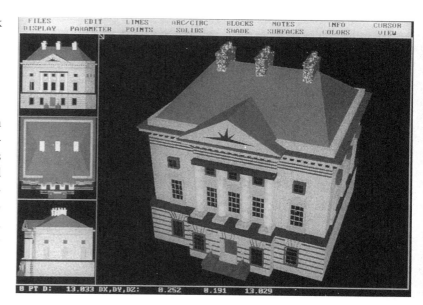

virus may also remain dormant until triggered. In 1992, the well-publicized Michelangelo virus was triggered on Michelangelo's birth date. Computers have internal clocks that keep track of the time and date. When the right date rolled around, if your system was infected, the virus could have become active.[34]

For the latter, certain parts of the United States, particularly California, have suffered from power outages. Besides inconveniencing people, they have a direct impact on a region's capability to support a growing computer-based society with its voracious energy consumption appetite. In response, some companies established their own power systems, among other measures.

Other potential risks range from the international disruption of computers systems, as exemplified by the Y2K scenario, to the numerous legal issues and entanglements born of computer systems and their applications. These subjects are explored in later chapters.

**Figure 3.6**
*An example of a house created with a 3-D CAD program. Please note the support for multiple views. (Courtesy of American Small Business Computers; DesignCad 3D.)*

FOUNDATION

## REFERENCES/NOTES

1. As defined by Anthony Ralston and Edwin D. Reilly, Jr., eds., *Encyclopedia of Computer Science and Engineering* (New York: Van Nostrand Reinhold Company, 1983), 969, a microprocessor is a "computer central processing unit (CPU) built as a single tiny semiconductor chip. . . . It contains the arithmetic and control logic circuitry necessary to perform the operations of a computer program." Microprocessors have been built into audio and video equipment for different functions.

2. This is due to their popularity. Other computer systems are also discussed when appropriate. These include mainframe and minicomputers.

3. Alan Freedman, *The Computer Glossary* (New York: AMACOM, 1991), 301.

4. There are also different types of RAM with different characteristics (e.g., speed).

5. Another system, which employs optical technology, is described in a later chapter.

6. Special cards are also used for specialized applications (e.g., video editing).

7. "Glossary," *Publish* 6 (October 1991): 124.

8. The footprint refers to the space a monitor may occupy on, for example, a desk; a 17-inch conventional monitor may only have a useable 16-inch (or less) viewing area in contrast to a 17-inch LCD monitor. Design/manufacturing factors play into this situation.

9. These include a spreadsheet program as described in a later section.

10. For example, an ad in a December 1982 issue of *Byte* listed an Atari 800 computer, equipped with 16K of RAM, for $689.95, excluding its $469.95 disk drive or other major components. In the early 2000s, even local computer stores were selling IBM PC compatibles, with 512 MB of memory, hard and CD-ROM drives, and a color monitor, for well under $1000.

11. This includes VR as described later in Chapter 14.

12. A keyboard—the way it feels and works—is also a personal matter.

13. Laurie Flynn, "Is Pen Computing for Real?" *InfoWorld* 13 (November 11, 1991), 75. *Note:* Think about your own handwriting. You might even have trouble, at a later date, recognizing what you wrote on a note.

14. Tablet PCs are suitable for various applications, including those in the medical field. Please see "Tablet PCs Could Boost Tablet Use in Health Care," by Joseph Goedert, for more information. Downloaded from www.healthdatamanagement.com/html/current/CurrentIssueStory.cfm?PostID= 13782.

15. Cell phones and wireless networking are covered in Chapter 8.

16. MS-DOS is Microsoft's DOS product for IBM PCs. It should also be noted that a PC may be able to support multitasking and other advanced functions. Multitasking enables a computer to run more than one program simultaneously. Other operating systems have included UNIX.

17. Frank Hayes and Nick Baran, "A Guide to GUIs," *Byte* 14 (July 1989), 250.

18. Other GUIs have included IBM's Presentation Manager.

19. Pascal is another language that many earlier PC programmers initially learned. While still used, it is not as popular as other languages (as of this writing).

20. Betty A. Toole, "Ada, Enchantress of Numbers," *Defense Science and Electronics* (Spring 1991), 32. *Note:* This issue has a number of articles about the Ada programming language and can be used to look at its roots.

21. There may, however, be some limitations.

22. Gene Apperson and Rick Doherty, "Displaying Images," in *CD-ROM Optical Publishing*, Vol. 2 (Redmond, WA: Microsoft Press, 1986), 134.

23. Corel Systems Corporation, Technical Reference (Ottawa, Ont: Corel Systems Corporation, 1990), 18.

24. At times, it may be advantageous to convert a bitmapped image into a vector-based one. There is also some cross-compatibility between software in that a drawing program

may, for instance, support bitmapped graphics.

25. The type may include spot and flood lights.

26. Impulse, Inc., Turbo Silver 3.0 User Manual, 1988, 17.

27. The transitions between slides can range from dissolves to wipes.

28. Phil Neray, "Visualizing the World and Beyond," *Photonics Spectra* 25 (March 1991), 93.

29. Jim Martin, "Supercomputing: Visualization and the Integration of Graphics," *Defense Science* (August 1989), 32.

30. Multiuser systems are run by a powerful PC, or more typically, a mainframe or a mini-computer. A mainframe can be categorized as a physically large, powerful, and expensive computer that can accommodate numerous users through a multiuser environment. A main-frame is equipped with enhanced processing and memory systems physically built into a central console, the "main" frame. A minicomputer can be considered a scaled-down version of a mainframe. It can also support a multiuser environment, as can suitably equipped PCs, albeit to a lesser degree than a mainframe.

31. Please see the *High Tech Dictionary* for more information (//www.computeruser.com/resources/dictionary). This is an excellent resource for identifying computer terms.

32. Because it is shielded, it is less susceptible to outside interference.

33. *Log on* and *log off* are two commands.

34. "Information Sheet on the Michelangelo Virus," compiled by J. M. Allen Creations/Michael A. Hotz, February 17, 1992.

## SUGGESTED READINGS

*Byte* on CD-ROM. Compilations of *Byte* articles on CD-ROMs.

Garcia, Emmanuel. "Desktop Visualization Tools for Designers." *CADENCE* (October 2000), 18–29; Mark Hodges. "Visualization Spoken Here." *Computer Graphics World* 21(April 1998), 55–62; Diana Phillips Mahoney. "Launching a Construction Simulation." *Computer Graphics World* 21 (August 1998), 60–62; NASA. "New Technologies Drive Changes in Auto Design and Production." *NASA Tech Briefs* (April 2001), 18–24; Richard Spohrer. "Making CAD Models Shine." *Computer Graphics World* 21 (January 1998), 51–56; J. R. Wilson. "Virtual Proto-typing Is Revolutionizing Aircraft System Design." *Military & Aerospace Electronics* 11 (February 2000), 1, 26. Different computer graphic techniques, including CAD, and their applications. The NASA article also covers complementary technologies and their role in the automotive industry.

Gilster, Ron, and Diane McMichael Gilster. *Build Your Own Home Network*. NY: McGraw-Hill, 2000; John Kincaid and Patrick McGowan. "When Faced with the Office Wiring Decision, 'Let the Buyer Beware.' " *Communication News* (February 1991), 59–61; Jim Quraishi. "The Technology of Connectivity." *Computer Shopper* 11 (May 1991), 187–198; Winn L. Rosch. "Net Gain." *Computer Shopper* 12 (April 1992), 534, 536–544. Overviews of LAN technology.

Glassner, Andrew S. "Ray Tracing for Realism." *Byte* 15 (December 1990), 263–271. A detailed examination of ray-tracing.

Hodges, Mark. "It Just Feels." *Computer Graphics World* 21 (October 1998), 48–56; Jeffrey R. Young. "Computer Devices Impart a Real Feel for the Work." *The Chronicle of Higher Education* XLV (February 26, 1999), A23–A24. Interesting looks at human–computer interfaces.

Mahoney, Diana Phillips. "The Picture of Uncertainty." *Computer Graphics World* 22 (November 1999), 44–50. Visualization and some applications.

Pournelle, Jerry. "User's column." *Byte*. A column in the now discontinued *Byte* magazine. Jerry Pournelle is a well-known science fiction writer and computer expert. His column covered a range of topics for the computer user. As Pournelle might have said, this column was and still is "highly recommended."

Rivlin, Robert. *The Algorithmic Image.* Redmond, WA: Microsoft Press, 1986. A richly illustrated history of computer graphics and applications.

Webster, Ed, and Ron Jones. "Computer-Aided Design in Facilities and System Integration." *SMPTE Journal* 98 (May 1989), 378–384. The use of CAD software in the television industry.

## GLOSSARY

*Disk Drive:* A data storage device.

*Electronic Mail (e-mail):* Electronic messages or mail that can be relayed all distances (for example, over a LAN or around the world).

*Graphical User Interface (GUI):* A visual, rather than text-based, interface.

*Graphics Programs:* The generic classification for computer graphics software. These range from paint to CAD software, and they excel at certain applications. Paint programs, for example, are used to create/manipulate bitmapped images. CAD programs are geared for architectural/technical drawings.

*Human-Computer Interface:* The tools we use to work with computers. They range from keyboards to touch screens.

*Local Area Network (LAN):* A dedicated data communications network. It can link computers, printers, and other peripherals for exchanging/sharing information, programs, and other resources.

*Monitor:* A computer's display component.

*Multiuser System:* A computer that can accommodate multiple users, typically through terminals. The computer provides central processing and data storage capabilities.

*Personal Computer (PC):* A computer typically designed to serve one user.

*Photorealistic:* A video display card/monitor that can produce realistic or lifelike images.

*Printer:* A device that produces a hardcopy of the computer's information.

*Programming Language:* A computer language used to create a computer program, the instructions that drive a computer to complete various tasks. Typical languages include C, Pascal, and Fortran.

*Random Access Memory (RAM):* A computer's working memory.

*Server:* A computer that manages and controls the flow of information through a network. It also stores program and data files.

# 4 Computer Technology: Legal Issues, Y2K, and Artificial Intelligence

The previous chapter introduced us to computer technology. This chapter focuses on related topics. These include legal issues, a potential millennium nightmare, and artificial intelligence. The latter has implications that cut across technical and philosophical fields.

## LEGAL ISSUES

### Software Piracy

A challenge facing the computer industry is software piracy, the illegal copying and distribution of software. It is an economic issue—potential revenues are lost or stolen—and a matter of intellectual property rights. A panel convened to examine this situation suggested that the U.S. government should increase its "antipiracy efforts" and "strengthen the enforcement of intellectual property rights abroad and in the United States."[1] The last sentence is a key one because pirating is not limited to other countries. Although there have been crackdowns, led by software industry organizations, piracy is still rampant in the United States.

Piracy presents a difficult and complex situation that points out a dilemma of the information age. The same tools that can create an information commodity can be used to steal it, in this case, software.[2] The situation is also exacerbated by the PC's ubiquitous presence at work and home.

Legislation can be enacted against piracy to protect intellectual property, but how do you enforce it? A few major "pirates" might be caught, but what of the thousands of individuals who may copy software either for sale or, more typically, for personal use?

In response, some companies have adopted software or hardware-based protection schemes. An example of the latter is bundling a *dongle* with a program. A dongle is a small hardware component that plugs

**Figure 4.1**
*A screen shot of an expert system. The window on the left shows the code. The window on the right shows one element of this expert system in action: a question and list of possible answers. (KnowledgePro screen shot by permission of Knowledge Garden, Inc.; KnowledgePro for Windows.)*

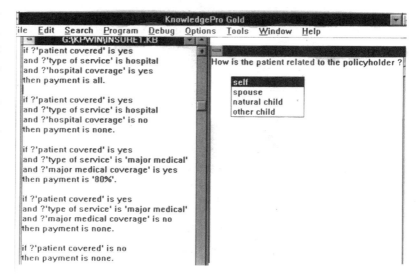

FOUNDATION

into a PC's USB or other data port. As designed, the software will not work without it. Other protection schemes also exist.

Legal issues have also been somewhat exacerbated by the language of licensing agreements, particularly when dealing with multiple computers in the workplace or a school. Depending on the situation, you sometimes require legal advice to determine what you can and cannot do. Even individual users may be faced with a dilemma. For example, if you own a desktop and a notebook PC, can you buy and use a single copy of a program on both computers? When upgrading an OS, can you use one copy for both PCs or do you have to buy two? Have you sat down and actually read an agreement before using a package?

Years ago, Borland International, a major software house, had adopted a clear cut licensing agreement for its products. It treated a program like a book. Multiple people could use it, but like a book, only by one person, on one PC, at a time.[3]

To sum up, software piracy is a major problem, especially in an information age, and manufacturers have taken steps to combat it. However, more can be done. As indicated by the panel mentioned earlier, the U.S. government can increase its efforts in this area.[4]

Education is one way. Intellectual property may not be viewed as a gold watch, money, or other *real property*. A common perception is that you should not steal money, but it is all right to copy a disk.

Part of this problem is *philosophical*. While the legal system may safeguard intellectual property, the philosophical element may lack focus. We have to recognize and accept the philosophical basis behind the idea of ownership before we follow the legal guidelines. Otherwise, the only preventive measures are legal, and if they cannot be fully enforced, the situation will continue.

Finally, note that other types of pirating

affect the communications industry. These include the illegal copying and distribution of movies and CDs and, as covered in Chapter 6, pirating of pay television services.

## L Is for Lawsuit

The computer industry has also been a fertile ground for lawsuits. The software arena has been especially hotly contested, and the focus has been on *patent* and *copyright* protections and violations.

In general, "a copyright provides long term conditional ownership rights in a specific expression of an underlying concept, without protecting the concept itself. A patent provides relatively short term . . . conditional ownership rights in an underlying concept (the patent's 'invention') without reference to the particular concept's embodiment."[5]

A highly publicized lawsuit was the dispute between Apple Computer and the Microsoft Corporation for copyright violations.[6] Apple contended that Microsoft's Windows borrowed heavily from Apple's GUI in regard to the interface's "look and feel."[7]

The issue came to a head in 1992. At that time, most of Apple's claims were thrown out by U.S. District Judge Vaughn Walker in a series of rulings. This included Apple's claim that Windows was "substantially similar to the look and feel of the Macintosh user interface."[8] If the ruling had gone the other way, Apple could have had a hammerlock on GUI rights and, possibly, future developments.

With respect to patents, a 1981 Supreme Court decision, *Diamond v. Diehr,* opened the software patent floodgate. A patent can provide broader and wider protection than a copyright and, as such, is a valuable legal and financial commodity.

Since that time, companies have filed

more patent applications, the U.S. patent office has been overburdened, and the hardware end of the computer industry experienced its own share of lawsuits. The patent process is also laborious and expensive, which may shut out individuals and small businesses from this form of protection.

A pessimist might say this escalation of patent applications and potential litigation has a negative impact. If, for instance, the developers of the first electronic spreadsheet had pursued and been granted a patent, they could have blocked the development of rival software products.[9] This action could have hampered the industry's growth.

An optimist might contend that legal protection can promote an industry's growth. There's a financial incentive for developers to continue their work, license agreements can be made with other companies, and new products can be introduced, to the industry's benefit.[10] It is, after all, a balancing act between protecting intellectual property while ensuring that a dynamic industry continues to grow.

Finally, copyright issues cut across the entire information and communications arena. The topic will be discussed at greater length in a later chapter. Suggested Readings for this topic will also be included in that chapter.

## Y2K

*For want of a nail the shoe is lost,*
*For want of a shoe the horse is lost,*
*For want of a horse the rider is lost,*
*For want of the rider the battle is lost,*
*For want of the battle the war is lost,*
*For want of the war the nation is lost,*
*All for the want of a horseshoe nail.*
　　　　　　　　—George Herbert[11]

As the millennium approached, the global community was faced with a potential time bomb—the Y2K bug. Early computer programmers worked with hardware and storage space limitations. To conserve resources, programmers used a shorthand method of designating dates. Instead of writing 1981, or any other such date, only the last two digits—81—were used. But what would happen in the year 2000? A computer could read the double zero (00) as 1900, thus potentially disrupting computer-based operations. In an information society, the results could be catastrophic.

Concerns ranged from air traffic control failures to massive power outages to banks losing financial data. The problem was exacerbated by embedded technology—hardware components in elevators, telecommunications equipment, and other systems that also relied on dates for operation. From a systems perspective, other key factors included

1. An enormous volume of date-sensitive transactions help drive our information society. What would happen if they were disrupted?
2. Programmers worked to solve and forestall potential problems. This included searching through computer programs and replacing old Y2K-sensitive code with new code. Programmers in older languages (e.g., COBOL) also became in demand for this operation.
3. Estimates to fix the global problem ranged from 1 to 2 trillion dollars.[12]
4. Organizations had to identify and replace affected components (for embedded technology). Banks, power companies, and other organizations also ran tests to evaluate potential problems and implemented Y2K corrections.
5. It was believed that some countries were not Y2K ready.
6. Legal issues surfaced. The Securities and Exchange Commission, for one, issued a statement about public companies and

their obligation to disclose potential problems.[13] Due to Y2K, the new millennium promised to be a financial boon for lawyers.

7. The Y2K problem could be likened to a row of dominos. Even if your company was Y2K ready, what would happen if a key supplier was not? It is also like an EMP—companies and countries were threatened by electronic paralysis.

8. People responded to the possible disruptions of services in different ways. Some individuals ignored the matter and believed the problem would be solved. Still others took a middle-of-the-road approach. They believed there would only be minor disruptions. A third group, though, looked back to the Cold War era and began to stockpile food, water, and other essential supplies. They believed services would be cut off, and unprepared national and global economies would be shattered.

The reality of the situation? The year 2000 dawned with only minor disruptions. The smooth transition surprised a lot of people. As they watched New Year's parties televised from around the world, they partly expected to see the lights go out in different countries at the stroke of midnight. It did not happen.

This led to some charges that the potential problems were actually exaggerated. The media were also blamed for "hyping" the situation. It made good news. This was countered by others who believed the investment in time and money actually paid off. In some cases, it led to the modernization and enhancement of computer operations, thus paying long-term dividends for this investment.[14]

Regardless of the viewpoint, it is interesting to note that all this chaos was caused by the placement of two numbers—seemingly insignificant—but with monumental, cascading implications. Much like the nail in the poem that opened this section.

## ARTIFICIAL INTELLIGENCE

This section of the chapter examines artificial intelligence (AI), another computer area relevant to our discussion of information and communications technologies.

The AI field is dedicated, in part, to developing computer-based systems that seemingly duplicates the most important of human traits, the ability to think or reason. This section serves as a brief introduction to the field and related topics.[15] More detailed information can be found in the Suggested Readings section.

### Natural Language Processing

Natural language processing can simplify the communication between humans and computers. For example, you may not have to learn specific command sequences because the computer is able to "understand" you. For a database program, a query for information can be phrased in an ordinary sentence.[16]

### Speech Recognition

Speech recognition, a subset of natural language processing, allows a computer to recognize human speech or words. You verbally instruct the computer to perform an operation instead of inputting the instructions with a keyboard. The system will initially digitize your voice, and then it must recognize various words before the instructions are carried out.

To complete the circle, the computer could answer you. At the PC level, a speech synthesis card can be used with special software, enabling a blind PC user to, for example, type on a keyboard, and have the words subsequently vocalized.[17]

This capability can be linked with an optical character recognition (OCR) system, as examined in the desktop publishing chapter, Chapter 10. An OCR configuration can recognize text from printed documents. The information can subsequently be reproduced in computer-generated speech, making the material available to visually impaired individuals.[18]

When this operation is viewed with the entire spectrum of speech research, it is an important achievement. Even at this developmental stage, our productivity is boosted, the human-computer interface becomes increasingly transparent, and working with computers turns into a natural process. These tools can also help individuals to communicate their thoughts and ideas in a manner that would have been impossible to achieve a few short years ago.

## Expert Systems

An expert system is a computer-based advisor. It is a computer program that can help in the medical, manufacturing, and other fields.

The heart of an expert system is knowledge derived from human experts and includes "rules of thumb," or the experts' real-world experiences. Additional information sources could encompass books and other documents.

This expertise can then be retrieved during a consultation. The computer leads you by posing different questions. Based on your answers, in conjunction with a series of internal rules, essentially the stored knowledge, the computer "reaches" a decision.[19] In one application, the Foundation Bergonie, a French research center, developed a system that would "help doctors in general hospitals in the management of breast cancer patients."[20] The expert system provided doctors with a level of advice that would not normally have been available.

Expert systems are valuable information tools having other applications and can

- help fill an information gap,
- support a field where there may be too few human experts, and
- preserve, in a sense, the knowledge and experience that would otherwise be lost when a human expert dies.

Yet despite their advantages, expert systems are not infallible. Their capabilities are limited by the quality of the information, the rules that govern the system, and by other criteria.[21] There's also the problem of working with a machine. A human expert is typically better equipped to ask the right questions to define a client's situation. Therefore, the human expert may arrive at a superior solution.

Another potential trouble area has been litigation. If a medical expert system made a mistake, who would be liable? This question, the high cost for liability insurance, and other factors, have held up product releases.[22] Other fields could be similarly affected, and in an article devoted to this topic, two pages outlined potential risk areas for expert system developers and designers.[23]

Finally, expert systems have been joined by a related field, *neural networks*. In brief, "neural network technology, which attempts to simulate electronically the way the brain processes information through its network of interconnecting neurons, is used to solve tasks that have stymied traditional computing approaches."[24] A network can, for example, be trained: It learns by example. A neural network can also work with incomplete data, much like humans, but unlike traditional computer systems.[25]

Neural networks have been trained to adjust a telescope to improve its performance, to handicap horse races, to make stock market predictions, and to recognize a face, even if the expression changes.[26] In

FOUNDATION

**Figure 4.2**
*Demeter. Demonstrates the "automation of terrestrial agricultural operations through the application of space robotics technology." (Courtesy of NASA/Space Telerobotics Program.)*

keeping with the communications industry, research has also been conducted to tap a neural network as part of a sophisticated tool for measuring television audiences' viewing habits.

### Computer Vision

Computer vision can be described as the field in which a picture, produced by a camera, is digitized and analyzed by a computer. In this setup, the camera and computer serve as machine equivalents of the human eye and brain, respectively.

During this operation, objects in a scene can be identified through template matching and other procedures.[27] In template matching, stored representations are compared to the objects in the image. The computer identifies the various objects when the templates match.

In one application, circuit boards can be examined for defects. If such a board is iden-

tified, it could be removed from the assembly line.[28]

Various technologies have also been harnessed to develop autonomous robotic vehicles. An autonomous vehicle would also process visual information, and in operation, could function without direct human intervention.

An example of an ambitious project is an autonomous rover for space exploration. Equipped with a robotic vision system, the vehicle would be activated on its arrival to explore Mars or some other planetary body's surface. The vision system would provide the onboard computer with information about the surrounding terrain. Based on an analysis and interpretation of the information, the computer could identify and avoid a potentially dangerous obstacle (e.g., a crater) without waiting for human intervention.[29]

This vehicle would be born from the union of hardware components and a sophisticated AI program that could process information. The integration of computer software and hardware may also serve as a model for vehicles built for use on Earth. These include experimental vehicles that already exist and those that may be placed in production.[30]

## Philosophical Implications

*I.  A robot may not injure a human being, or through inaction, allow a human being to come to harm.*

*II.  A robot must obey the orders given it by human beings, except where such orders would conflict with the first law.*

*III.  A robot must protect its own existence, as long as such protection does not conflict with the first or second law.*

—Isaac Asimov, *The Robots of Dawn*[31]

The development and integration of AI systems in society has philosophical implications. Some AI opponents believe, for example, that the new generation of intelligent machines will eliminate millions of jobs. There is also the fear that AI systems, including robots, diminish us as humans. Other AI-based systems have also raised the specter of machines wreaking havoc, as presented in *The Terminator, Creation of the Humanoids, Colossus: The Forbin Project*, and other movies. Although these concerns may be legitimate, AI proponents have offered counterarguments:

1. The tools of the AI field are just that, tools. An autonomous vehicle will not, for example, necessarily replace the human

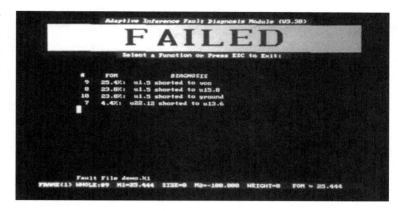

exploration of the planets, just as expert systems did not replace doctors.

2. Instead of diminishing our humanity, AI systems may enhance our lives. Cancer patients have benefited, and speech recognition and synthesis systems can provide individuals with better control over their environments.

3. The last concern was partly addressed by Isaac Asimov in his "Three Laws of Robotics" quoted earlier. The laws cut across scientific and science fiction boundaries and could help guide us as we start implementing advanced AI products.

To wrap up this discussion, it's also important to remember that while the arguments presented by both camps may have some merit, we are ultimately responsible for AI systems and their impact in our lives. If used correctly, AI and other technologies may continue to enhance the human exploration of our own world, of other worlds, and, just as important, the exploration of the human condition.

**Figure 4.3**

*A screen shot from the ExperTest(R) general purpose tester. The ExperTest has been used as a test development bed for PC motherboards, among other applications. By tapping artificial intelligence technology, the system can deliver an enhanced diagnostic capability, much like a human expert. (Courtesy of Array Analysis.)*

## REFERENCES/NOTES

1. Gary M. Hoffman, *Curbing International Piracy of Intellectual Property* (Washington, DC: The Annenberg Washington Program, 1989), 7. *Note:* The panel was not solely concerned with software issues.

2. Ibid., 10.

3. Borland International, *Turbo Prolog Reference Guide* (Scotts Valley, CA: Borland International, 1988), C2.

4. See Hoffman, pp. 19–24 of *Curbing International Piracy of Intellectual Property,* for a detailed discussion of the panel's recommendations.

5. Steve Gibson, "U.S. Patent Office's Softening Opens Floodgates for Lawsuits," *InfoWorld* 14 (August 17, 1992), 36.

6. Thanks to A. Jason Mirabito, a patent attorney, for his suggestions for this section.

7. Beth Freedman, "Look-and-Feel Lawsuit Expected to Go to Trial," *PC Week* 9 (February 24, 1992), 168. *Note:* The case also involved the Hewlett-Packard Company and its NewWave product.

8. Jane Morrissey, "Ruling Dashes Apple's Interface Hopes," *PC Week* 9 (August 17, 1992), 117.

9. Brian Kahin, "Software Patents: Franchising the Information Infrastructure," *Change* 21 (May/June 1989), 24. This is a sidebar in Steven W. Gilbert and Peter Lyman, "Intellectual Property in the Information Age," *Change* 21 (May/June 1989), 23–28.

10. "Patents: Protecting Intellectual Property," *OE Reports* 95 (November 1991), 1. *Note:* This interview with a patent lawyer provides a good overview of patent law and what can and cannot be patented.

11. Taken from *For Want of a Nail* by Robert Sobel, (London: Greenhill Books, 1997), foreword.

12. Barnaby J. Feder and Andrew Pollack, "Computers and 2000: Race for Security," *New York Times* CXlVIII (December 27, 1998), 22.

13. Jeff Jinnett, "Legal Issues Concerning the Year 2000 Computer Problem: An Awareness Article of the Private Sector," (Internet; downloaded March 16, 1999).

14. Mel Duvall, "Y2K Payoff: Systems Poised for New Projects." *Inter@active Week* 7 (January 10, 2000), 8.

15. For an introduction to AI's important elements, see Barry A. McConnell and Nancy J. McConnell, "A Starter's Guide to Artificial Intelligence," *Collegiate Microcomputer* 6 (August 1988), 243; and John Gilmore, "Artificial Intelligence in the Modern World," *OE Reports* (May 1987): 4A.

16. The Q&A program used such an interface. It also had the capability to add words to its vocabulary.

17. A program may even be compatible with a GUI. For example, icons can be described.

Please see Richard S. Schwerdtfeger, "Making the GUI Talk," *Byte* 16 (December 1991), 118, for more information.

18. Joseph J. Lazzaro, "Opening Doors for the Disabled," *Byte* 15 (August 1990), 258. *Note:* The article also provides an excellent overview of PC-based systems for the blind, deaf, and motor disabled.

19. The representation of knowledge by rules is only one expert system development tool. An expert system can also be created with a conventional programming language. A rule can take the form of:

IF the car doesn't start AND
the lights do not turn on
THEN the battery needs charging.

This example is very simplistic, and in a real-world situation, multiple rules would be employed.

20. Texas Instruments, "French Expert System Aids in Cancer Treatment," Personal Consultant Series Applications, product information release.

21. This includes novel situations where a human expert, in contrast, may be able to adapt to the new condition.

22. Edward Warner, "Expert Systems and the Law," *High Technology Business* (October 1988), 32.

23. G. Steven Tuthill, "Legal Liabilities and Expert Systems," *AI Expert* 6 (March 1991), 46–47.

24. Michael G. Buffa, "Neural Network Technology Comes to Imaging," *Advanced Imaging* 3 (November 1988), 47.

25. Gary Entsminger, "Neural-Networking Creativity," *AI Expert* 6 (May 1991), 19.

26. Maureen Caudill, *Neural Networks Primer* (San Francisco: Miller Freeman Publications, 1989), 4. See also Andrew Stevenson, "Bookshelf," *PC AI* 7 (March/April 1993), 30, 38, 57, 58, for an in-depth review of *In Our Own Image*, by Maureen Caudill (Oxford: Oxford University Press, 1992), a book relevant to this overall discussion.

27. Louis E. Frenzel, Jr., *Crash Course in Artificial Intelligence and Expert Systems* (Indianapolis, IN: Howard W. Sams & Co., 1987), 201.

28. Ibid., 205.

29. A neural network could be very appropriate in this type of situation, especially since it could be trained to avoid numerous obstacles and deal with novel situations.

30. As described in "Machine Vision," by J.R. Wilson, *Military & Aerospace Electronics* 12 (August 2001), 14–18, however, one problem is determining what the machine might be seeing.

As Wilson states as an example, "AI for years has been trying to look at what is a cup—does it have a handle . . . What makes a cup? It holds liquid, but that is hard to tell by just looking at it. That's a problem with AI that hasn't been solved yet." p. 16.

31. Isaac Asimov, *The Robots of Dawn* (New York: Doubleday and Company, Inc., 1983), back cover.

## SUGGESTED READINGS

Benfer, Robert A., and Louanna Furbee. "Knowledge Acquisition in the Peruvian Andes." *AI Expert* 6 (November 1991), 22–27; Jonathan K. Gable. "An Overview of the Legal Liabilities Facing Manufacturers of Medical Information Systems." *Quinnipiac Health Law Journal* 127, downloaded from LEXIS; Fiona Harvey. "A Key Role in Detecting Fraud Patterns: Neural Networks." *The Financial Times Limited,* January 23, 2002, downloaded from Nexis; Colin R. Johnson. "Neural-Network Expert-System Technologies Tapped to Build Smart-Vision Engine— AI Techniques Automate PC-Board Inspection." *Electronic Engineering Times,* downloaded from Nexis; Jessica Keyes. "AI in the Big Six." *AI Expert* 5 (May 1990), 37–42; Dan Shafer. "Ask the Expert." *PC AI* 3 (November/December 1989), 40, 49; J.C. Smith. "The Charles Green Lecture: Machine Intelligence and Legal Reasoning." *Chicago-Kent Law Review* 73 (1998), downloaded from LEXIS. Early to more recent looks at expert systems and neural networks.

Blackwell, Mike, and Susan Verrecchia. "Mobile Robot." *Advanced Imaging* 2 (November 1987), A18–A21; Maureen Caudill. "Driving Solo." *AI Expert* 6 (September 1991), 26–30. A look at autonomous vehicles.

Cleveland, Harlan. "How Can 'Intellectual Property' Be Protected." *Change* 21 (May/June 1989), 10–11; Francis Drummer Fisher. "The Electronic Lumberyard and Builders'

Rights." *Change* 21 (May/June 1989), 13–21. Earlier looks about intellectual property.

Curran, Lawrence J. "Vision-Guided Robots Assemble Wheels Parts." *Vision Systems Design* 7 (September 2002), 17–21. A vision system application.

Frenzel, Louis E., Jr. *Crash Course in Artificial Intelligence and Expert Systems.* Indianapolis, IN: Howard W. Sams & Co., 1987. A guide to AI technology and applications.

Godwin, Mike. *Cyber Rights.* New York: Times Books, 1998. Comprehensive coverage of free speech issues, including intellectual property issues in an electronic age. Also see Kenneth C. Creech, *Electronic Media Law and Regulation.* Boston: Focal Press, 2000, for additional information about copyright/communications issues.

The January 10, 2000, issue of *Inter@active Week* carried a series of Y2K articles and various implications. Two include: Doug Brown, "Y2K Overseas: What Went Right," 9 and Charles Babcock, "Spending on Y2K: Waste of Time or Prudent Investment," 8–9.

Kavoussi, Dr. Louis R. "NY Doctors Use Robotics in Baltimore Surgery." *Communications Industries Report* 15 (October 1998), 23. An example of using robotics in telemedicine; i.e., from a distance.

Markowitz, Judith. "Talking to Machines." *Byte* 20 (December 1995), 97–104; Jeffrey Rowe. "Look Who's Talking." *Computer Graphics World.* 24 (February 2001), 42–45. Speech recognition technology.

NASA. "Remote Agent Experiment," downloaded from NASA web site; John Rhea. "Report from Washington and Elsewhere." *Military & Aerospace Electronics* 9 (October 1998), 8, 19. AI and space applications.

United States Government. The Library of Congress (www.loc.gov/copyright) and the U.S. Patent and Trademark Office (www.uspto.gov) are valuable information resources about copyrights and patents and how to file for each.

## GLOSSARY

*Artificial Intelligence (AI):* The field dedicated, in part, to developing machines that can seemingly think.

*Computer Vision:* The field that duplicates human vision through a computer and a video camera system.

*Copyright and Patent:* A copyright provides protection in the expression of a concept. A patent protects the underlying concept.

*Expert Systems and Neural Networks:* An expert system manipulates knowledge and can serve as an in-house expert in a specific field. It combines "book information" with the knowledge and experience of human experts. A neural network, for its part, simulates the way the brain processes information through its network of interconnecting neurons.

*Natural Language Processing:* Natural language processing focuses on simplifying the human-computer interface. Rather than using special keywords to initiate computer functions, you use conventional word sequences.

*Optical Character Recognition (OCR):* After a document is scanned, an OCR system can recognize the text.

*Software Piracy:* The illegal copying and distribution of software.

*Speech Recognition:* A subset of natural language processing, it allows a computer to recognize human speech or words.

*Y2K:* A potential, adverse computer problem caused by using shorthand notation to represent a year (e.g., 98 for 1998) in computer programming. Massive financial and personnel time investments helped rectify the situation.

# II

## INFORMATION TRANSMISSION

# 5 The Magic Light: Fiber-Optic Systems

The concept of harnessing light as a modern communications tool stretches back to the late nineteenth century. Alexander Graham Bell, the inventor of the telephone, patented an invention in 1880 that used light to transmit sound.[1] Bell's invention, the photophone, employed sunlight and a special light-sensitive device in the receiver to relay and subsequently reproduce a human voice.

It wasn't until the twentieth century, though, that the idea for such a tool could be transformed into a practical communications system. Two developments helped bring about this advance: the perfection of the laser and the manufacturing of hair-thin glass lines called *fiber-optic* (FO) lines. When combined, they helped create a lightwave communications system, a system in which modulated beams of light are used to carry or transmit information.

## OVERVIEW

### Light-Emitting Diodes and Laser Diodes

FO technology has contributed to the development of high-capacity communications lines. A transmission is conducted with optical energy, beams of light, produced by a transmitter equipped with either a light-emitting diode (LED) or a laser diode (LD). The light is then confined to and carried by a highly pure glass fiber.

Both LEDs and LDs are used in different communications configurations. An LED is less expensive and has generally supported lower volume, short-distance relays. An LD, like an LED, is a semiconductor, but in the form of a laser on a chip. It is a small, powerful, and rugged semiconductor laser that is well suited for high volume and medium- to long-distance relays. The LD family has also served optical storage systems.

### The Fiber-Optic Transmission

In an FO transmission, a beam of light, an optical signal, serves as the information-carrying vehicle. Both analog and digital information are supported.

In operation, the light is launched or fed into the fiber. The fiber itself is composed of two layers, the cladding and the core. Due to their different physical properties, light can travel down the fiber by a process called total internal reflection. In essence, the light travels through the fiber via a series of reflections that take place where the cladding and core meet, the cladding-core interface. When the light reaches the end of the line, it is picked up by a light-sensitive receiver, and after a series of steps, the original signal is reproduced.

To sum up, a video camera's output or other such signal is converted into an optical signal in an FO system. It is subsequently transmitted down the line and converted back following its reception.

INFORMATION TRANSMISSION

**Figure 5.1**
*How a fiber-optic system works. The electrical signal is converted into an optical signal, and back, following the relay. (Courtesy of Corning Inc.)*

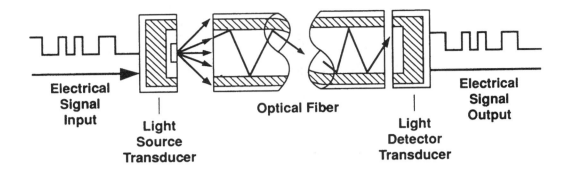

Electrical Signal Input    Light Source Transducer    Optical Fiber    Light Detector Transducer    Electrical Signal Output

Finally, the fiber, which may be made out of plastic in short-distance runs, is covered by a protective layer or jacket. This layer insulates the fiber from sharp objects and other hazards, and it can range from a light protective coating to an armored surface designed for military applications. The now-protected fiber is called an FO cable, and it consists of one or more fiber strands within the single cable enclosure.

***Advantages.*** An FO line has distinct advantages as a communications channel:

1. In comparison with some other communications systems, light—the information-carrying vehicle in an FO system—can accommodate an enormous volume of information. Transmissions are in the gigabit-plus range (billions of bits per second), and a single 0.75-inch fiber cable can replace 20 conventional 3.5-inch coaxial cables.[2]
2. Fiber-optic lines are immune to electromagnetic and radio interference. Because light is used to convey the information, adjacent communications lines cannot adversely affect the transmission. An FO line can also be installed in a potentially explosive environment where gas fumes may build up over time.
3. An FO line offers a higher degree of data security than conventional systems. It is more difficult to tap, and FO lines do not

radiate. A signal cannot be picked up by instruments unless the line is physically compromised.[3]
4. Information can be relayed a great distance without repeaters. A new generation of LDs and complementary fiber, as well as sensitive receivers, have made it possible to relay signals great distances without repeaters. In practical terms, if fewer repeaters are used, building and maintenance costs are reduced.
5. An FO line is a valuable asset where space is at a premium, such as a building's duct space for carrying cable. Because fiber is comparatively narrow, it can usually fit in a space that may preclude the use of conventional cable.

***Disadvantages.*** Some factors do, however, limit an FO line's effectiveness. Like other communications systems, there may be a loss of signal strength, which in this case, may be caused by physical and material properties and impurities.[4] In another example, *dispersion* can affect the volume of information a line can accommodate in a given time period, that is, its channel capacity.

The latter can be addressed by using single-mode rather than multimode fiber. The single-mode fiber is constructed with a very narrow core, and the light essentially travels straight down the fiber in a single path, thus preventing the smearing of pulses that comprise the optical signal.[5] A single-mode fiber can also accommodate a high information rate and has been the backbone of the telephone industry and other high-capacity long-distance systems.

Fiber is also harder to splice than conventional lines, and the fiber ends must be accurately mated to ensure a clean transmission. But special connectors, among other devices and techniques, have facilitated this process.

Another element is the current state of the communications industry. Despite the inroads made by FO technology, the overall system is still dominated, in certain settings, by a copper standard. This includes the local telephone industry.

FO components can also be more expensive. An FO cable hookup, for example, may be more expensive than a conventional one. The actual FO line may likewise cost more in specific applications, even though this price differential has diminished.

Thus, FO technology is somewhat analogous to digital technology. It has been integrated in the current communications structure.

The prices for LDs and other components are also dropping, and the growing information torrent our communications systems must handle may mandate the use of such high-capacity channels. They may be joined by fiber/coaxial hybrid systems. Note also that by using compression techniques, the traditional copper plant is still very much alive and well.

## APPLICATIONS

An FO line is an attractive medium for the video production industry. Its lightweight design and transmission characteristics can make it a valuable in-field production tool. Several hundred feet of line, for instance, may weigh only several pounds, and a camera/FO combination can be used at a greater distance from a remote van than a conventional configuration.[6]

Fiber's information capacity also makes it an ideal candidate for an all-digital television studio. The digitization of broadcast-quality signals can generate millions of bits per second, and an FO system could handle this information volume. The line would also fit in a studio's duct space, which is usually crammed with other cables, and its freedom from electromagnetic and radio frequency interference would be valuable assets in this environment.

Besides these roles, FO systems have been used to create efficient and high-capacity local area networks (LANs). Businesses, hospitals, schools, and other organizations are also designing and implementing their own FO lines that can accommodate computer data as well as voice and video.

The telephone and cable industries also have a vital interest in this technology. Long-distance carriers have developed extensive FO networks, and eventually, fiber may be extended to all our homes. Cable companies, for their part, are converting their systems to fiber or, as covered in Chapter 15, fiber/copper hybrid systems.

Fiber can also provide us with high-speed data highways. As we generate not only more information, but more complex information with each passing year, the communications infrastructure may be hard pressed to handle this data flow. But FO technology will continue to play a key role in solving this problem. It can also support an array of new television and information options.

### Underwater Lines
Beyond landline configurations, AT&T headed an international consortium that developed the first transatlantic undersea FO link between the United States and Europe. The system, TAT-8, is more than 3000 nautical miles in length and was the first trans-

oceanic FO cable. TAT-8 was designed to handle a mix of information. When inaugurated, it had an estimated lifetime in excess of 20 years.[7] Even though fiber had already been used in long-distance land and short-distance undersea operations, TAT-8 was the first of a new class of cables. Its installation was preceded by extensive deep-water experiments and trials conducted in the early 1980s to demonstrate the project's feasibility. Once completed, the findings confirmed the designers' expectations: The FO cable and repeater relayed the information within acceptable error rates and transmission losses. This held true even when the system was subjected to the ocean's cold temperature, tremendous water pressure, and other extreme environmental conditions.

Like other new applications, though, FO lines have had some problems. Certain TAT-8 users, for instance, experienced some outages from cable damage caused by fishing trawlers.[8] But despite some problems, the technology proved its value. In fact, a series of planned and existing FO systems may span the globe, ringing it with a band of light.

## Fiber-Optic Lines and Satellites

At first glance, undersea cables and long-distance landlines may appear to be obsolete in light of satellite communication. Yet an FO system has a wide channel capacity, an extended lifetime, and can be cost effective for long-distance applications.

An FO line also has certain other advantages. A satellite transmission may be susceptible to atmospheric conditions, and a slight time delay is inherent in satellite traffic.[9] An FO link is not subject to these constraints.

Fiber may also be a more secure conduit for sensitive material. A satellite's broad transmission area makes it possible for unauthorized individuals to receive its signal.

Although it could be encrypted for protection, the encryption scheme could be broken.

Fiber is also well suited for intracity relays and point-to-point communication (e.g., New York to Washington, DC). It has been used for television network news, by the telephone industry, and in teleconferencing applications. Various processing techniques have also enhanced an FO system's relay.

A satellite, for its part, can support a flexible point-to-multipoint operation that can readily accommodate additional receiving sites. This may not be the case with an FO system where a special line may have to be laid to reach the new location.

Thus, FO lines and satellites actually supplement and complement each other on national and international levels. They also have their own particular strengths, and both will continue to support our communications needs.

## Other Applications

Fiber-optic technology has also been adopted for other applications. In the medical field, an FO line can be used in certain types of laser surgery. The fiber can act as a vehicle to transport the intense beam of light to the operating site. In a related application, fibers serve as a visual inspection tool in the form of a fiberscope. This device "consists of two bundles of optical fibers. One, the illuminating bundle, carries light to the tissues, and the other, the imaging bundle, transmits the image to the observer."[10] Doctors have used such devices to peer inside the human body.

Scientists have also experimented with and have used FO sensors to monitor the physical condition of different structures. In one application, fibers have been embedded in composite materials, which can be used to create items ranging from airplane parts to, potentially, a space station. Through testing, it would be possible to determine if a structure was subject to damage and undue stress.[11]

Although the fiber may not be relaying computer data, it is still providing observers with information. As discussed in the book's opening, this is one of the key features of the communication revolution. Information will take new forms and shapes, and in this setting, the data about a structure's physical condition does, in fact, convey an intelligence.

All the applications described thus far only touch on FO technology's different roles. It has even been tapped by the aerospace industry to supplement and to potentially replace fly-by-wire systems with fly-by-light configurations. In this set-up, conventional, heavier wiring would be replaced by fiber for command-and-control operations. The fiber would save weight, and the freedom from electromagnetic interference would lend itself to this environment.[12] The automobile industry has also explored this option.

## CONCLUSION

Fiber-optic lines have had an impact on our communications system. On one front, they have already supported sophisticated national and international telephone connections. On another, they promise to change the way we view television and, potentially, the way we receive and use information.

The growth of FO lines can also be viewed as an evolutionary rather than a revolutionary development. For example, the infrastructure is modernized as FO lines and components are integrated, but the current system is not immediately abandoned. This follows the pattern of digital and other technologies and applications.

Current FO systems will also be complemented and supplemented by newer, higher

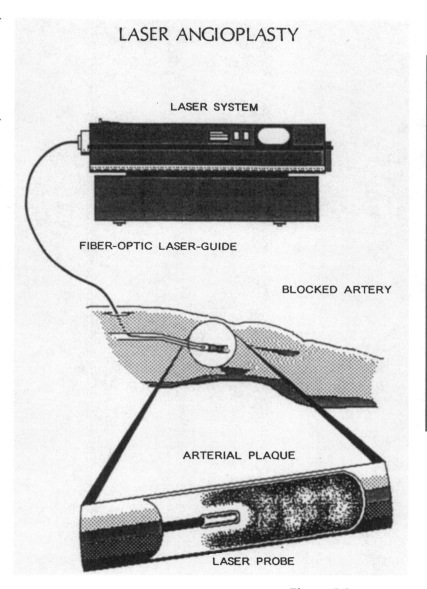

**LASER ANGIOPLASTY**

LASER SYSTEM

FIBER-OPTIC LASER-GUIDE

BLOCKED ARTERY

ARTERIAL PLAQUE

LASER PROBE

**Figure 5.2**
*A medical application of fiber-optic technology. (Courtesy of KMI Corp., Newport, RI.)*

INFORMATION TRANSMISSION

information capacity systems. Project Oxygen, for instance, called for the design of a transoceanic global system that would "have a maximum throughput of 2.56 terabits (2560 gigabits) per second on [its] undersea cable segments."[13] Besides this high capacity capability, the system would support a sophisticated network design that would provide users with enhanced connection options. On land, a pan-European project in 17 countries was designed to

**Figure 5.3**

*Using fiber-optic technology, surgeons can examine, in this case, a torn meniscus (knee) for repair. (Courtesy of Dr. J. Henzes.)*

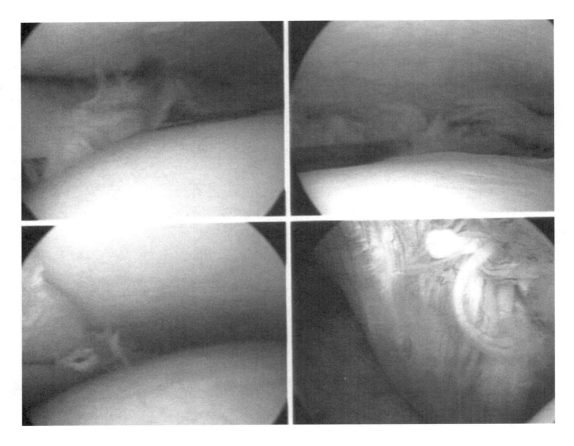

"reach more than 70 cities with 8 million km of . . . optical fiber—enough to circle the globe 200 times."[14]

Fiber-optic capabilities will also improve. One driving force is the marriage between optical and electronic components (opto-electronics) and the use of light alone. In one example, optical amplifiers can improve an FO system's performance. The signal could remain in the optical domain for longer distances, thus reducing the number of regenerative repeaters.[15]

A similar development is taking place in the switching field, where an optical configuration would be advantageous.[16] In one application, data traffic may be managed by microelectromechanical systems (MEMS). One such system could consist of "tiny moving mirrors [which] are used to deflect beams of light" to different fibers. Mounted on an integrated circuit, an individual mirror may be "smaller than the diameter of a human hair."[17] In essence, optical networking can lead to the creation of faster networks with increased capacity.[18]

In a related application, light may also be used to develop newer, faster computers. Analogous to an optical network, optical computing will make it possible to produce computers with an enhanced processing capability.

In sum, these developments are exciting—and we have only just begun to tap light's potential as a communications tool. Its role in our information society cannot and should not be underestimated.

INFORMATION TRANSMISSION

## REFERENCES/NOTES

1. Richard S. Shuford, "An Introduction to Fiber Optics," *Byte* 9 (December 1984), 121.

2. Arthur Parsons, "Why Light Pulses Are Replacing Electrical Pulses in Creating Higher-Speed Transmission Systems," *Communications News* (August 1987), 25.

3. "Optical Fiber Technology: Providing Solutions to Military Requirements," *Guidelines* (1991), 6.

4. The latter factor, though, has been greatly reduced.

5. Ronald Ohlhaber and David Watson, "Fiber Optic Technology and Applications," *Electronic Imaging* 2 (August 1983), 29.

6. Richard Cerny, "Fiber Optic Systems Improve Broadcast ENG/EFP Results," *Communications News* (April 1983), 60.

7. R.E. Wagner, S.M. Abbott, R.F. Gleason, et al., "Lightwave Undersea Cable System," reprint of a paper presented at the IEEE International Conference on Communications, June 13–17, 1982, Philadelphia, PA, 7D.6.5.

8. Barton Crockett, "Problems Plaguing Undersea Fiber Links Raise Concern Among Users," *Network World* 20 (August 21, 1989), 5.

9. As described in Chapter 6, a communications satellite is typically placed in an orbit some 22,000 miles above the earth. The round-trip to and from the satellite results in the slight time delay.

10. Abraham Katzir, "Optical Fibers in Medicine," *Scientific American* (May 1989), 120. *Note:* A fiberscope may also be incorporated in an endoscope, which provides physicians with remote access to the body regions under observation. The article presents an excellent overview of this field.

11. Richard Mack, "Fiber Sensors Provide Key for Monitoring Stresses in Composite Materials," *Laser Focus/Electro-Optics* 23 (May 1987), 122.

12. Luis Figueroa, C.S. Hong, Glen E. Miller, et al., "Photonics Technology for Aerospace Applications," *Photonics Spectra* 25 (July 1991), 117.

13. "Frequently Asked Questions About Project OXYGEN™," downloaded from Project Oxygen's web site.

14. Robert Pease, "Largest Pan-European Network Project to Light Later This Year," *Lightwave* (May 2000), 1.

15. Robert G. Winch, *Telecommunication Transmission Systems* (New York: McGraw-Hill, 1998), 456.

16. Paul R. Prucnal and Philippe A. Perrier, "Self-Routing Photonic Switching with Optically Processed Control," *Optical Engineering* 29 (March 1990), 181.

17. Robert Pease, "Microscopic Mirrors May Manage Future Optical Networks," *Lightwave* 16 (May 1999), 1.

18. Jeff Hecht, "An Introduction to Optical Networking," *Laser Focus World* 37 (January 2001), 115.

## SUGGESTED READINGS

Belanger, Alain. "Broadband Video Goes Mainstream." *Lightwave* 15 (November 1998), 49–52, 56; Richard Cerny. "Using Fiber in the Field." *Broadcast Engineering* 38 (January 1996), 52–60, 64. Coverage of fiber-based video applications; the first article also examines fiber systems in other countries; the second article also covers video/field production applications.

Branst, Lee. "Optical Switches Are Coming—But When." *Lightwave* 15 (July 1998), 63–64, 68; Jones-Bey Hassaun. "Ecological Metaphors Color Future MEMS Perspectives." *Laser Focus World* 37 (January 2001), 122–230; Jeff Hecht. "An Introduction to Optical Networking." *Laser Focus World* 37 (January 2001), 115–120; Jeff Hecht. "Optical Regeneration Will Be Key for 40 Gbit/s Success."

**Figure 6.1**

*Satellites play a major role in the communication revolution. This artist illustration depicts a satellite that delivers television programs directly to viewers, as described in this chapter. (Courtesy of Hughes Space and Communications Company.)*

fixed. Today's communications satellites have generally followed suit and are placed in *geostationary* orbital positions or slots. Simply stated, a satellite in a geostationary orbital position appears to be fixed over one portion of the Earth. At an altitude of 22,300 miles above the equator, a satellite travels at the same speed at which the Earth rotates, and its motion is synchronized with the Earth's rotation. Even though the satellite is moving at an enormous rate of speed, it is stationary in the sky relative to an observer on the Earth.

The primary value of a satellite in this orbit is its ability to maintain communica-

tion with ground stations in its coverage area. The orbital slot also simplifies this link: Once a station's antenna is aligned, it may only have to be repositioned to a significant degree when contact is established with a different satellite. Prior to this time, a ground station's antenna had to physically track a satellite as it moved across the sky.

Based on these principles, three satellites placed in equidistant positions around the Earth can create a worldwide communications system. This concept was the basis of Arthur Clarke's original vision of a globe-spanning communications network.

Finally, note that while geostationary slots are important, low and medium Earth orbits are still used. As will be discussed, these may range from remote sensing to specialized communications satellites.

***Uplinks and Downlinks.*** According to the FCC, an *uplink* is the "transmission power that carries a signal . . . from its Earth station source up to a satellite"; a *downlink . . .* "includes the satellite itself, the receiving Earth station, and the signal transmitted downward between the two.[1] To simplify our discussion, the uplink refers to the transmission from the Earth station to the satellite, and the downlink is the transmission from the satellite to the Earth station. This two-way information stream is made possible by special equipment. The station relays the signal via an antenna or dish and a transmitter that produces a high-frequency microwave signal.

The communications satellite, for its part, operates as a repeater in the sky. After a signal is received, the satellite relays a signal back to Earth. This is analogous to an Earth-based or terrestrial repeater, but now it is located more than 22,000 miles above the Earth. The satellite

1. receives a signal,
2. the signal is amplified,

3. the satellite changes the signal's frequency to avoid interference between the uplink and downlink, and

4. the signal is relayed to Earth where it is received by one or more Earth stations.

To create this link, the satellite uses transponders, equipment that conduct the two-way relays. A communications satellite carries multiple transponders.

As illustrated by the Intelsat family, the number of transponders per satellite class has increased over the years. For example, the original Intelsat satellite, Early Bird, was equipped with 2 transponders that supported either a single television channel or 240 voice (telephone) circuits. The later Intelsat IV satellites, launched in the early to mid-1970s, carried 12 transponders that generally accommodated 4000 voice circuits and two television channels.

The newer Intelsat VIII spacecraft are equipped with 44 transponders. Intelsat VIII can accommodate, on the average, "22,500 two-way telephone circuits and three television channels."[2] Using digital technology, a satellite could potentially handle over 100,000 simultaneous two-way telephone circuits.

Intelsat VIII is also a hybrid satellite. It supports both the C- and Ku-bands. As will be covered, C- and Ku-band satellites employ the C- and Ku-band communications frequencies, respectively. Consequently, while some satellites employ only one band, other satellites support both, providing for a more flexible communications platform.

This design concept was similarly adopted by later satellites, including Intelsat 906, another hybrid spacecraft, but one equipped enhanced C-band and Ku-band capabilities. Slated to serve Asia, Africa and other regions, the satellite supports the relaying of data streams, among other information types.[3]

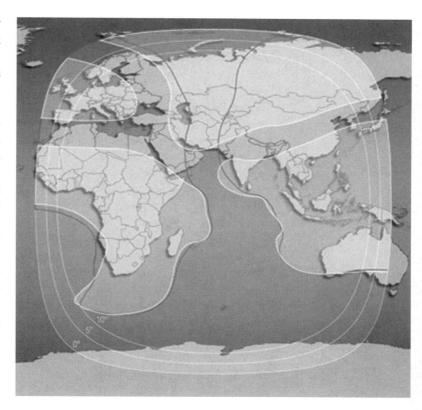

INFORMATION TRANSMISSION

**Figure 6.2**
*Example of a satellite transmission's coverage area (global). (Courtesy of Intelsat.)*

### Parts of a Satellite

*Satellite Antennas.*    An important factor that influences the satellite communications network is a satellite's antenna design. A satellite's transmission is focused and falls on a specific region of the Earth. This reception area, the footprint, can vary depending on the satellite and its projected applications. The footprints can range from global to spot beams, in descending order of coverage.[4]

A global beam provides the most coverage and is used for international relays. A spot beam falls on a narrowly defined geographical zone, making it particularly effective for major metropolitan areas. Because this signal is concentrated in a relatively small area, it is also stronger than one distributed to the entire country for cable television relays. A smaller and less expensive dish could subsequently be used.

***Satellite Spacing and Antennas.*** Geostationary communications satellites had traditionally been spaced four degrees or well over 1000 miles apart in the orbital arc to create buffer zones. The zones reduce the chance of cross-interference during transmissions. If eliminated, an uplink, for example, which was not tightly focused, could unintentionally spill over and affect another satellite.

In the 1980s, the FCC decided to situate communications satellites closer together to open up additional orbital slots. This action was complemented by specifications for Earth-based antennas that satisfied the technical demands imposed by this arrangement.[5]

The plan was devised to address the orbital scarcity problem. As stated, a geostationary slot is an advantageous orbital position. But the number of available slots is finite. The region of space that supports this type of orbit is confined to a narrow belt above the Earth.

A political consideration also plays a role in the allocation process. A nation cannot launch and place a satellite in any slot it chooses because slots are assigned on an international basis through the auspices of the International Telecommunications Union.

The FCC's program was implemented to maximize the United States' orbital allocations. It was also a reflection of the increased pressure to launch more satellites, which demanded the availability of additional orbital assignments.

It should also be noted that orbital slots have economic connotations, analogous to the use of the electromagnetic spectrum as described in Chapter 2. Companies may spend hundreds of millions of dollars to enhance orbital slots through the deployment of satellites. As covered in a later section, satellites support numerous revenue generating applications. Consequently, a satellite placed in a desirable slot, that is, one in which it can communicate with a specified geographical area for a given service (e.g., relay television programming), would be a valuable space-based economic asset.

***Power System.*** A satellite's electrical power is supplied through the conversion of sunlight into electricity by solar cells and ancillary equipment. Cylindrically shaped satellites (spin-stabilized) are covered with the cells, whereas a three-axis stabilized satellite uses wings or extended solar panels.[6] A communications satellite is also equipped with a battery system.[7]

Besides solar cells, satellites use another power source during their 12+ year approximate lifetime. A satellite is equipped with external thrusters and a fuel supply. The thrusters, when activated by ground station controllers, emit small jets of gas to help maintain the satellite's station, which is its position in its slot. Even though a satellite in a geostationary orbit appears to be fixed, it actually moves slightly or drifts. The thrusters correct this drifting.

Once the fuel is expended, however, the satellite may be lost to ground operators.[8] The satellite's drifting can no longer be corrected, and the satellite may become inoperable even though its other systems may still be functional. In view of this situation, a plan has been devised to extend a satellite's life: Let the satellite drift in a controlled fashion. Because the thrusters are now fired less frequently, the remaining fuel can be stretched.

The system's primary drawback is antenna modifications: The satellite must be tracked. But recent designs have made this process simpler and less expensive.[9] The upgrade would also be compensated by the satellite's longer life.

Finally, while communications satellites rely on solar energy, spacecraft designed for deep space and long-term missions have used radioisotope thermoelectric generators

(RTGs), that is, nuclear power. The RTGs have extended lifetimes and have provided power on missions where a spacecraft may be too far from the sun to tap its energy as a power source.[10]

### Transmission Methods

A satellite's information capacity is limited by different factors, as are other communications systems. For a satellite, these include the number of transponders and the power supplied to the transmission system. A typical 36-MHz transponder carried by a C-band satellite, for example, can accommodate a television channel and a number of voice/data channels or subcarriers.

Like terrestrial systems, satellite relays can be either analog or digital. Various transmission schemes have also been implemented, including multiple-access systems. Multiple ground stations can gain access to and "share" a satellite's transponder.[11]

Satellite traffic is also increasingly digital, and processing techniques have decreased the bandwidth requirement and lowered transmission costs. An important implication of this development, digital compression, is making satellite communication available to a broader user group. By employing compression, an organization can use a portion of a transponder for a video relay.[12] Although the receiving sites must be equipped to handle this information, the transmission costs are reduced and the satellite can carry additional channels.

Supported applications include those outside of the traditional cable and broadcast industries. In education, a school could start a distance learning network, because satellite time would be less expensive. Or a school could produce a series of satellite-distributed educational programs. If properly implemented, it could be an efficient and cost-effective way to share resources and to support students.

While it is a valuable tool in the satellite industry, compression is not limited to this field. It is also a powerful tool that has reshaped whole segments of the communications industry. These include the teleconferencing, multimedia, and video production markets, as will be explored throughout the book.[13]

***C-Band and Ku-Band.*** Until the early 1980s, transmissions in the commercial satellite communications field were primarily conducted in the 4/6 Gigahertz (GHz) range, the C-band. The 4/6 notation indicates the downlink and the uplink, respectively. Ground stations that receive their signals are generally equipped with 10-foot to about 15-foot-diameter dishes.

Besides C-band spacecraft, a newer generation of satellite that employs a higher frequency range, the K-band, has become operational. The first class of this type of spacecraft uses the Ku-band (12/14 GHz). This has an advantage. C-band satellite transmissions have been limited in power "to avoid interference with terrestrial microwave systems."[14] A Ku-band satellite is not similarly restricted, and the power of its downlink can be increased. This higher power also translates into smaller receiving dishes and points out a generalization between a satellite's transmission and a dish's size. As the power increases, the dish's size can decrease.

The Ku-band also offers a user more flexibility. A dish's smaller size and a Ku-band system's freedom from terrestrial operations simplifies finding a suitable dish site. C-band systems are not afforded this same luxury. The possible joint interference and a dish's size may make it harder to find a location.[15]

The Ka-band for satellites, which has more recently been adopted, is described in the next chapter in the context of a satellite that helped pioneer this new communications tool.

**Figure 6.3**
*NASA's ACTS Spacecraft showing, as originally envisioned, its various components. This advanced communications vehicle is described in the next chapter. (Courtesy of NASA; Lewis Research Center.)*

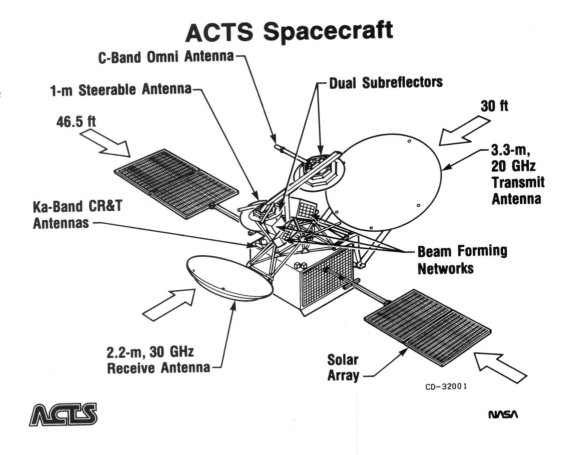

## ACTS Spacecraft

C-Band Omni Antenna

1-m Steerable Antenna

46.5 ft

Dual Subreflectors

30 ft

3.3-m, 20 GHz Transmit Antenna

Ka-Band CR&T Antennas

Beam Forming Networks

2.2-m, 30 GHz Receive Antenna

Solar Array

CD-32001

ACTS

NASA

***Very Small Aperture Terminals.*** Ku-band technology spurred the growth of a satellite system that employs very small aperture terminals (VSATs). A VSAT is a compact dish mated with the necessary electronic hardware to create a cost-effective communications system composed of a few or numerous sites.

VSAT technology has grown in popularity. The small dish can be mounted in a confined area, and a complete station can be cost effective. A VSAT setup also supports various network configurations, information, and can even service geographically remote sites.

A VSAT's multipoint distribution capability also highlights a satellite's key strength.

Unlike point-to-point configurations, where a dedicated line may have to be laid to connect a new site, with a VSAT, you set up a dish and its companion electronics. It also allows an organization to create its own communications network.

In one example, Wal-Mart tapped this technology to process credit card purchase transactions. It is faster and less expensive than traditional systems.[16] VSATs have also been used to distribute press pictures to news organizations and to deliver video to K-Mart stores.[17] In essence, VSATs help large and small organizations take advantage of satellite technology.

There are, however, some disadvantages of Ku-band systems. This includes a greater susceptibility to interference created by rain. A severe rainstorm could disrupt a transmission to varying degrees.[18]

## GENERAL SATELLITE SERVICES

The U.S. communications satellite fleet consists of government and commercial spacecraft. Government satellites are used for military and nonmilitary applications. These range from the creation of a worldwide communications net for military installations to serving as a testbed for new technologies.

On the commercial front, Western Union's Westar I satellite, launched in 1974, was the country's first commercial satellite designed to serve domestic communication needs. The FCC, which oversees the private element of the satellite fleet, also helped spur the industry's growth through its open skies policy. Articulated in the early 1970s, the policy promoted the commercialization of outer space with regard to commercial satellite communication.

A company that plans to launch and operate a commercial communications satellite must also seek FCC approval. Other details must then be resolved, including contracting with a launch agency and obtaining insurance against possible losses. The latter could include the destruction of the launch vehicle (and the satellite) and the satellite's failure to activate once it reaches its assigned station.

The satellite's emergence as a dominant player in the communications field is a reflection of its general reliability and its

- wide channel capacity,
- capability to handle a mixed bag of information,
- capability to simultaneously reach multiple sites, and
- capability to reach areas not serviced by terrestrial lines.

The last characteristic may be salient for countries that do not have a developed telecommunications infrastructure or whose terrestrial telecommunications expansion is hindered by deserts, mountainous regions or other geographical obstructions.

Organizations have tapped this tool by leasing satellite channel space or by contracting for a turnkey communications system for a specified fee and time period. Another option calls for leasing transponder time for occasional use.

### Intelsat, Inmarsat, and Comsat

The international market had been dominated by the International Telecommunications Satellite Organization (Intelsat) and other organizations. Intelsat was founded in 1964 as an international satellite consortium (intergovernmental). Intelsat still owns and operates a fleet of satellites that supports the international distribution of television and telephone signals and other services.

The Communications Satellite Corporation (Comsat) served as the U.S. representative to Intelsat. This made Comsat a powerful and far reaching satellite organization that supported an array of services.

The International Maritime Satellite Organization (Inmarsat) for its part, extended satellite communication to ships at sea, oil drilling rigs, and even remote land sites (mobile communication). Ships, for example, have established satellite links with land bases through stabilized antennas.

Intelsat and other intergovernmental organizations helped foster the growth of satellite communication. Nevertheless, independent organizations, such as PanAmSat, have stepped forward to compete in this international market. While other satellite networks served various regions of the Earth, they were generally not viewed as Intelsat's competitors. The new private systems were, however, competitors for satellite communication traffic.

This development was fueled by new regulatory developments and the rapid integration of satellite technology and equipment on the world market.[19] When Intelsat and other organizations were founded, the satellite industry was in its infancy, and consortia were partly established to foster satellite communication for developed and developing nations. Today, the tools of satellite technology have reached a level of maturity and cost effectiveness where it is possible to launch private operations.

In fact, the wave of privatization also extended to the intergovernmental satellite arena. For example, Intelsat was privatized in 2001. One goal was to eliminate certain restrictions to enable the organization to engage more effectively in the international market.[20] Inmarsat, for its part, was set up as a limited company in 1999.[21] This trend, where private industry rather than government-controlled organizations or agencies offer specified services, is also not unique to the satellite industry.

### Teleports

Satellites can work with terrestrial communications systems. A company's data may be transmitted over a high-speed landline prior to an uplink and after the downlink. Consequently, satellite and terrestrial systems can be interdependent. The satellite transmission can serve as the long-distance connection, while terrestrial lines provide the intracity hookup.

This integration of systems has been exemplified by New York City's teleport. The Port Authority, Western Union, and Merrill Lynch joined forces to develop a sophisticated communications center on Staten Island. A series of ground stations tied the teleport to national and international satellites and high-speed lines provided the local connection.

Teleports also helped spur the satellite industry's growth. Organizations could share satellite facilities to reduce each participant's financial burden. This could make satellite communication more economically attractive for organizations that have not yet entered the field.

Similarly, a company's geographical location may preclude the building of an Earth station.[22] The establishment of a nearby teleport would provide the company with the capability to establish a satellite link.

### Satellites and the Broadcast and Cable Industries

Satellite-distributed television programming is the backbone of the U.S. cable television industry. HBO and other companies uplink programming to a satellite, where it is subsequently downlinked and received by cable companies. The programming is then locally distributed to individual subscribers.[23]

WPIX in New York City, WTBS in Atlanta, and other independent television stations also tapped this resource. They have used satellites to distribute programming, much the same way as HBO. The television networks also turned toward satellite communication. NBC, for example, the first Ku-band network, established a national satellite system that linked affiliate television stations. The same technology also helped give birth to new networks by providing organizations with a national distribution vehicle. These developments, and others, follow up on the success enjoyed by the Public Broadcasting System, a pioneering satellite organization.

***Satellites and Scrambling.*** As just described, television programming is distributed by satellite to U.S. cable companies. Once the signals are received, they are routed to individual subscribers. This distribution system is efficient, but creates a problem for cable companies and program providers. Other satellite dish owners could bypass the local cable company to gain access to this programming.

The situation developed into a serious problem by the mid-1980s. The Television Receive-Only (TVRO) industry, partly composed of companies that manufacture and supply Earth stations, experienced a rapid growth of its consumer business. This was a result of the FCC's 1979 deregulation of receive-only stations.

Consumer-based TVRO systems, consisting of small backyard receive-only dishes and the complementary electronic components, were purchased by more than a million Americans. These configurations, which can be called home satellite dishes (HSDs), sprouted up across the country, and people were able to watch pay television and other programming for free.

In response, *scrambling* was inaugurated in the mid-1980s by HBO—a television signal was rendered unintelligible unless you had access to a special descrambling device. For cable companies, each site would be equipped with a descrambling unit so their subscribers would continue to receive uninterrupted programming. Other services joined the bandwagon, creating a furor in the TVRO industry. Ultimately, dish owners would have to pay to gain access to programming or use illegally altered decoders.[24]

One individual became so angry at this situation that he illegally interrupted HBO's programming on April 27, 1986, and relayed his own antiscrambling and antisubscription fee message:

Good evening HBO
From Captain Midnight
$12.95/month
No way!
Showtime/Movie Channel beware

This satellite pirate, Captain Midnight, was eventually identified as a part-time employee at a teleport and was subsequently convicted.[25] Captain Midnight used the teleport's facilities to override HBO's signal, and the satellite distributed his signal instead.

Eventually, a compromise was reached between HSD owners and the cable and television industries. The owners were generally treated more equitably, on a par with standard cable subscribers, in terms of monthly fees. This initiative, among others, helped to somewhat defuse the situation, but the problem with illegal descramblers continued.[26]

The controversy between these groups also clouded a key issue—the vulnerability of the commercial satellite fleet. If Captain Midnight could disrupt HBO's transmission, other individuals with access to the proper facilities could follow suit. Even though this group's size may be somewhat limited, the situation could change as new facilities, both permanent and portable, are brought online.

Beyond television programming, financial data and other information vital to the world community is exchanged daily. An ongoing disruption of these services would be disastrous. Ultimately, the same technology that advanced our communications system could potentially harm us. In response, the FCC indicated that security systems had to be implemented to protect the integrity of satellite transmissions.[27]

## DIRECT BROADCAST SATELLITES

As conceived, a direct broadcast satellite (DBS) was a class of spacecraft that would initiate a new communications service. Equipped with a very powerful transmission system, a satellite would operate in the K-band and would bypass conventional media outlets to relay programming directly to consumers.[28] But as will be discussed, this original concept evolved over the years.

It is also important to note that two DBS concepts were and still are important: a dish's

INFORMATION TRANSMISSION

**Figure 6.4**
*Satellites have played a crucial role in monitoring the weather. In this shot, Hurricane Fran as viewed by the GOES-8 satellite. (Courtesy of NASA.)*

size and programming choices. Thanks to a powerful transmission and to technical advancements, a dish less than 2 feet in diameter could be used. This is in contrast to the typical 7-foot or more C-band HSD configuration.

Dish size is a vital concern for DBS companies since a small dish is unobtrusive, compact, fairly inexpensive, and easy to set up. It is also permanently aligned toward a specific satellite, unlike an HSD dish, which can be moved.

Viewing choices are also important. A DBS company can provide subscribers with movies as well as television and sports programming. Current systems, as covered in a later section, can also compete with cable operations in the delivery of a mixed bag of program options.

### Early History

During the early 1980s, various companies floated DBS proposals, some of which shifted over time. For example, the Satellite

Television Corporation (STC), a subsidiary of Comsat, planned to use four satellites to cover the United States.[29] Because each satellite would target only a sector of the country (for example, the Eastern time zone), its "focused" signal would help make it possible to use the smaller dish. But in one modification, this geographical zone was extended so the entire country could be serviced in an accelerated time frame.[30] Proposed DBS systems were also somewhat handicapped by a limited channel capacity. Weight and power demands had an impact on the number of transponders and channels a satellite could support.[31] Consequently, five- or six-channel offerings were not uncommon.

Regardless of the scheme, none of the high-power DBS ventures became operational. Different factors contributed to this situation.

1. The development of a national system demanded a large capital investment. Beyond the millions of dollars to build, launch, and maintain the satellites, a terrestrial support network had to be created. The latter ranged from local sales and repair offices to an advertising campaign to program licensing fees. For some organizations, the investment was too high for an untested and potentially risky business.[32]

2. The rapid expansion of the TVRO/HSD and VCR industries exacerbated this situation since the consumer market was already served by these applications. In fact, more than 40 million households were already equipped with VCRs by the mid-1980s.[33] This was an unfortunate development for DBS companies since movies were slated to be a staple feature.

3. As stated, subscribers would have received only a limited number of channels. Although this may have been acceptable to consumers who lived in areas with few

programming options, would consumers in areas served by cable follow suit?

4. The DBS industry could not sustain a sufficient level of financial support. It was also dealt a severe blow in the 1980s when STC suspended its plans. Other companies, including CBS, had previously bowed out of the field. Consequently, a high-power DBS system did not materialize in the United States.

In contrast, a low-power service was actually created by United Satellite Communications, Inc. (USCI). Instead of constructing a fleet of expensive and untested high-power satellites, a more proven medium-power Ku-band satellite was used. Launched in 1983, USCI offered subscribers five channels of entertainment programming. Future options included specialized information services and bilingual programming.

But despite the advantages of using a less expensive spacecraft and beating high-power systems to the punch, financial pressures forced USCI to close its doors in 1985.

## The Digital Option

Other DBS ventures followed suit, including one proposed by NBC, Hughes Communications Inc., Cablevision Systems Corporation, and the News Corporation Limited. The plan collapsed, though, in the early 1990s.[34]

Nevertheless, as exemplified by DirecTV, a new generation of DBS systems finally became a reality in the same era. Initiated in 1994 and owned by the Hughes Electronics Corporation, DirecTV had already attracted more than a million subscribers by 1996.[35]

Digital technology and processing techniques have been tapped to make DirecTV an efficient and comprehensive service. The several channel limitation has been eliminated and enhanced audio and video signals can be relayed.

As of this writing, high-power Ku-band satellites deliver well over 100 digital channels. Movies, pay-per-view options, and standard programming fare are supported, and this programming depth, as well as comparable pricing structure, makes DirecTV directly competitive with cable systems.[36] You could also opt for a high-speed data relay to tap into the Internet. Thus, your satellite service could serve as an integrated entertainment and information utility.

## Summary

The attraction of a DBS system is a powerful one that fuels continued interest in the field. Consumers can receive an array of programs, including ones that may not be otherwise available. The prerequisite technology base has also matured since the 1980s, making DBS operations more feasible and, ultimately, fully integrated in the U.S. communications infrastructure. They support individuals in rural areas who do not have broadcast or cable options and offer traditional cable subscribers another choice.

It will be interesting to watch the overall communications field as satellite, cable, and telephone companies compete for subscribers. As the latter two industries upgrade their physical plants, they will be better positioned to contend with DBS systems.

Note also that on the international front, other countries are well versed in DBS technology and continue to draft plans for sophisticated systems. Japan and various Euro-pean nations are the major contenders in this field.

DirecTV-type services also targeted Latin America during the late 1990s. In one case, Brazil represented a marketing prize. It held a good portion of Latin America's total households, while only "2.5 percent [were] reached via cable television lines."[37]

This same environment would also offer satellite companies new challenges. In the

**Figure 6.5**
*An SNG vehicle. Note the portable dish size. (Courtesy of Intelsat.)*

United States, telephone lines are used for the return loop and for relaying billing information. In countries where universal telephone service do not exist, other options have to be adopted.[38]

Finally, like other fields, the DBS industry had to contend with auctions. Bids for available DBS slots surpassed several hundred million dollars.[39] Similar auctions were held for spectrum allocations for other communications services.

## SATELLITES, JOURNALISTS, AND THE NEWS

### Satellite Newsgathering

Satellite communication has revolutionized another facet of the television industry—television news. Besides using satellites for story distribution, television stations can participate in satellite newsgathering (SNG), a newer production form.

Wider satellite availability, lower costs, and portable equipment have prompted stations to create their own remote setups. A station buys either a van or truck and a portable satellite dish. This rig is taken on the road, and an uplink to a satellite is established when the reporters reach the story's site. The

transmission is subsequently picked up by the home television station. This capability makes it possible for the station to conduct relays from distant sites. For example, a station from Seattle, Washington, could send its satellite rig to Washington, DC, to establish a link between Seattle's congressional representatives and their constituents.

On the international front, flyaway systems have extended satellite communication to regions where standard satellite links may not be available or accessible. Accommodated by a commercial airliner, the system is stored in trunks and reassembled on arrival. The Cable News Network (CNN) has pioneered the use of such systems to cover world events. These include 1989s Tiananmen Square student occupation in Beijing, China, and Operation Desert Storm, 1991s Persian Gulf conflict.[40]

### Operation Desert Storm

Satellite technology made Operation Desert Storm the first "real-time" war.[41] People witnessed missile attacks and other events as they actually occurred. Satellite links also provided reporters with the ability to relay voice, video, and computer data (news stories) to their home offices in a timely fashion. In Vietnam, for instance, satellites were generally not available, and footage, shot on film, had to be shipped and processed before airing. This led to a time delay.[42]

But this new found immediacy triggered negative and positive U.S. responses. In the most publicized case, Peter Arnett, a CNN reporter, continued to file stories from Baghdad, Iraq, during the war.[43] He was criticized for this action by the Victory Committee, a coalition that included the Accuracy in Media organization.[44] The criticism centered on his reporting while under Iraqi censorship. On the flip side, tight restrictions were placed on the press by the U.S. government and its allies.

## Operation Iraqi Freedom

The same region as Operation Desert Storm erupted in conflict in early 2003. But in this instance, reporters were "embedded"—integrated—with military units. Using satellites and other technologies, they could file reports from the field as they moved with the units. At times, proponents and opponents used this information to try to sway world opinion as to the war's justice or injustice. For the media, Ted Koppel had another observation:

What's totally unpredictable, of course, is the impact that all this coverage will have back at home and around the world. If the campaign doesn't go as quickly or as well as anticipated, if friendly casualties are high, if some of the reporting is deemed too critical, or if some of the information proves inadvertently helpful to the Iraqis, the military may quickly rethink the value of having us journalists along.[45]

These sentiments reflected the mixed access the media had in previous military actions. But General Tommy Franks, the U.S. commander, indicated he supported this concept since it provided, in a sense, a snapshot of the truth—it opened-up a picture window to events as they actually unfolded. Franks also stated it was important in light of the First Amendment.[46]

This conflict was also the first "Internet" war. Live reports streamed in from reporters to the world's television sets and computers. Personal diaries and journals on the Internet, web logs or *blogs*, were popular Internet fare. The media also used videophones, compact systems that supported video relays, to send in live reports.[47] Live video from the Persian Gulf and information from U.S. and international sources were also accessible on the Internet.

As footnotes to this event,

- Peter Arnett was fired from NBC and National Geographic for personal comments made on Iraqi television in 2003.[48]

Arnett subsequently apologized for his remarks.

- In the field, Geraldo Rivera, another reporter, left the Middle East for revealing inappropriate information in a report.[49]
- The BBC singled out U.S.–Iraq coverage as so "unquestioningly patriotic and so lacking in impartiality that it threatened the credibility of America's electronic media . . ."[50] The BBC was, for its part, criticized for being "soft" on the Iraqi government.[51]
- The Internet provided a global audience with multiple news perspectives from multiple news sources. While the U.S. and British media, among others, were charged with certain biases, you could seek out alternate news pools via the Internet. One potential pitfall, however, was and still is a news source's reliability. Almost anybody with access to a computer and the Internet could post information to a web site—but how credible is it?

*Questions.* The Persian Gulf conflicts highlighted the media's capability to deliver the news "as it happened." Nevertheless, questions were raised about the collision between technology and what should be good reporting. For example,

- What is technology's impact?
- Does good reporting transcend technology?
- It may have taken more than a month to receive a news report if you lived some 200 years ago. Has the recent immediacy of news altered the news process itself?
- Is a good reporter a good reporter regardless of the technology that is used?

The answers, or part of them, may be found in a report that explored the media's role in Operation Desert Storm. It also applied in

2003 and, potentially, for the foreseeable future. In brief, "technology cannot be an end in and of itself in making sense of war-related events. That, as always, remains the job of the journalists themselves, with the new technology facilitating but not replacing the task."[52]

### Remote Sensing and the News

Pioneered in part by Mark Brender and ABC, the news media have adopted another satellite-based system, remote sensing satellites, to support their reports. A remote sensing satellite is a sophisticated spacecraft equipped with high-resolution cameras and an array of scientific instruments. Instead of being locked in a geostationary orbit, a satellite can cover the Earth in successive orbits. The pattern is then repeated.

Remote sensing satellites were originally designed to examine and explore the Earth. They have helped document the Earth's physical characteristics as well as environmental changes. The latter include the impact of deforestation and pollution.

The United States and France initially led the world in this field through their Landsat and SPOT satellites, respectively, and news organizations subsequently tapped this tool. For example, this type of satellite was an invaluable resource in the aftermath of the Chernobyl nuclear reactor accident. Satellite pictures of the site were obtained and released by the media. The pictures highlighted the facility's damage and helped prevent a potential cover-up.

ABC News also used remote sensing images in a special program televised in July 1987. The pictures documented various facets of the Iran–Iraq war. The same region was also under scrutiny during Operation Desert Storm and other operations.

Despite this service's benefits, all governments, including the U.S. government, were not happy with this newfound capability.

Since the media could order a photograph of a region of the Earth covered by the spacecraft, military maneuvers and other situations, which a government may want to keep hidden, could be revealed. In the United States, the government could have imposed restrictions on private, domestic remote sensing licenses, via ambiguous licensing procedures. If implemented, the media's access to specific images may have been limited.[53]

This stance, and others, triggered a reaction from the Radio-Television News Directors Association (RTNDA) and other media groups. It was argued that restrictions violated First Amendment rights, and the government's concern over potential national security breaches was unfounded. The press had generally been responsible in the use of sensitive information in the past and would follow suit with the satellite pictures.

Besides this initial government intervention, media organizations were and still are faced with other problems that may hinder this investigative tool's effectiveness. In one case, a satellite may not be in the correct orbital position to deliver a requested picture immediately. This delay could hamper the coverage of fast-breaking events.

Nevertheless, the remote sensing field still holds great promise. New domestic and international satellites have been developed, including the deployment of Ikonos in 1999, the "first commercial imaging satellite."[54] It can produce high-resolution images, and this class of satellite should reduce the response time for pictures. Simplified image manipulation software has also contributed to enhanced operations. ABC News, for one, employed such a system during Operation Desert Storm to generate graphics for its programs.

The U.S. government continues to play a pivotal role in this field. In one example, the Clinton administration released a policy statement about remote sensing licenses. While it somewhat relaxed restrictions, they could still be implemented.[55]

In April 2003, the Bush administration redefined this role. The new Policy, which superceded a 1994 Presidential Directive, provided for more government support for commercial ventures. Other goals were to streamline various licensing procedures, to enhance the government's relationship with the commercial U.S. remote sensing industry, and to give the field a general "boost."[56]

However, while implementation details were still under development as of this writing, it appears the new Policy was still somewhat vague in regard to potential government restrictions and the news media. This topic is discussed more fully in the Conclusion.

## CONCLUSION

Satellites helped transform our communications system. In one example, distance is less of a factor than it was in the past. Satellites enable us to relay information rapidly around the world and to view news and human events, such as Tiananmen Square, in real time. Other applications support remote sensing and the delivery of entertainment programming through DBS companies.

In essence, satellites have provided us with a new set of communications tools. It is also important to remember that we have only been using these tools for a relatively short time. We have only begun to tap their potential and, as introduced in Chapter 8, their potential to promote a personal communication revolution and, ultimately, their potential to better explore our own world.

But this optimistic view could be dampened. The remote sensing arena serves as an example of the impact of potential government intervention. Barbara Cochran, RTNDA President, wrote an article about

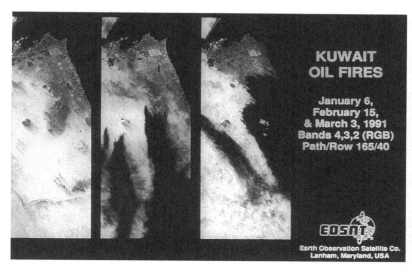

such a scenario, "Fighting the Feds Over Shutter Control." (Copyright by Radio–Television News Directors Association. Reprinted with Permission of the RTNDA.)[57] Her words clearly define this issue, and just as important, the satellite images' inherent power, particularly for a free press.

While written before the Bush Policy, the article's main points may still be valid. If, however, the scenario does evolve, the article still provides valuable insights about the collision between a new technology, journalistic privileges, and the government.

Proposed government restrictions on the use of commercial satellite images are unconstitutional.

History was made two months ago when the first image from a U.S.-owned commercial satellite was beamed back to earth and released to the general public. That image was remarkable because, although it was made by a satellite orbiting 423 miles above Earth, it showed detail around the Washington Monument so clearly that you could pick out automobiles on the street.

The technology is nifty, but why should you care? Because these images can help television journalists tell stories with a graphic reality never before possible and can provide images from places where cameras are forbidden. Satellite imagery can help you report on major storm

**Figure 6.6**
*Satellite systems have enhanced our communications capabilities and our ability to cover world events. In this photo, the impact of the Kuwait oil fires (e.g., smoke), set during operation Desert Storm, is evident in this sequence of photos. (Courtesy of the Earth Observation Satellite Company, Lanham, MD.)*

INFORMATION TRANSMISSION

systems, ecological change or urban sprawl in your community. Soon you will be able to give viewers a "fly-through" of your city or the surrounding area. With satellite imagery, producers will be able to show terrain in forbidden territory or a gravesite indicating a massacre.

There's one more reason you should care: The U.S. government has adopted a policy for these commercial satellites that would allow the government to cut off access to images whenever the State Department or Pentagon deems it necessary. The policy is unconstitutional, a violation of the First Amendment right of the press to publish or broadcast without government interference, and RTNDA is leading the fight against this policy.

Until recently, government satellites collected pictures from space, which were released to the public at the government's discretion. For years, journalists have used weather satellite imagery in their newscasts to inform viewers and even save lives of those threatened by hurricanes and tropical storms. Now Earth imagery is available from Russia, France and India. When these images are combined with computer technology, producers can create much more realistic 3-D graphics that replace artists' sketches and give the viewer the sense of "flying through" a landscape.

Dan Dubno of CBS News says the network has used satellite imagery in recent months to report on "everything from the crisis in Kosovo to the Gulf War; North Korean nuclear development; the atomic bomb tests in Pakistan and India; the assault on Osama bin Laden's hideaway in Afghanistan; the 1999 hurricane season; El Nino; and the shuttle launch."

This evolution in journalists' use of satellite imagery brings enormous benefits. Stories can be more accurate and truthful and can give the public access to geographic areas that are politically inaccessible or too expensive to get to.

Now, with the advent of more commercial satellite companies that will create a business out of making these images widely available, journalists will have more opportunities than ever to use this imagery.

But the choices will exist only to the extent that the U.S. government permits the imagery to be made available. And contained in the government policy for licensing of commercial satellites is some disturbing language on "shutter control."

Shutter control is the term for cutting off imaging over a given geographic area for a given period of time. If a satellite is really nothing more than a camera in the sky, government exercise of shutter control constitutes prior restraint of publication of images.

U.S. constitutional law puts a heavy burden on the government if it wants to prevent publication. The First Amendment ensures that the press is free of government control and restraint. The Supreme Court has insisted that the government must take its case to a court of law and present evidence to show why publication should be forbidden before the fact. In the case of government arguing that publication would endanger national security, the court has said the government must show that publication presents a clear and present danger to the national security.

The policy for commercial satellite licensing is much broader and more vague than constitutional law prescribes. In addition, the policy puts the power to decide to exercise shutter control solely in the hands of the executive branch, leaving out the judiciary entirely. The Secretary of Commerce may invoke shutter control after being informed by either the Secretary of State or Defense that a period exists when national security, international obligations, or foreign policy interests may be compromised.

No judicial review. No presentation of evidence in court. No "clear and present danger" test. Rather, the policy makes the executive branch of government both judge and jury in the matter. And the causes that could trigger shutter control are so broad as to be meaningless.

These conditions for shutter control were included in new rules proposed by the federal government in 1998. RTNDA and the National Association of Broadcasters filed comments strongly objecting to the rules. I also met personally with Secretary of Commerce William Daley to explain our position. So far, the government has not issued a final version.

But in 1998 the dispute was theoretical. No satellite had been successfully launched, no company was providing imagery to the commercial market. That all changed on September 24, when IKONOS successfully launched and three weeks later when it began distributing images with a resolution of one meter, or approximately three feet. Now, journalists and other users have access to imagery with much more detail than ever before of countries, weather systems, ecological change, and a host of other subjects. Other companies with licenses have plans to launch in the future.

So the question is now a practical one. What if the conflict in Kosovo were going on now, with severe restrictions on reporting and deep involvement of U.S. military forces? What would the U.S. government do? Would it prevent the commercial satellite from photographing Kosovo and the surrounding area? Would it prove a national security impact, or would the reasons be "foreign policy interests"? And how would news organizations react? Would they try to win the right to argue the case with the executive branch in court?

At RTNDA, we believe this issue is essential to preserving the right to broadcast material in the public interest without prior restraint. A test of the government's role in shutter control is almost certain to occur in the near future. When it does, RTNDA will be in the forefront on the issue, protecting the ability of journalists to use the best tools available—including satellite imagery—to tell stories that are accurate, independent, and in the public interest.

Some observers have noted this issue may actually be a nonissue at this point in time.

In one case, during U.S. operations in Afghanistan, the Department of Defense entered an "'assured access' agreement with . . . Space Imaging, buying all of its satellite images of Afghanistan to prevent other users from obtaining them. . . . In this way, formal shutter control never had to be exercised . . ."[58] This action prevented the triggering of shutter control and could set a precedent for similar situations.

Other factors include the potential impact of the Bush Policy and the support for shutter control-type initiatives in light of the time sensitive nature of certain government actions. For the latter, some proponents believe the government should have the freedom, when appropriate, to rapidly initiate security measures without prior approval from a federal judge, as some journalists have advocated.[59]

On the flip side, even if the government does enter into an "assured access" agreement and does not implement shutter control, could this action be viewed as a different form of information control? The Policy's actual implementation must also be fully mapped out to gauge its impact. But from an initial reading of the Policy's mandates, Barbara Cochran's concerns may still hold true.

## REFERENCES/NOTES

1. *Telecommunications: A Glossary of Telecommunications Terms* (Washington, DC: Federal Communications Commission, April 1987), 4, 14.

2. Intelsat web site, downloaded January 2000.

3. Intelsat, "Intelsat 906 @ 64 Degrees E: Expanding the Network of Major Carriers in Europe, Asia, Africa, and Australia," downloaded from www.intelsat.com/globalnetwork/satelites_launches_906.asp.

4. Mark Long, *World Satellite Almanac* (Boise, ID: Comm Tek Publishing Company, 1985), 73.

5. For example, to produce a more focused signal.

6. *Stabilization* refers to the way a satellite maintains its stability while in orbit. A spin-stabilized satellite rapidly rotates around an axis while the antenna is situated on a despun platform so it continues to point at the Earth. Three-axis-stabilized spacecraft use gyros to maintain their positions. Other systems also play a role in this process. NASA has used a three-axis-stabilized concept for many of its outer space probes. See Andrew F. Inglis, *Satellite Technology* (Boston: Focal Press, 1991), 32; and

P.R.K. Chetty, *Satellite Technology and Its Implications* (Blue Ridge Summit, PA: TAB Books, 1991), 174–179, for more detailed information.

7. It can be used during eclipses.

8. Another vital element in the transmission between a satellite and a ground station is telemetry data. This is essentially housekeeping data that are relayed by the satellite to indicate its current operational status. In addition, the ground station can uplink commands, such as the aforementioned activation of the thrusters and a possible command to move the satellite to a different orbit. Thus, depending on its design, a spacecraft is not necessarily locked into a position once it achieves its assigned slot.

9. Phil Dubs, "How to 'Recycle' a Dying Bird," *TV Technology* 12 (May 1994), 20.

10. A controversy erupted about RTGs prior to the launching of the *Galileo* spacecraft. Designed to investigate Jupiter and its moons during an extended mission, *Galileo's* systems are powered by RTGs. Protesters sought to block its launch legally. There was a concern over the possible contamination of the Earth if the RTGs' fuel was scattered in a launch disaster. NASA and space advocates replied the nuclear fuel was in protective containers. Even if there were an explosion, the fuel would not be scattered. Ultimately, the spacecraft was launched in October 1989. Robert Nichols provides an excellent overview of this issue in his article "Showdown at Pad 39-B," *Ad Astra* 1 (November 1989), 8–15. Other, subsequent spacecraft raised similar concerns.

11. Long, *World Satellite Almanac*, 68. See also Chetty, *Satellite Technology and Its Implications*, 402–403, for details. *Note:* Two such systems have been frequency- and time-division multiple access (FDMA/TDMA). They also created a more efficient communications operation.

12. Conversation with Scott Bergstrom, Ph.D., director, Technology-Based Instruction Research Lab, Center for Aerospace Sciences, University of North Dakota, August 6, 1992. For additional information, see Peter Lambert, "Digital Compression; Now Arriving on the Fast Track," *Broadcasting* (July 27, 1992), 40–46.

13. Other areas are also discussed at appropriate points in the book.

14. Andrew F. Inglis, *Satellite Technology* (Boston: Focal Press, 1991), 30. *Note:* They share a frequency range.

15. Satellite Communication Research, *Satellite Earth Station Use in Business and Education* (Tulsa, OK: Satellite Communication Research), 18.

16. George Lawton, "Deploying VSATs for Specialized Business Applications," *Telecommunications* 28 (June 1994), 28.

17. Ibid., 30.

18. Long, *World Satellite Almanac*, 99. *Note:* At these higher frequencies, the signal can be weakened, that is, absorbed or scattered by raindrops. This is analogous to the way the light from a car's headlights is dispersed and reduced in intensity by fog.

19. Please see "Satellite Communications," by Carolyn A. Lin, p. 289, which appears in *Communication Technology Update*, 8th Ed. (Boston: Focal Press, 2002) for further information about the regulatory developments.

20. Intelsat, "Four Decades of Inspiring Achievement," downloaded from www.intelsat.com/company/history.asp.

21. Inmarsat, "About Inmarsat," downloaded from www.inmarsat.com/about_inm.cfm.

22. For example, a downtown district.

23. The placement of HBO on satellite also helped spur the growth of urban cable systems.

24. Illegal decoders became available when the supposedly unbreakable scrambling system was cracked.

25. William Sheets and Rudolf Graf, "The Raid on HBO," *Radio-Electronics* 10 (October 1986), 49. *Note:* HBO provided the impetus for the development of the VideoCipher scrambling system, originally developed by M/A-Com, Inc. This concept was described in a 1983 HBO brochure, "Satellite Security," for its affiliates.

26. Please see the Satellite Home Viewer Improvement Act of 1999, available from www.fcc.gov, for more information about HSD initiatives.

27. This included the adoption of an automatic transmitter identification system (ATIS). Please see Hughes Communications, Inc. "Staying Clean," *Uplink* (spring 1992), 6, for details.

28. The transmission system was on the order of 150 to 200 watts of power.

29. David L. Price, "The Satellite," *COMSAT* 11 (1983), 14.

30. "STC Asks for Modifications in DBS Plans," *Broadcasting* (July 23, 1984), 99.

31. Andrew F. Inglis, "Direct Broadcast Satellites," *Satellite TV* (October 1983), 33.

32. "DBS Ranks Cut in Half," *Broadcasting* (October 15, 1984), 75.

33. "Commerce Department Sees Bright Future for Advertising," *Broadcasting* (January 12, 1987), 70.

34. "USSB, Hughes Revive DBS in $100 Million + Deal," *Broadcasting* (June 10, 1991), 36.

35. "The Growing World of Satellite TV," *Cable & Broadcasting* (February 5, 1996), 59. *Note:* Stanley Hubbard has been one of the most vocal DBS advocates in the United States and started such a service.

36. Hughes Communications, Inc., "Direc TV," information flyer.

37. William H. Boyer, "Across the Americas, 1996 Is the Year When DBS Consumers Benefit from More Choices," *Satellite Communications* 20 (April 1996), 24.

38. Ibid., 26. *Note:* One option was to use local institutions, such as banks, as billing centers.

39. Rich Brown, "DBS Auctions Yield $735 Million," *Broadcasting & Cable* (January 29, 1996), 6.

40. Marc S. Axelrod, "Transmitting Live from Beijing," *InView* (summer 1989).

41. John Pavlik and Mark Thalhimer, "The Charge of the E-Mail Brigade: News Technology Comes of Age," in *The Media at War: The Press and the Persian Gulf Conflict* (New York: Gannett Foundation Media Center), 1991, 35.

42. Ted Koppel, "Journalists overseas rely on their experience, wits and a host of new technologies," Report aired on March 17, 2003; downloaded from ABCnews.com.

43. Arnett was a CNN reporter at the time.

44. "Group Launches Campaign to 'Pull Plug' on CNN's Arnett," *Broadcasting* (February 18, 1991), 61.

45. Ted Koppel, "Journalists overseas rely on their experience," downloaded from ABCnews.com.

46. TV interview between Tony Snow (FOX) and Gen. Tommy Franks; broadcast on FOX News, April 13, 2003.

47. Videoconferencing is discussed in a later chapter.

48. National Geographic News, "National Geographic Fires Peter Arnett," March 31, 2003, downloaded from http://news.national geographic.com/news/2003/03/0331_030331 _arnettfired.html.

49. Rivera drew a rough map in the sand as to his unit's location.

50. Merissa Marr, "BBC Chief Attacks U.S. Media War Coverage," Reuters Industry, April 24, 2003, downloaded from http://story.news. yahoo.com/news?tmpl=story&u=/nm/200304 25/media_nm/iraq_media_bbc_dc_4.

51. Ibid.

52. Pavlik and Thalhimer, "The Charge of the E-Mail Brigade," 37.

53. Jay Peterzell, "Eye in the Sky," *Columbia Journalism Review* (September/October 1987), 46.

54. Karen Anderson, "Eagle Eye in the Sky," *Broadcasting & Cable* (October 12, 1999), 72. *Note:* In one case, a 10-meter resolution limitation was lifted (Ikonos has a 1 meter capability). Other countries have also gotten on the remote sensing bandwagon, including Russia for high-resolution images. See The Commercial Space Act of 1997, H.R. 1702, 105th Congress, 2nd session and the Land Remote Sensing Policy Act of 1992 for information about U.S. policy issues.

55. U.S. Congress, Office of Technology Assessment, *Civilian Satellite Remote Sensing: A Strategic Approach*, OTA-ISS-607 (Washington, DC: U.S. Government Printing Office, September 1994), 114.

56. "U.S. Commercial Remote Sensing Policy," Fact Sheet, April 25, 2003, downloaded from www.fas.org/irp/offdocs/nspd/ remsens.html.

57. "Fighting the Feds Over Shutter Control," by Barbara Cochran, RTNDA President. Copyright by Radio-Television News Directors Association. Reprinted with permission of the RTNDA.

58. Jefferson Morris, "Rand: Satellite 'Shutter Control' Not as Big an Issue as Expected," downloaded from Nexis.

INFORMATION TRANSMISSION

59. Captain Michael R. Hoversten, "U.S. National Security and Government Regulation of Commercial Remote Sensing from Outer Space," 2001 Air Force Judge Advocate General School; *The Air Force Law Review;* downloaded from LEXIS. *Note:* The article also provides an excellent review of the remote sensing field and related issues.

## SUGGESTED READINGS

Boyer, William. "Across the Americas, 1996 Is the Year When DBS Consumers Benefit from More Choices." *Satellite Communications* 20 (April 1996), 23–30; Robert M. Frieden. "Satellites in the Global Information Infrastructure: Opportunities and Handicaps." *Telecommunications* 30 (February 1996), 29–33; David Hartshorn. "Conjuring the Cure for Asian Flu." *Satellite Communications* 23 (January 1999), 24–31; *Via Satellite* extensively covers the international satellite market with articles such as: Peter J. Brown. "It Takes Two to Tango . . . Latin America." *Via Satellite* XVI (July 2001), 32–43; Nick Mitsis. "Africa Uniting the Continent Through Satellite Technology." *Via Satellite* XVI (November 2001), 36–45; Nick Mitsis. "Asian Economic Tiger Awakens." *Via Satellite* XVI (April 2001), 20–28. Articles that cover various international satellite issues.

Brender, Mark E. "Remote Sensing and the First Amendment." *Space Policy* (November 1987), 293–297. An excellent review of remote sensing, journalism, and First Amendment issues, prior to the Clinton administration.

Careless, James. "Databroadcasting: Pushing Boundaries." *Via Satellite* XVII (October 2002), 28–33; James Careless. "Ka-Band VSATs: Blazing the Next Great Frontier." *Via Satellite* XVI (February 2001), 40–48; Patrick Flanagan. "VSAT: A Market and Technology Overview." *Telecommunications* 27 (March 1993), 19–24; George Lawton. "Deploying VSATs for Specialized Business Applications." *Telecommunications* 28 (June 1994), 28–32; Gino Picasso. "VSAT's: Continuing Improvements for a Workhorse Technology." *Telecommunications* 32 (September 1998), 56–57; John

Sweitzer. "The VSAT Ka-Band Configuration." *Satellite Communications* 23 (October 1999), 42–44, 48. Early VSAT developments, applications, the players, and Ka-band systems.

Chetty, P.R.K. *Satellite Technology and Its Applications.* Blue Ridge Summit, PA: TAB Professional and Reference Books, 1991; Andrew F. Inglis. *Satellite Technology: An Introduction.* Boston: Focal Press, 1991; Donald Martin. *Communication Satellites,* 4th Ed. El Segundo, CA: The Aerospace Press, 2000. Satellites and their applications.

*Communications News.* From late 1987 through 1988, *Communications News* ran a series of articles devoted to VSATs. These articles provide an interesting perspective of this field. Includes David Wilkerson, "VSAT Technology for Today and the Future—Part 3: Use Private Networks or Leased Services?" (November 1987), 60–63.

Dorr, Les, Jr. "PanAmSat Takes on a Giant." *Space World* W-12–276 (December 1986), 14–17; "Anselmo, Landman Team Up to Tackle Intelsat." *Broadcasting* (August 5, 1991), 48. Early look at private, international satellite networks.

Dubs, Phil. "How to 'Recycle' a Dying Bird." *TV Technology* 12 (May 1994), 20. An interesting look at how to extend a satellite's useful lifetime.

"EOSAT Operations Underway." *The Photogrammetric Coyote* 9 (March 1986), 8, 17; "SPOT to Fly in October." *The Photogrammetric Coyote* 8 (September 1985), 2, 4; David L. Glackin and Gerard R. Peltzer. *Civil, Commercial, and International Remote Sensing Systems and Geoprocessing.* El Segundo, CA: The Aerospace Press, 1999; NASA. "The Landsat Satellites: Unique National Assets." (FS-1999(03)-004-GSFC), downloaded; U.S. Congress, Office of Technology Assessment. *Civilian Satellite Remote Sensing: A Strategic Approach,* OTA-ISS-607. Washington, DC: U.S. Government Printing Office, September 1994. Earlier to more recent examinations of remote sensing satellites, policies, and applications.

Gildea, Kerry. "Lawmakers Put Pressure on DoD to Devise Commercial Imagery Strategy," downloaded from Nexis; "RTNDA Protests Imaging Satellite Constraints." *Broadcasting & Cable* (August 12, 1996), 84. Domestic remote sensing restrictions and security concerns (on the part of the U.S. government).

Nelson, Robert A. "What Is the Radius of the Geostationary Orbit." *Via Satellite* XVI (September 2001), 80. Brief but excellent description of the geostationary orbit and factors that necessitate the use of thrusters to maintain a satellite's station.

Niekamp, Raymond A. "Satellite Newsgathering and Its Effect on Network-Affiliate Rela-

tions." An AEJMC Convention paper, Radio–TV Journalism Division, August 1990. Network affiliates and "their policies in sharing news video with their networks and other stations."

U.S. Chamber of Commerce. "National Space Policy Review Remote Sensing," downloaded from www.uschamber.com/space/policy/remotesensing2.htm; U.S. Chamber of Commerce. "U.S. Chamber's Space Enterprise Council Welcomes New National Remote Sensing Policy," downloaded from www.uschamber.com/press/releases/2003/may/03-82.htm. The U.S. Chamber of Commerce's recommendations for the government's role in this field (first article) and its response to the Bush Policy (second article). While the response is positive, there may still be gaps that have potential First Amendment implications.

## GLOSSARY

*Active Satellite:* A satellite equipped to receive signals and to relay its own signal back to Earth.

*C-Band:* A satellite communication frequency band and satellite class. Commercial C-band satellites are the older of the contemporary communications satellite fleet.

*Direct Broadcast Satellite (DBS):* A communications satellite that delivers movies and other offerings to subscribers equipped with compact satellite dishes.

*Earth Station:* An Earth station establishes a communication link with a satellite. Some Earth stations, also called ground stations, transmit and receive signals; others only receive signals.

*Footprint:* The shape of a satellite transmission's reception area on the Earth.

*Geostationary Orbit:* A desirable orbital position/slot for a communications satellite. The satellite's motion is synchronized with the Earth's rotation and appears, to ground observers, to be stationary. This has technical advantages for maintaining a communications link.

*International Telecommunications Satellite Organization (Intelsat):* A pioneering international satellite consortium. Intelsat supports a broad range of satellite services.

*Ku-Band:* A newer satellite communication band and class. Ku-band satellites also support more powerful downlinks and have news (media) applications.

*Orbital Spacing:* Buffer zones physically separate the satellites to help eliminate interference.

*Passive Satellite:* A satellite that does not relay its own signal back to Earth.

*Remote Sensing Satellite:* A remote sensing satellite scans and explores the Earth with different instruments, including cameras. Images can highlight the Earth's physical character-

INFORMATION TRANSMISSION

and the continuation of satellite communications research in the Ka-band."[3] Thus, another of the mission's original goals—the fostering of a collaborative environment between governmental and academic/other nongovernmental agencies—continued.

More pointedly for this chapter, the success of the ACTS mission helped open-up the Ka-band for commercial satellite communication. For example, Ka-band satellites will benefit from NASA's work with spot beam and internal signal switching capabilities. Two targeted applications are the support of high-speed VSAT configurations and Internet service.[4]

Satellites may also be built with more autonomous capabilities. One goal is to develop systems that can process data with less human intervention.

NASA has tapped this capability for terrestrial and space-based applications. One project, the Three Corner Sat (3CS) mission, would reduce the input from ground controllers. The onboard software would, in one operation, have "the ability to make real-time decisions based on the images it acquires and send back only those it deems important . . . Less time will be needed to transmit the data, freeing up power and allowing the spacecraft to concentrate on other important tasks."[5]

As the technology base matures, this principle could be adopted for other space-based projects, including autonomous planetary rovers. It could similarly be used in the design of intelligent communications satellites and imaging spacecraft with enhanced operational capabilities.

Yet, while these developments have advantages, there are some pitfalls. The implementation of any new technology, for instance, generally entails a financial investment and, potentially, costs that may be passed on to the customer. Ka-band satellites are also more vulnerable to rain fade—rain can disrupt a transmission. More sophisticated systems also dictate the construction of a more complex satellite. This may have reliability implications.[6]

Finally, it is important to note the commercial satellite field is a dynamic one. Research continues to enhance current spacecraft while companies, sometimes in league with the government, continue to explore new communications options. One such case is the potential broadening of the commercial use of another satellite band, the X-band. Primarily relegated to government-based activities (e.g., military use), the X-band represents another frequency that may be tapped, in one example, to support governmental initiatives through commercial satellite systems.[7]

## Smallsats and Space Platforms

Besides an ACTS-type spacecraft, other developments have and will continue to advance satellite technology. These include the launching of *smallsats* and space platforms.

Smallsats are small, cost-effective satellites. They can be used for remote sensing, creating personal communications networks, and for other applications.[8] The bottom-line figure for countries and organizations is that satellite technology has become more affordable. A smallsat is less expensive than a conventional satellite and can be designed and assembled in an accelerated time frame.

These factors are crucial for new and developing applications. As covered in Chapter 8, satellite-based personal communication could be supported, in one configuration, by a satellite series or constellation placed in low Earth orbit.

Basically, if you have to build and launch a number of satellites to start a service, you cannot spend two or more years to manufacture a single spacecraft. This is where smallsats step in. Instead of building customized satellites, you reuse existing technologies and products. Modular satellite systems, which can accommodate different space-based applications, are also employed.[9]

The process is somewhat analogous to mass production, but there may be a trade-off. The smallsat may not be as sophisticated or capable of handling as many tasks as its larger and more costly counterpart. But this may not be the design goal in creating a smallsat in the first place.

Similarly, the reuse of components or design concepts has been extended to outer space exploration. To save mission costs, commercially produced systems, or a design implemented on a satellite, for instance, could be used in the production of an outer space probe.

Finally, a space platform would be a large structure placed in orbit. It would route a high volume of information while occupying only a single slot. A single platform could potentially replace several contemporary spacecraft.

INFORMATION TRANSMISSION

**Figure 7.1**
*The* Challenger *at liftoff. (Courtesy of NSSDC.)*

## Space Weather and Space Debris

It is important to note that scientists believe some satellites may be more susceptible to space weather—for our discussion, adverse solar activity that could damage a spacecraft.[10] As we move toward more mass produced systems, which may employ commercial, off-the-shelf components, they may not be adequately shielded from radiation. Other solar activities may also harm "traditional" satellites as well. In fact, the National Oceanic and Atmospheric Administration (NOAA) has developed a "space weather scale" that tracks the potential damage that could be caused by solar originated "geomagnetic storms."[11]

The upshot of these effects? If communications satellites are damaged, our communications system may be affected, and by extension, certain communications capabilities.

Besides natural phenomena, satellites, particularly in low and medium orbits, face another hazard. The Earth is surrounded by a "cloud" composed of millions of pieces of debris. This space junk, which can vary from paint particles to radioactive droplets leaked from other satellites, poses new challenges for satellite designers.[12] While the possibility for a fatal impact is slight, precautions have to be taken. Organizations have also drafted plans to produce less "polluting" spacecraft and rockets.[13] Much like environmental conditions on certain regions of the Earth, our lack of foresight has had unforeseen consequences. This time, it is in space.

The problem may also be magnified if an armed conflict or terrorist attack is extended to space. Commercial and military satellites could be potential targets, particularly the newer generation remote sensing and imaging satellites. As discussed, these spacecraft could produce detailed images of a country that may want to shield its activities from view. If a satellite were destroyed in space, the resulting debris would create another, potentially harmful, space hazard beyond the service disruptions and monetary costs.

## LAUNCH VEHICLES

New launch vehicles and organizations will fuel the growth of the satellite system. In the past, companies and most nations signed with NASA for this job, placing a heavy demand on launch vehicles and facilities. This situation looked as if it was going to be altered in the 1980s. The development of NASA's space shuttle and the entry Arianespace, a private consortium, promised to facilitate satellite launch operations.

### The Space Shuttle

The space shuttle is the world's first refurbishable manned or piloted spacecraft. After a mission, the shuttle returns to Earth and is refurbished for its next flight. The space shuttle can carry a self-contained laboratory, scientific experiments, and satellites in its hold, the cargo bay. In its latter role, the shuttle initially carries a satellite to a low Earth orbit where it is subsequently released. If the satellite's final destination is a geostationary orbit, an attached rocket booster propels the satellite to a specified altitude, and it eventually reaches a preassigned orbital slot after maneuvers.

The shuttle's original promise and premise was to make space accessible and operations cost effective. Yet it was plagued by mechanical and structural problems. This was partly a reflection of the spacecraft's heritage:

- As a piloted vehicle, it had to support a crew;
- As a multipurpose vehicle, it had to support a range of missions; and
- It was refurbishable.

Consequently, the shuttle was a complex spacecraft. While it proved successful in many ways, its complex design and other factors, including refurbishing delays, led to setbacks. A somewhat burdensome organizational hierarchy also hampered the shuttle program.

### The Challenger Explosion.

In February 1984, the space shuttle's credibility as a launch vehicle received a blow. After its release, the Westar VI satellite's booster malfunctioned, and the satellite was stuck in a useless orbit. Indonesia's Palapa-B2, the mission's second satellite, suffered a similar fate.[14] A little less than 2 years later, the explosion of the space shuttle *Challenger* shocked the world. The entire crew was lost in the most devastating tragedy in NASA's history. (As discussed in a later section, this was followed by the loss of the space shuttle *Columbia*, in 2003.)

In the wake of the *Challenger* disaster, and a report released by an investigatory commission, President Ronald Reagan announced that NASA would generally withdraw from the commercial satellite launch industry. The frequency of future flights would also be scaled down, and the shuttle's primary role would be to support scientific and military missions.

This directive reflected the president's attitude toward the government's role in private enterprises and the realization that an overly ambitious launch schedule contributed to *Challenger's* destruction. As stated by the commission, "The nation's reliance on the Shuttle as its principal space launch capability created a relentless pressure on NASA to increase the flight rate."[15] This pressure played a role in the decision to launch *Challenger* under adverse weather conditions. Other contributing elements, which led to *Challenger's* destruction, were design flaws in the shuttle's rocket boosters, as well as possible flaws in the spacecraft's overall design and the booster refurbishing process.[16]

Various companies subsequently stepped forward to fill the void in the commercial launch industry. Even though there was some prior activity, the list of interested parties has grown at an accelerated rate.[17]

Companies adopted existing rockets and developed new unpiloted rockets, expendable launch vehicles (ELVs). An example of the latter was Orbital Sciences Corporation's *Pegasus*. Instead of a typical ground launch, *Pegasus* was carried by a jet, released, and then proceeded on its own power.

This class of ELV was developed to lift small payloads into low Earth orbits. They also complemented smallsats since they were less expensive than conventional ELVs.[18] But depending on the circumstances, a smallsat could even hitch a ride on a conventional ELV, as part of another payload.[19]

These innovations will help fuel the smallsat industry. An organization may have access to a smallsat, but without a cost-effective ELV, it may not be able to launch it. Launch costs are also factors in the creation and maintenance of satellite constellations.

The future U.S. satellite launch fleet will include the space shuttle and ELVs. The commercial sector will use ELVs while the government will also rely on the shuttle.

NASA has helped support this industry by leasing its facilities to private companies and through other programs. This support has been timely in light of the stiff competition American companies will continue to face in the international satellite launch market.

This mixed fleet has also provided the United States with a more balanced launch capability. When the shuttle program was in full swing, ELVs were delegated to a secondary role. But they have reemerged from the background.

The space shuttle, for its part, will continue to fulfill the role for which it is best suited, that of a special utility and research vehicle. An example of the former was a dramatic and televised salvage operation.

In 1992, the shuttle *Endeavor* rendezvoused with an Intelsat VI satellite stuck in a low orbit. After some difficulties, astronauts retrieved the satellite, brought it into the shuttle's bay, and attached a motor for a subsequent boost to its final geostationary position.

Besides contributing to the knowledge base for recovery missions, some of the shortcomings of simulations were revealed. A simulation, which attempts to duplicate the conditions of an actual event, was inaccurate in this case. An alternate plan had to be devised and implemented.

This type of practical experience in dealing with novel situations is important for the future of extravehicular space-based activities. It also highlighted, at this stage of our technology base, the value of the human presence in space. Humans, unlike current robotic devices, can adapt to unique circumstances.[20]

## Arianespace

Arianespace is a private commercial enterprise and an offshoot of another European organization, the European Space Agency (ESA). Arianespace was created in response to NASA's earlier domination of the satellite launch industry.[21] Arianespace has aggressively promoted and marketed ELVs and a sophisticated launch and support operation. Its rockets have a flexible payload capability and can accommodate heavy payloads.

Arianespace has also maintained a competitive price structure, and its launch site in

INFORMATION TRANSMISSION

**Figure 7.2**
Pegasus *launch vehicle.*
*The diagrams highlight*
*its features. (Courtesy of*
*Orbital Sciences Corp.)*

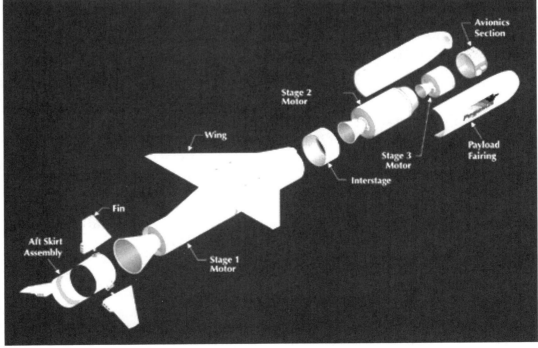

INFORMATION TRANSMISSION

Kourou, French Guyana, is particularly well situated to place satellites in geostationary and other orbital positions. These factors and others contributed to its growing share of the international launch market when NASA was still a participant in the field.

Despite its successes, Arianespace has suffered some of its own setbacks. Satellites have been destroyed by rocket failures, and the organization must face a host of new and potential competitors, including China, private U.S. companies, and Japan.

## The Current and Future State of the Satellite Launch Industry

The commercial launch industry, as indicated, experienced a series of upheavals. While it is true that NASA had left the field, companies and other nations have filled the void. The increased competition triggered by NASA's decision may actually make it easier, in the long run, for an organization to launch a satellite. In fact, a report issued by the Commission on the Future of the United States Aerospace Industry, noted that the supply of potential launch vehicles had actually outstripped the potential demand.[22]

On a bleaker note for the United States, its preeminent position in the satellite manufacturing field has been eroding. During the late 1980s to early 1990s, the United States manufactured 36 communications satellites and Europe and Japan built 23.[23] Prior to this time, the United States had dominated the industry.

This erosion was somewhat mirrored by other U.S. space initiatives. Mixed signals about the viability of the then-proposed space station were sent to the public and international community. The situation was further exacerbated by mishaps, including the loss of Mars probes and a technical problem with the *Galileo,* Jupiter mission.[24] The overall sentiment was summed up in

the early 2000s in the same *Aerospace Industry* report. It was noted "that a 'sense of lethargy' has taken over the U.S. space industry. Instead of the excitement and exuberance that dominated our early ventures into space, we at times seems almost apologetic about our continued investments in the space program."[25]

Nevertheless, there were and are some positive signs. These range from the extended ACTS mission to the development of enhanced satellite systems. Congress also recognized the necessity of supporting a

**Figure 7.3**

*The Ariane 42P, with two solid strap-on boosters. (Courtesy of Arianespace.)*

**Figure 7.4**
*Lockheed Martin's X-33 from different views. (Courtesy of NASA.)*

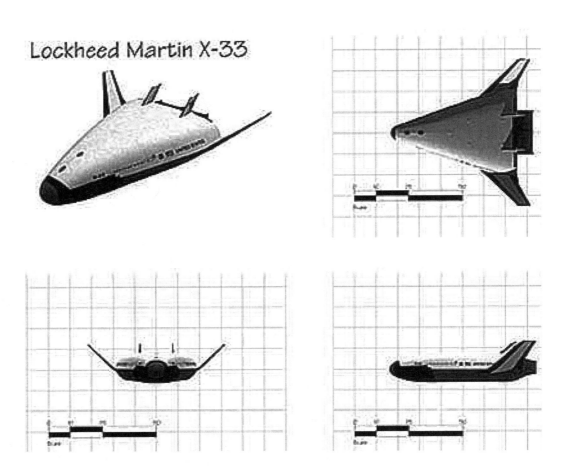

strong satellite and launch industry. As stated, "this industry contributed to the U.S. economy, strengthens U.S. scientific interests, and supports foreign policy and security interests."[26]

***New Ventures.*** The United States and other countries have also tried to make space more accessible and affordable. One proposal was the National AeroSpace Plane Program (NASP) with its experimental X-30 vehicle. The NASP was slated to pave the way for aerospace planes that could take off and land on conventional runways, attain a low Earth orbit, and be reusable instead of refurbishable.[27]

Much like an airline, the term *reusable* implies a quick turnaround time. Unlike the shuttle, a vehicle could be prepared for its next flight without major refurbishing. This capability would save time and money.

The X-30 is also a *hypersonic flight vehicle*, as are other, more recent NASA technology and flight demonstrators, including the Hyper-X series. As envisioned, this new generation of vehicle would "routinely fly about 100,000 feet above Earth's surface and reach sustained travel speeds in excess of Mach 5, or 3750 mph—the point which 'supersonic' flight becomes 'hypersonic' flight."[28] Projected applications for such vehicles, if and when operational, could include retrieving low orbit satellites and servicing the space station.[29] Another spinoff is more down to Earth. New airliners based on this concept could carry passengers between cities faster, for example, in one route from Los Angeles to Sydney, Australia, in 2.5 rather than 13.5 hours.[30]

The goal of making space more accessible, and the search to replace the space shuttle, has similarly influenced other potential systems. One program's objective, for instance, was the development of a reusable launch system through a single-stage to orbit rocket. The Reusable Launch Vehicle Technology Program's objective was to create "technologies and new operational concepts that can radically reduce the cost of access to space. The program will combine ground and flight demonstrations. An important aspect . . . will be the use of experimental flight vehicles—the X-33 and X-34—to verify full-up systems performance. . . ."[31]

Consequently, the goal has been to replace current launch vehicles, where practicable, with cost-effective reusable systems. The Delta-Clipper Experimental (DC-X) rocket already tested key concepts at the White Sands Missile range in the early to mid-1990s.[32] This was followed by NASA's cooperative agreement with Lockheed Martin to work on the X-33 program. One stated goal was to "cut the cost of a pound of payload to orbit from $10,000 to $1000," the magical number that would help make space realistically affordable.[33]

It must also be noted that new designs and plans will inevitably be brought forward. Based on technological, programmatic, and budgetary needs, they will either remain images on a computer screen or possibly enter the prototype and production phases. New initiatives will also include ELVs as designs are generated and existing vehicles are updated and modified to carry heavier payloads.

Finally, Xs made another appearance in 1996 in the guise of the X-prize. Borrowing a tactic from aviation's early days when monetary prizes were offered to advance flying, the modern X-prize seeks to "accel- erate the development of low-cost, reusable vehicles and thereby jump-start the creation of a space tourism industry."[34]

As has also been indicated, satellite launches are not infallible. Accidents do oc- cur, and a launch vehicle and accompanying satellite(s) can be lost. Future launches may also be postponed until the accident's underlying cause has been determined. This waiting period may be unacceptable to an organization with a tight timetable. Thus, an organization may sign contracts with multi- ple vendors to launch its satellites. If one ELV is "grounded," other satellites could continue to be launched.[35] It is hoped, though, that as new ELVs come on line, the number of accidents will diminish and the decades old dream of making space access affordable will be realized—not only for organizations and governments, but also for the rest of us.

## SPACE EXPLORATION

In closing this chapter, it is appropriate to examine an area related to satellite commu- nication: space exploration. Both fields coin- cide to a certain extent, and space probes are sophisticated communications tools in their own right. More important, developments in this field have had an impact on the infor- mation and communications industries.

As will be discussed in Chapter 14, NASA helped pioneer image processing techniques to enhance and correct pictures transmitted by outer space probes. Similar techniques have been applied on Earth in the computer graphics and medical fields, among other application areas. The media has also used remote sensing satellites, originally designed to explore the Earth. Consequently, cross- fertilization can occur between what can be called outer and inner space operations. As such, outer space developments at least merit an overview.

INFORMATION TRANSMISSION

**Figure 7.5**
*The Hubble Space Telescope being refurbished during a shuttle mission. (Courtesy of NSSDC.)*

This section also provides a brief history of NASA and highlights some of the forces that have shaped and continue to shape the space program. These include social and political issues and pressures. The section concludes with an overview of legal implications governing space-based activities.

### History

The 1950s witnessed the birth of the modern era of space exploration.[36] A milestone was the inauguration of NASA on October 1, 1958, as the successor to the National Advisory Committee for Aeronautics (NACA). NACA, founded in 1915, helped advance the nation's aeronautical industry through research and related activities. The new agency was given the same mandate and oversight of the civilian space program.

Some of the space program's major events and influencing factors are as follows:

1. The Soviet Union launches the first artificial satellite, *Sputnik 1*, on October 4, 1957.
2. On April 12, 1961, Soviet cosmonaut Yuri Gagarin becomes the first human in space. The U.S. space program centers about Project Mercury and its seven astronauts.
3. President John F. Kennedy commits the nation to landing an astronaut on the moon before the end of the decade (Project Apollo).
4. On January 27, 1967, a fire in the Apollo command module kills three astronauts.[37] A Russian cosmonaut loses his life in the same year.[38]
5. On July 20, 1969, Neil Armstrong and Edwin ("Buzz") Aldrin of Apollo 11 become the first humans to walk on the moon, while their comrade, Michael Collins, orbits overhead.
6. The Apollo program comes to a halt after Apollo 17 in December 1972, owing to financial considerations and the changing U.S. social and political climate (for example, the Vietnam War).
7. The 1970s and early 1980s witness other missions, including the Apollo-Soyuz Test Project (1975), in which a Soviet and U.S. spacecraft dock in orbit, and the development of the space shuttle.
8. Budgetary constraints and the loss of the space shuttle *Challenger* contribute to the dearth of planetary missions from

the late 1970s until the late 1980s.[39] The potential lack of appropriations for ACTS and other projects could have similarly derailed advanced satellite/planetary spacecraft missions.

9. The *Galileo* probe investigates Jupiter and its moons in the 1990s and the Hubble Space Telescope continues its exploration of the universe from a low Earth orbit. Both systems, though, have suffered from performance problems. A rover, controlled from Earth, explores Mars.

10. Space probes explore the Solar System, including Mercury, Venus, Mars, Jupiter, Saturn, Uranus, Neptune, asteroids, the Moon, and the Sun. Only Pluto remains to be visited by a probe, which may come to fruition at some future date, if a proposed mission is launched.

11. The Earth is explored by different satellites. For example, remote sensing satellites image the Earth for geological, and more recently, media applications.

12. A new series of spacecraft explore Mars, while an international space station orbits the Earth.

Finally, as stated, the space shuttle *Columbia* was lost in early 2003. Engaged in a scientific mission, the shuttle broke-up during re-entry in the Earth's atmosphere. Shuttle pieces and sections were subsequently recovered and reconstructed. As of this writing, the best data indicate a piece of foam insulation, which ripped off the external fuel tank during launch, slammed into the shuttle's wing. The impact damaged a section of the wing's leading edge, which was ultimately breached by hot gas during the re-entry.[40] This triggered a series of events and, ultimately, led to *Columbia's* destruction.

A follow-up investigation indicated that some shortfalls at NASA might have contributed to the loss. They may have included the following:

- ignoring safety warnings by experts,[41] and
- if NASA should have used satellite imagery or a spacewalk to investigate the wing prior to re-entry since the foam impact was observed during launch.[42]

As was the case with *Challenger*, some individuals called for the cessation of the human exploration of space since it was dangerous. Others, however, believed the flights should and must continue. They extend our knowledge and contribute to the centuries-old human exploration of our world, and now, of other worlds as well.[43]

## Legal Implications

As our satellites' capabilities increase and we develop commercial space-based enterprises, legal issues become more important. Relevant topics already discussed are orbital assignments, signal piracy, and the role of Intelsat and competing organizations in the international arena. Additional subjects include the following:

- The UN's concern about the international free flow of information and its balance; not solely from the developed to the developing world.
- The international dissemination of data by remote sensing satellites.
- The U.S. media's use of such images and the impact of government restrictions and regulations.
- The sociopolitical impact of DBS relays if the signals spill over to a neighboring country.[44]
- The "upward extent" of a country's national sovereignty—is it 100 miles, or could it be lower or even higher?[45] How high is high? What are the political implications for satellites with respect to their orbital slots?
- Ownership/property rights in space. Various legal questions remain as to

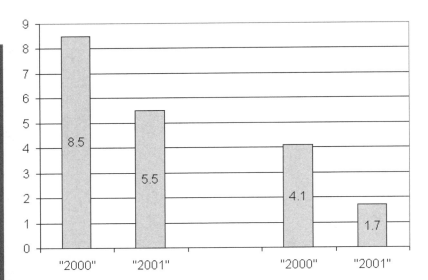

**Figure 7.6**
*The chart highlights the decline in the U.S. satellite manufacturing (8.5 and 5.5) and launch industries (4.1 and 1.7). The figures represent revenues and are measured in the billions of dollars. (Source:* Via Satellite, *"Strategic Planning and Resource Guide," 2002.)*

property rights on the moon and other celestial bodies.[46] In one case, would a company or consortium invest hundreds of millions of dollars to develop a mining operation if its rights were not clearly defined?

• Individual rights for space explorers and/or colonists. The U.S. Constitution

Bicentennial Committee covered this issue in a subcommittee. An outcome, the "Declaration of First Principles for the Governance of Outer Space Societies," declared that the U.S. Constitution should also apply to "individuals living in outer space societies under United States jurisdiction."[47] The document's drafters believed that individual rights, such as freedom of speech, assembly, and *media* and *communications* [my emphasis], are fundamental principles that would extend to U.S. space societies, balanced against the unique environment afforded by outer space.[48]

Finally, there are other legal issues beyond the scope of this book, such as licensing policies and procedures for communications satellites. For the realm of outer space, there's a growing body of space law. It is a fascinating field, and one that will continue to evolve as we begin to take our first outward steps in space.

## REFERENCES/NOTES

1. NASA, "Advanced Communications Technology Satellite (ACTS) Hardware," Information Sheet, 3.

2. Frank Gedney and Frank Gargione, "ACTS: New Services for Communications," *Satellite Communications* (September 1994), 48.

3. OCACT, "Background Information Sheet," downloaded from www.csm.ohiou.edu. *Note:* Ohio University plays a major role in the consortium.

4. Peter J. Brown, "Ka–Band Services: Available in Different Flavors," *Via Satellite* XVI (February 2001), 22.

5. NASA, "Artificial Intelligence Software to Command Mission," Press Release, May 30, 2001, downloaded from http://solarsystem. nasa.gov/whatsnew/pr/010530C.html.

6. James Careless, "Ka–Band Satellites," *Via Satellite* XVI (February 2001), 41.

7. Nick Mitsis, "X–BAND: How Interested Should the Commercial Sector Be?" *Via Satellite* XVIII (August 2003), 33.

8. Brian J. Horais, "Small Satellites Prove Capable for Low-Cost Imaging," *Laser Focus World* 27 (September 1991), 148.

9. Amy Cosper, "Crank Up the Assembly Line," *Satellite Communications* (February 1996), 29. *Note:* NASA has adopted a similar philosophy for some of its outer space missions.

10. Sten Odenwald, "Solar Storms: The Silent Menace," *Sky and Telescope* 99 (March 2000), 54. *Note:* The article provides an excellent overview of solar storms and

their potential impact, including terrestrial implications.

11. Ibid., 55.

12. William J. Broad, "Radioactive Debris in Space Threatens Satellites in Use," *New York Times* (February 26, 1995), 12.

13. Leonard Davis, "Lethal Litter," *Satellite Communications* (January 1996), 24.

14. Following this mission, the satellites were subsequently recovered and returned to Earth via a shuttle.

15. William P. Rogers, Neil Armstrong, David C. Acheson, et al., *Report of the Presidential Commission on the Space Shuttle Challenger Accident* (Washington, DC: U.S. Government Printing Office, 1986), 201.

16. Yale Jay Lubkin, "What Really Happened," *Defense Science* 9 (October/November 1990), 10.

17. The Reagan administration had been a proponent of the commercialization of outer space, especially in the area of the launch industry. See Edward Ridley Finch, Jr. and Amanda Lee Moore, *AstroBusiness* (Stamford, CT: Walden Book Company, 1984), 56–63, for more information.

18. A *Pegasus* launch would have cost approximately $7 to 10 million versus millions of more dollars for a standard ELV.

19. Rick Fleeter, "The Smallsat Invasion," *Satellite Communications* (November 1994), 29.

20. Note that salvage missions of this nature are limited, at least at this time, to low Earth orbits.

21. The ESA and its member states support a wide range of space activities (for example, ELV developmental work and space exploration).

22. Jeff Foust, "Recommendations Issues for Revitalizing U.S. Space Industry," *Spaceflight Now* (November 18, 2002), downloaded from www.spaceflightnow.com/news/n0211/19com mission/.

23. R. T. Gedney, "Foreign Competition in Communications Satellites Is Real," *ACTS Quarterly* 91/1 (February 1992), 1.

24. Despite the mission's successful deployment of a probe into Jupiter's atmosphere, an antenna mishap reduced its transmission rate (for example, of photos).

25. Foust, "Recommendations Issues for Revitalizing U.S. Space Industry," *Spaceflight Now*.

26. The Commercial Space Act of 1997, H.R. 1702, 105th Congress, 2nd session.

27. U.S. General Accounting Office, "National Aero-Space Plane; A Technology Development and Demonstration Program to Build the X-30," GAO/NSIAD-88-122, April 1988, 14.

28. NASA, "NASA Developing Hypersonic Technologies; Flight Vehicles only Decades Away," Press Release, July 22, 2002. Release # 02-182, downloaded from http://www. spacelink.msfc.nasa.gov/NASA.News/NASA. News.Releases/Previous.News.Releases/02. News.Releases/02-07.News.Releases/02-07-22.Hypersonic.Technologies.Developed.

29. See Jim Martin, "Creating the Platform of the Future: NASP," *Defense Science* (September 1988), 55, 57, 60, for more information about the NASP program, including potential military applications. NASP-based vehicles would not eliminate ELVs since one vehicle may not be able to perform all functions equally well. The payload weight and final orbital placement (e.g., geostationary) will also help determine the type of launch vehicle that will be used for a given mission.

30. U.S. General Accounting Office, National Aero-Space Plane, 50.

31. NASA Facts OnLine, Marshall Space Flight Center, "The Reusable Launch Vehicle Technology Program."

32. A witness to a DC-X flight (launch/hovering/flight periods), indicated "the landing was the way God and Robert Heinlein intended." For details, see Marianne J. Dyson, "A Sign in the Heavens," *Ad Astra* 5 (November/December 1993), 17.

33. "Lockheed Martin Selected to Build the X-33," NASA Press Release, July 2, 1996.

34. X-Prize brochure, 1996.

35. James M. Gifford, "Going Up," *Satellite Communications* 20 (February 1996), 33.

36. Parts of this section are taken from Michael Mirabito, "Space Program," in *The Reader's Companion to American History* (New York: Houghton Mifflin Company, 1991),

INFORMATION TRANSMISSION

1013–1014. Houghton Mifflin kindly extended permission for its use.

37. Virgil Grissom, Edward White, and Roger Chaffee.

38. Vladimir Komarov. See J. K. Davies, *Space Exploration* (New York: Chambers, 1992), 193. Three other cosmonauts, Georgi Dobrovolsky, Vladislav Volkov, and Victor Patseysev, died in 1971.

39. Dick Scobee, Mike Smith, Ellison Onizuka, Judy Resnik, Ron McNair, Gregory Jarvis, and Christa MacAuliffe.

40. William Harwood, "Foam Impact Centered on Panel 6 of Wing's Edge," *Spaceflight Now* (March 26, 2003), downloaded from http://spaceflightnow.com/shuttle/sts107/030 326hearing.

41. "Report: NASA Removed Advisers Who Warned on Safety," Reuters (February 3, 2003), downloaded from Yahoo.com.

42. William Harwood, "Shuttle Columbia and Crew Lost," *Spaceflight Now* (February 2, 2003), downloaded from http://spaceflightnow.com/shuttle/sts17/030202columbialost.

43. Space flight proponents supported the continued human presence in space, and in one instance, adopted a quote from Teddy Roosevelt: "Life belongs to those who know the great enthusiasms, the great devotions; who spend themselves in a worthy cause; who at the best know in the end the triumph of high achievement, and who at the worst, if they fail, at least fail while daring greatly, so that their place shall never be with those cold and timid souls who have never known neither victory nor defeat." From Nathan Miller, *Theodore Roosevelt* (NY: William Morrow and Co., Inc., 1992), 507. Thus, as supporters stated, while space exploration can be dangerous, it can also be a worthy cause. *Columbia's* crew: Rick Husband, William C. McCool, David M. Brown, Kalpana Chawla, Michael P. Anderson, Laurel Clark, and Ilan Ramon.

44. Stephen Gorove, "The 1980 Session of the UN Committee of the Peaceful Uses of Outer Space: Highlights of Positions on Outstanding Legal Issues," *Journal of Space Law* 8 (spring/fall 1980), 179.

45. S. Houston Lay and Howard J. Taubenfeld, *The Law Relating to Activities of Man in Space* (Chicago: The University of Chicago Press, 1970), 41.

46. Ty S. Twibell, "Legal Restraints on the Commercialization and Development of Outer Space," *University of Missouri at Kansas City Law Review*, (spring, 1997), downloaded from LEXIS.

47. Nathan C. Goldman, "Space Colonies: Rights in Space, Obligations to Earth," in Jill Steele Meyer, ed., *Proceedings of the Seventh Annual International Space Development Conference* (San Diego, CA: Univelt, Inc., 1991), 220.

48. Rights versus the space environment include the right to bear arms, an important U.S. concept. But in space, where a weapon could physically compromise the integrity of a colony's protective shielding, this issue becomes more complex. The same question applies to the freedom of assembly and the press, among others. This general concept has also been used as the plot in various works, including Robert A. Heinlein's science fiction book, *The Moon Is a Harsh Mistress* (New York: Berkeley Publishing Corp., 1968). For more information, see William F. Wu, "Taking Liberties in Space," *Ad Astra* 3 (November 1991), 36.

## SUGGESTED READINGS

Abutaha, Ali F. *The Space Shuttle: A Basic Problem.* This videotape was produced by George Washington University, Washington, DC. This is a taped lecture conducted by Ali Abutaha for George Washington's Continuing Engineering Education program. The tape covers shuttle design problems discovered by Abutaha after the *Challenger* disaster.

Banke, Jim. "Ticket to Ride." *Ad Astra* 8 (January/February 1996), 24–26; Dr. Peter

Diamandis. "The 'X' Prize." *Ad Astra* 7 (May/June 1995), 46–49. Space tourism development and a look at the X-prize and aviation examples.

Boeke, Cynthia. "On the Road." *Via Satellite* XVI (September 2001), 76–79; John Kross. "Fields of Dreams." *Ad Astra* 8 (January/February 1996), 27–31; Justin Ray. "Delta 4 Fleet Goes from 'Medium' to 'Heavy'." *Spaceflight Now* (November 12, 2002), downloaded from www.spaceflightnow.com; Renee Saunders. "Rocket Industry Agenda for the Next Millennium." *Satellite Communications* 19 (July 1995), 22–24. The U.S. and international outlooks for spaceports and rockets.

Fleeter, Rick. "The Smallsat Invasion," *Satellite Communications* 18 (November 1994), 27–30. Smallsats.

Gedney, Frank, and Frank Gargione. "ACTS: New Services for Communications," *Satellite Communications* 18 (September 1994), 48–54; NASA. "Advanced Communications Technology Satellite (ACTS)," downloaded from http://acts.grc.nasa.gov. The ACTS satellite, a review of its first year of operation, and its influence on other satellite designs. The second article includes retirement information.

Hardin, R. Winn. "Solid-State Lasers Join the Space Race." *Photonics Spectra* 32 (June 1998), 114–118; Walter L. Morgan. "Pass It Along." *Satellite Communications* 22 (May 1998), 50–53. Laser applications in satellite communication.

Kerrod, Robin. *The Illustrated History of NASA: Anniversary Edition.* New York: Gallery Books, 1988. A comprehensive and richly illustrated history of NASA and the U.S. program.

Lewis Research Center, NASA. *ACTS Quarterly.* A Lewis Research Center newsletter that traces the development, launching, and testing of the ACTS satellite. As a collection, it is an interesting overview of the birth and life of a satellite, as well as the impact of external forces (for example, budgetary appropriations).

*Military & Aerospace Electronics* covers new launch vehicle/propulsion developments, among other topics. Three articles include: John Keller. "Avionics Innovation Marks New Space Shuttle." (April 1997), 1, 7; John McHale. "Electrical Xenon Ion Engine to Power New Millennium Spacecraft." (June 1997), 1, 33; John Rhea. "Avionics Key in Drive to Cut Space Launch Costs." (December 1998), 1, 8.

NASA/JPL. CASPER. (Artificial Intelligence Group). Overview of autonomous space-based missions and other information, http://www.aig.jpl.nasa.gov/public/planning/casper/.

*Sky and Telescope.* The November 1993 issue had a series of articles/sidebars that focused on the Hubble Telescope and its repair. An example is Richard Tresch Fienberg, "Hubble's Road to Recovery," 16–22.

*Space Law.* Numerous articles and books cover space law (the last article covers space debris) and satellite regulatory issues. These include: Carl J. Cangelosi. "Satellites: Regulatory Summary," in Andrew F. Inglis, ed. *Electronic Communications Handbook.* New York: McGraw-Hill Book Company, 1988, 6.1–6.9; Ty S. Twibell. "Legal Restraints on Commercialization and Development of Outer Space." *University of Missouri at Kansas City Law Review.* (spring 1997). (65 UMKC L. Rev. 589), downloaded from LEXIS; Alan Wasser. "A New Law Could Make Privately Funded Space Settlement Profitable." *Ad Astra* 9 (July/August, 1997), 32–35; Christopher D. Williams. "Space: The Cluttered Frontier." *1995 Southern Methodist University Journal of Air Law and Commerce.* May/June 1995 (60 J. Air & Com. 1139), downloaded from LEXIS.

## GLOSSARY

*Advanced Communications Technology Satellite (ACTS):* A NASA satellite that is a prototype for the future commercial fleet.

*Arianespace:* A commercial satellite launch agency.

*Challenger:* The space shuttle that was lost to equipment and systems failures.

*Launch Vehicles:* Both expendable (ELV) and piloted vehicles used to launch satellites and space probes. Two launch organizations have included NASA and Arianespace, and other countries have either developed, or are developing, their own capabilities.

*NASA:* The U.S. space agency given the mandate to oversee the civilian space and aeronautical programs. Also the successor to the NACA.

*Reusable Launch Vehicle Technology Program:* Program designed to create the next-generation U.S. launch vehicle.

*Smallsat:* A small, relatively inexpensive satellite.

*Space Shuttle:* The world's first refurbishable piloted spacecraft. A shuttle can carry a variety of payloads to a low Earth orbit.

*X-Prize:* Echoing back to an earlier era, a proposed monetary prize to advance reusable launch vehicle technology (for example, to develop space tourism).

# 8 Wireless Technology and Mobile Communication

Wireless technology comes in different flavors. As implied by the name, you are not physically connected with or tied to a communications line. One application supports data exchanges in a building while another extends this relay to another country. In essence, the wireless industry embraces an assortment of technologies and applications you can tap to solve your communications needs.

For our discussion, we also focus on systems that support mobile business and personal communication. These include cellular telephones (cell phones), personal communications services (PCS), certain satellite relays, the virtual office, and wireless local area networks (WLANS). But before we cover these topics, other wireless systems are quickly reviewed for a full coverage of this universe.

## WIRELESS SYSTEMS

### Microwave, Laser, and Cable

Microwave systems have handled long, short, and intracity relays. A microwave system is cost effective and can accommodate a range of information with a wide channel capacity. Microwave applications may also be easier to implement than fiber-optic or copper-based ones. You do not, for instance, have the potential problem of obtaining clearances to lay the cable.

On a negative note, a microwave transmission could be affected by heavy rain. An FCC license has also been required and, as a line-of-sight medium, the transmitter and receiver must be in each other's line of view.

Another wireless system uses an infrared laser to relay voice, video, and computer information through the air. It is cost effective and can support a wide and secure communications channel. An FCC license is not required, and operations could be set up in areas where microwave communication may not be feasible.[1] A laser relay is, however, line of sight. Smog and other atmospheric conditions could also variably affect it.

Wireless technology has also been used by the cable industry. Wireless relays existed for a number of years under the aegis of the Multichannel Multipoint Distribution Service (MMDS).[2]

Finally, the communication transactions supported by some of these wireless systems have been conducted as bypass operations. An organization bypasses, or goes around, the traditional public communications networks.[3]

Thanks, in part, to the divestiture of AT&T in the early 1980s, a new dawn greeted the telecommunications world. Industry segments once dominated by AT&T

**Figure 8.1**
*An over-the-air (atmospheric) optical communications system. It can accommodate a range of information. (Courtesy of ICS, Inc.)*

became open markets for sales and leasing opportunities.

Companies also became more responsible for their own communications systems. Various intracity and intercity links were established that could be more responsive to an organization's unique communication needs. Thus, a company could more readily react to new communication demands since it used private or leased links.[4]

The latter concept is a key one. Communications systems will conform to our communication needs and not the other way around.

## MOBILE WIRELESS SERVICES

In the context of our present discussion, mobile wireless systems could be viewed as personal bypass systems—the important concept for us is mobility. Various technological and social changes have combined to make us a society on the go, a mobile society. Our information and communications tools are following suit.

- Powerful notebook computers are smaller and weigh less.

- We conduct business from our cars and can use the same telephone to call home.
- Cell phones and pagers are no longer relegated to only a few professions.
- Satellite-based relays can support everyday communication.

A goal is to free us from physical wires—constraints. As stated by one author when describing such an application, "PCS is a new wireless mobile technology for voice and data communication to and from *people*, not *locations* [my emphasis]."[5] Consequently, we may no longer be bound to a physical space to communicate or to exchange information. Communication would be centered on us rather than an office or other site.

## CELLULAR TELEPHONE AND PERSONAL COMMUNICATION SERVICES

The cellular telephone industry, a key wireless player, is a communications fixture. A specified geographical region is divided into small physical areas called *cells*, each of which is equipped with a low-power transmission system.

In a typical operation, as you approach a cell's boundary while driving, the signal between the phone and the transmitter becomes weaker. At this point, the new cell the car is approaching basically picks up the connection. The telephone is then switched over to a different frequency, to avoid potential interference with adjacent cells. This procedure is automatically completed by a sophisticated control network.

Cellular technology has also been integrated with portable PCs for remote data relays. In a related development, digital operations support a cleaner signal, for both data and voice, and provide for a more secure (private) relay.[6]

INFORMATION TRANSMISSION

The cellular technology universe exploded in the late 1990s and early 2000s. New price breaks made it possible to replace your conventional home telephone service with a cell phone. Features also abound. These range from color screens to Internet search/e-mail capabilities to functioning as a worldwide phone. For the latter, different international standards exist. With the appropriate phone, you could use the same unit in more than one country or continent. As briefly covered in Chapter 3, cell phone and PDA capabilities have also been married in some units.

Like the computing field, these technological enhancements did *not* come with a physical price tag. Earlier phones were very bulky—you could not fit one in your pocket unless you wore a trench coat. Newer models, which are also more sophisticated, are compact, fairly lightweight, and have a longer battery life than their predecessors.

The last point is important. Taking a page from a systems approach, battery enhancements have helped fuel the cell phone, notebook computer, and other markets. Batteries are now more efficient with enhanced service capabilities. As such, they contribute to the continued growth of the portable communications market.

While the cell phone picture is rosy, there are limiting factors. Some may also affect other communications systems and include interference created by terrain and human-made obstructions (e.g., buildings). In another example, if a company is creating a new network, its initial coverage area may not be as extensive as an older, but more established, competing system.

Another factor is a potential health concern. Analogous to the situation with earlier computer monitors, some experts believe the cell phone is a potential health hazard. As a transmitting device, the cell phone produces radio frequency (RF) energy. Critics contend that with extended use, a cell phone could cause a biological and, potentially, a health effect. Proponents argue that no such link exists.

A key measurement, which could help determine a cell phone's potential impact, is the *Specific Absorption Rate* (SAR). According to the FCC, the "SAR is a value that corresponds to the relative amount of RF energy absorbed in the head of a user of a wireless handset. The FCC limit from public exposure from cellular telephone [as of this writing] is an SAR level of 1.6 watts per kilogram . . ."[7] Cell phones are tested and are assigned an SAR rating—as the SAR value decreases, so too does the possibility of a potential adverse effect.[8]

It is important to repeat that, as of this writing, a direct link *has not* been established between cell phone use and health effects. But since the data are also not conclusive, it makes sense to take some precautions until the facts become clearer. You can, for instance, select a cell phone with a lower SAR value. These figures are available on different Internet sites.[9] The FCC web site (www.fcc.gov) is another valuable information resource. Besides offering information about RF safety in general, you can look-up your phone's SAR value by its FCC I.D. number.[10] A second option is to keep the phone unit away from your head by using a headset or an earpiece, much like a component you may use with a portable CD player.[11] Manufacturers are also designing safer phones, including two piece units where the transmitting component is at a greater distance from your head.

In a related health issue, cell phone use has caused numerous car accidents. Some drivers focus on holding/talking on the phone rather than on road conditions. In response, new regulations have gradually been adopted. In some locales, you must use a "hands free" accessory if you want to talk while driving.[12]

INFORMATION TRANSMISSION

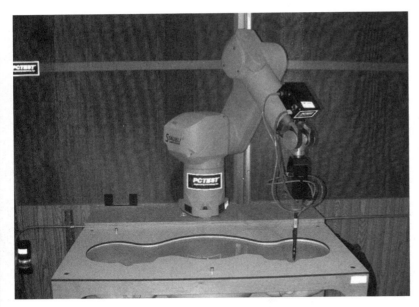

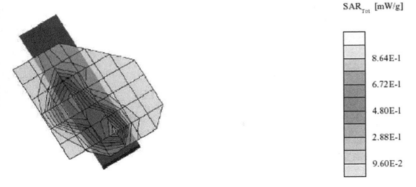

SAR$_{Tot}$ [mW/g]

8.64E-1

6.72E-1

4.80E-1

2.88E-1

9.60E-2

**Figure 8.2**

*A set-up to test a mobile phone's SAR characteristics and a plot that highlights this information. (Courtesy PCTEST Engineering Lab., Inc.)*

lular carriers must comply with more detailed technical and operational requirements, such as rules regarding mandatory provision of analog service, licensing, and interference criteria, that PCS licensees are not subject to.[14]

PCS companies have another advantage. As the new kid on the block, technologically speaking, a company could use the "newest network and digital technologies."[15] As such, the company was better positioned to tap technological advances and their complementary applications since it was not bound by an existing infrastructure. In fact, one early PCS promotional campaign highlighted the service's digital roots, which would make for a more secure relay, in contrast with its analog cellular competitors.[16] But over time, cellular telephone companies extensively upgraded and enhanced their own networks to level out parts of this playing field.

As of this writing, the average user who opts for a mobile communication service may not even know the type of technology he or she is using. In many instances, the term "cell phone" had become a universal moniker regardless of the technology base.

Personal communication services (PCS), for its part, has been defined by the FCC as "a family of mobile or portable radio communications services" designed to meet the "communications requirements of people on the move."[13] A digital service, the premise, much like the cell phone, has been to support mobile, wireless communication.

PCS companies also have some advantages over their cell phone competition. For example,

PCS licensees have greater leeway to choose the types of technologies and services they may provide than do cellular carriers. . . . Although cellular licensees may also provide alternative technologies as well as wireless fixed services, cel-

## Other Developments

The PCS industry experienced another, somewhat new development. Traditionally, segments of the communications industry have used the airwaves for free. But to help defray the budget deficit, auctions were initiated for specific U.S. PCS licenses. Companies paid for the right to use the spectrum allocations.

The international mobile communications industry was also kicked into higher gear by the development of advanced mobile wireless—*third generation (3G)*—systems. Such systems support enhanced services, including high speed data access and communications devices that could tap terrestrial and satellite networks.[17]

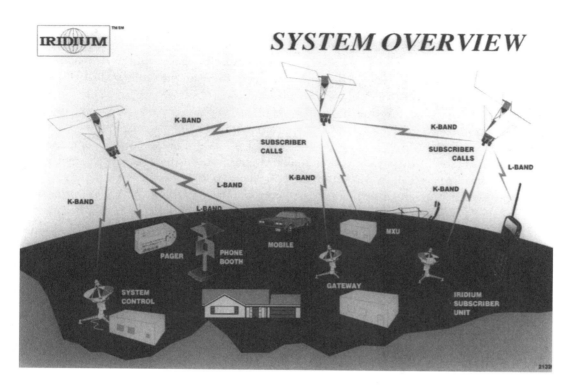

*Figure 8.3*

*A space and earth-based communication operation can support a sophisticated personal communications network. The various elements are highlighted in this diagram, including the satellites and personal communications device (subscriber unit; right bottom hand corner). (Courtesy of Motorola Inc.; Iridium.)*

INFORMATION TRANSMISSION

While spectrum allocation issues have delayed U.S. operations, they have been deployed on a broader level in the international market. In one example, "the launch of a 3G service . . . by NTT DoCoMo in Japan . . . allows users to access the Internet at speeds of up to 384 kbps, transmit and download video clips, and send large data files quickly."[18] The U.S. market actually received a boost for enhanced wireless systems in 2003 following FCC action.[19]

When fully implemented, we may be able to retrieve a range of data types with international-compatible mobile devices. But will consumers become willing buyers since enhanced services typically carry enhanced price tags? Or like cell phones, will it take a number of years and reduced rates to promote its broadbased adoption? Or could two markets evolve, basic and enhanced mobile services, analogous to dial-up modem and DSL/cable modem access? These questions are also pertinent for digital television and other application areas.

## Satellite Communications

Satellites are powerful long distance and point-to-multipoint communications tools. Satellites also support mobile and personal relays, and this development is a space-based extension of mobile communications technology.

In one configuration, a series of small satellites are launched and placed in low Earth orbits (LEOs).[20] This constellation of satellites can provide global coverage while their low altitudes support portable transmitters and receivers on Earth. Applications include data/voice communication over small terminals and service for regions that are not covered by terrestrial communications networks.[21]

INFORMATION TRANSMISSION

Different companies have stepped forward in this field. Iridium, for instance, was set-up as a multisatellite network that would function as an interconnected Earth- and space-based venture—you could talk with a friend half a world away through portable, mobile telephones.[22]

All in all, it is an interesting concept, especially in light of the history of our satellite system. As you may recall, LEO satellites were generally abandoned in favor of their geostationary counterparts for communications applications. Yet the new generation of LEO satellites, when properly configured and coordinated, can deliver certain satellite services right into our hands. It is almost like Dick Tracy and the wrist radio come to life.

But there have been some problems. Competition with established terrestrial systems as well as higher service costs in some quarters (e.g., for a telephone call) have caused some financial turbulence in this market.[23] Nevertheless, satellite-based mobile services remain a viable communication option for various reasons, including the following:

- system enhancements have improved and expanded satellite services;
- a satellite can reach remote and undeveloped regions;
- by extension, a satellite can support highly specialized and targeted market groups (e.g., maritime industry); and
- other initiatives, including the Memorandum of Understanding regarding Global Mobile Personal Communications by Satellite (GMPCS-MoU), have promised to facilitate international satellite-based mobile services.[24]

The upshot of these developments? Over time, it may become easier and less expensive to use a satellite system for mobile and global communications needs.

## Virtual Office and Wireless Local Area Networks

Organizations have either experimented with or have adopted the virtual office for elements of their workforce.[25] If you are a salesperson, you may no longer be confined to an office. You may be equipped with a portable computer and other information and communications tools so you can work on the road. The virtual office has also been considered a "variant of telecommuting," discussed in a later chapter.[26] But because of some of the virtual office's wireless possibilities, it has been placed in this chapter.

Through the virtual office, a salesperson can work directly with clients in the field and, depending on the application, could tap either wireless or wire-based communications systems. The former will also become more important as they are refined. You can be untethered, that is, free from constraints imposed by physical wires and connections. In one sense, your communications tools become as mobile as you.

Besides serving the client base better, a virtual office can help cut real estate costs because fewer offices have to be set up and maintained.[27] If properly implemented, a company could enhance its productivity while cutting costs. The key to success, though, is proper implementation and addressing key issues. These include securing a suitable communications channel, offering employee training, and providing clerical and other logistical support.

Companies have also established special work sites to provide "an office away from the office for transient workers."[28] For Xerox, this meant creating a large workspace where employees could meet, exchange ideas, and plug-in their PCs when they're not on the road.

A wireless local area network (WLAN), for its part, is just what it sounds like. You can tap a LAN's resources without a direct, physical connection. In one application, doctors and nurses working in a hospital can readily gain access to information.[29] In another application, you can go to a meeting and retrieve pertinent data, or record notes, without physically connecting your PC or other device. Educational institutions are also using the technology to provide their students and faculty with access to information throughout a campus.

As of this writing, wireless components are more expensive than their wired counterparts. But the afforded flexibility is valuable, especially for applications that demand mobility and for facilities that cannot be readily wired for conventional LANs.[30] It may also be used for locations where you can only install a limited number of data ports, yet must support numerous individuals.

Another potential concern is security. Since data are relayed through the air, the data are easier to intercept. While special encryption techniques may make the data unintelligible to the average user, other individuals could use software to "sniff" your network and, subsequently, to break the encryption—the data are intercepted and stolen.[31]

It is important to remember that a wireless network's advantages also make it more vulnerable. But as is the case with other databased applications, stronger security measures are in development.

Wireless technology also became a household tool in the early 2000s. In a typical operation, you could use a wireless system to connect your household computer to your data network. Like a commercial operation, you could then use your notebook or other computer untethered.

The same era saw the broader introduction of *Bluetooth* technology. Designed as a cost-effective personal area network, Bluetooth supports comparatively low data rate/short distance communication, potentially between an assortment of devices. The

standard has broad industry support, which should lead to equipment interoperability.[32] Bluetooth technology, embedded in communications and computer equipment, could also make these transactions transparent to users.

Finally, you may run across the term *Wi-Fi* or wireless fidelity. Wi-Fi is a more recent, and now popularized name, for a wireless network that taps the IEEE 802.11 family of standards.[33] There is also a Wi-Fi Alliance, which helps promote and certify equipment interoperability.[34] One goal is to make wireless networking ubiquitous in the workplace and at home.

## CONCLUSION

Wireless systems may free you from having to plug your computer or telephone into an outlet for certain communications applications. They can also work with traditional communications systems, as may be the case with a virtual office. Even though you may be untethered, a conventional telephone line could be used for data relay. In many ways, the wireless and wired worlds are complementary.

Wireless technology may also allow you to receive e-mail while in the field and to relay data to your office. You could potentially tap into a network, dial a specific number, and reach another person, wherever he or she may be, via a portable telephone. It is like super call forwarding, free of the constraints imposed by wires, distance, and location.

New developments will also continue to refine our wireless systems. In one example, very small cells, called microcells, "are being designed to take care of downtown users on city streets, and picocells, for inside office buildings."[35] Essentially, our communications world could be subdivided into smaller, physical regions (e.g., a building or floor in a building) to handle our increasing wireless traffic.

On the flip side, wireless communication carries a price. Wireless components can be more expensive, connections can be sporadic, and security concerns continue to surface.[36] Much like earlier computer monitors, questions have also been raised about cell phones and potential health risks.

From a nontechnical standpoint, wireless technology has another implication. If a call can be routed to you, and if your home becomes a workplace, what impact will this have on your individual privacy? Will the "electronic clutter" become too pervasive?[37] Is it necessary to have access to information at every waking moment? Or do we also require quiet, reflective periods, free from distractions, even in a business setting? How do you strike a balance between the two?

Finally, the wireless field also illustrates the interdependence of technological developments. The launching of an Iridium-type network serves as an example. The technology now exists to build sophisticated small satellites that are matched by portable telephones and other communications devices. This entire system, in turn, is influenced by the satellite launch industry. Without cost-effective launch vehicles, the deployment of a constellation of satellites could be prohibitively expensive.

Thus, as introduced in Chapter 1, although you can examine individual applications in isolation, it may also be important to explore related areas. If you do not, you might miss key connections that could have a major impact on an industry's success. In this example, cost-effective launches may play a role in an Iridium-type system's future. So too does its acceptance by the targeted user group(s).

## REFERENCES/NOTES

1. This has included high-density urban areas where physical obstacles and the lack of licenses have restricted microwave communication.

2. Mark Hallinger, "Wireless Cable Ready for the World," *TV Technology* 13 (July 1995), 1.

3. Dwight B. Davis, "Making Sense of the Telecommunications Circus," *High Technology* (September 1985), 20.

4. These also include fiber and copper lines.

5. Lou Manuta, "PCS's Promise Is in Satellites," *PCS* (September 1995), 14.

6. Robert Corn-Revere, "Cellular Phones: Only the Illusion of Privacy," *Network World* 6 (August 28, 1989), 36. *Note*: Data relays could also be intercepted.

7. FCC. "Radio Frequency Safety-Cellular Telephone Specific Absorption Rate (SAR)," downloaded from www.fcc.gov.

8. Health Council of the Netherlands, "Mobile Telephones," January 28, 2002. The report notes the importance of differentiating between a *biological* versus a *health effect*. It also covers this issue in a comprehensive fashion. The report supports mobile communication proponents who believe the phones are essentially safe.

9. Cnet has kept such a list on its site.

10. As of this writing, the URL is: www.fcc.gov/oet/fccid/. The FCC I.D. # is on the telephone; you may have to remove the battery to find it.

11. This may, in turn, subject that part of the body to higher RF energy—but the head has been the primary concern in the typical report.

12. These include headsets with mics for talking.

13. From an August 1992 FCC Notice of Proposed Rule Making and Tentative Decision; GEN Docket #90–314, prepared by Scott Loftesness, August 25, 1992, downloaded from CompuServe.

14. FCC, "Cellular Operations," November 19, 2002, downloaded from http://wireless.fcc.gov/services/cellular/operations.

15. Maxine Carter-Lome, "Cellular in the Lead, But Can't Stay Ahead," *PCS Today* (September 1995): 8.

16. Harry A. Jessell, "Sprint Unveils U.S. PCS Service," *Broadcasting & Cable* (November 20, 1995): 47.

17. FCC. "Third Generation Wireless, 3G Information," downloaded from www.fcc.gov/3G/.

18. FCC, "In the Matter of Service Rules for Advanced Wireless Services in the 1.7 GHz and 2.1 GHz Bands," WT Docket No. 02-353, Notice of Proposed Rulemaking, adopted: November 7, 2002, Released: November 22, 2002, downloaded from http://hraunfoss.fcc.gov/edocs_public/attachmatch/FCC-02-305A1.txt.

19. "FCC Reallocates Spectrum for New Wireless Services," Press Release, January 30, 2003, downloaded from http://hraunfoss.fcc.gov/edocs_public/attachmatch/DOC-230754A1.pdf.

20. There are also different categories of LEOs. Part of the distinction has to do with a satellite's size and relay capabilities (e.g., signal strength).

21. Orbital Sciences Corporation, "Orbcomm Signs Marketing Agreement in Five More Countries," news release.

22. Jim Foley, "Iridium: Key to Worldwide Cellular Communications," *Telecommunications* 25 (October 1991): 23. *Note:* The ground-based sector would handle, in part, billing and the necessary authorization to use the system. This would be provided through gateways.

23. Alan Pearce, "Is Satellite Telephony Worth Saving," *Wireless Integration* (September/October, 1999): 19. Iridium, for one, went through such a financial period.

24. "FCC Proposes Steps to Implement 'GMPCS-MOU,' Facilitating Satellite Services while Protecting Against Interference to Radionavigation Services," Report No. IN 99-9, February 25, 1999, downloaded from www.fcc.gov/Bureaus/International/News_Releases/1999/nrin9010.html.

25. Michael Nadeua, "Not Lost in Space," *Byte* 20 (June 1995): 50.

26. Osman Eldib and Daniel Minoli, *Telecommuting* (Norwood, MA: Artech House, 1995), 1.

27. Patrick Flanagan, "Wireless Data: Closing the Gap Between Promise and Reality," *Telecommunications* 28 (March 1994): 28.

28. Nadeua, "Not Lost in Space," 52. *Note:* Much like telecommuting, interpersonal communication between employees was still important.

29. Elisabeth Horwitt, "What's Wrong with Wireless," *Network World* 12 (November 13, 1995): 65.

30. Flanagan, "Wireless Data," 30.

31. Sniff refers to monitoring network traffic. Please see www.webopedia.com/TERM/s/sniffer.html for more information. The webopedia is also an excellent information source about computer/information technologies.

32. For more information, please see Formulasys, "The Basics of Wireless," downloaded from www.formulasys.com/Whitepapers/Basics-of-Wireless_Formulasys.pdf.

33. Wi-Fi Alliance, "Wi-Fi Overview," downloaded from www.wi-fi.org/OpenSection/why_Wi-Fi.asp?TID=2.

34. For more information, you can go to the Alliance's web site, www.wi-fi.org/OpenSection/index.asp.

35. Robert G. Winch, *Telecommunication Transmission Systems* (New York: McGraw-Hill, 1998), 363.

36. Horwitt, "What's Wrong with Wireless," 64.

37. Patrick M. Reilly, "The Dark Side," *Wall Street Journal* (November 16, 1992): technology supplement on "Going Portable," R12.

INFORMATION TRANSMISSION

## SUGGESTED READINGS

Bernard, Josef. "Inside Cellular Telephone." *Radio-Electronics* 11 (September 1987), 53–55, 93; Jim Rendon. "It's Not Just a Phone, It's a Lifestyle." *Mbusiness* (May 2001), 52–53. An earlier overview of this field and one of a series of articles about cell phone use/development in Finland, respectively.

Brewin, Bob. "Watch Out for Wireless Rogues." *ComputerWorld* 36 (July 15, 2002): 36; John McHale. "Wireless Devices Link Soldiers on the Digital Battlefield." *Military & Aerospace Electronics* (January 2001): 22–25, 29; Ellen Messmer. "Feds to Clamp Down on Wireless LANS." *Network World* 19 (August 19, 2002): 12; Florence Olsen. "The Wireless Revolution." *The Chronicles of Higher Education* (October 13, 2000): A59–A62; Marisa Picker. "It's a Wireless World." *Mobile Computing & Communications* 10 (May 1999): 95–101; Jeffrey R. Young. "Are Wireless Networks the Wave of the Future?" *Chronicle of Higher Education* XLV (February 5, 1999): A25–A26. Wireless technologies, applications, and implications.

Caldwell, Bruce, and Bruce Gambon. "The Virtual Office." *Information Week* (January 22, 1996): 33–36, 40. Telecommuting, requirements, and is it appropriate for you?

CIT Publications. "A Wireless Decade: A European Survey." *Telecommunications* 32 (September 1998): 70–79. Comprehensive examination of wireless systems in Europe. Individual countries are highlighted.

Dornan, Andy. "Wireless Optics: Fiber Is Cheap, but Space Is Free." *Network Magazine* 17 (September 2002): 28–32; Michael Fink. "Lasers over Manhattan." *Communications News* (May 1994): 30; Dick Guttendorf. "Flash Gordon Meets Ma Bell." *Communications Industries Report* (September 1995): 26. Over-the-air laser communications systems.

Eldib, Osman, and Daniel Minoli. *Telecommuting.* Norwood, MA: Artech House, 1995. Excellent and comprehensive examination of telecommuting.

Flanagan, Patrick. "Personal Communications Services: The Long Road Ahead." *Telecommunications* 30 (February 1996): 23–28; Rob Frieden. "Satellite-Based Personal Communication Services." Telecommunication 27 (December 1993): 25–28; Susan O'Keefe. "The Wireless Boom." *Telecommunications* 32 (November 1998): 30–36. Personal communications services: terrestrial and satellite-based.

Formulasys. "The Basic of Wireless; A White Paper," downloaded from www.formulasys.com.; Tim Kridel and Meg McGinity. "The Trouble with Bluetooth." *The Net Economy* (January 8, 2001): 44–46; Verizon. "How Bluetooth Short Range Radio Systems Work," downloaded from www.verizon.com. Information about Bluetooth. The first document also presents an overview of wireless technology/systems.

Foster, Kenneth R, and John E. Moulder. "Are Mobile Phones Safe." IEEE Spectrum Online, downloaded from www.spectrum. ieee.org/publicfeature/aug00/prad.html; FCC. "Information on Human Exposure to Radio Frequency Fields from Cellular and PCS Radio Transmitters," downloaded from www.fcc.gov/oet/rfsafety/cellpcs.html; Reuters. "Italian Study Raises Concerns About Mobile Phones." October 24, 2002, downloaded from http://story.new.yahoo. com/news?tmpl=tory&u=/nm/20021023/ sc_nmhealth_mobilephon. Mobile phones and potential health issues.

Grambs, Peter, and Patrick Zerbib. "Caring for Customers in a Global Marketplace." *Satellite Communications* (October 1998): 24–30; Corporate Information News Center. Iridium Press Release: "Iridium LLC Investors Commit Funding to Ensure Service During Restructuring Discussions." (December 9, 1999); Iridium Press Release, "About Iridium-Our Story" (Corporate Fact Sheet), downloaded from www.iridium.com/corp/ iri_corp-story.asp?storyid=2; "Iridium Constellation Finishes Launch Deployment." *Ad*

*Astra* 10 (July/August 1998): 12. A look at the Iridium system; from coverage of its operation to its financial restructuring.

Meeks, Brock N. "Spectrum Auctions Pull in $7.7 Billion." *Inter@ctive Week* 2 (March 1995): 7. PCS auctions.

NTIA, "An Assessment of the Viability of Accommodating Advanced Mobile Wireless (3G) Systems in the 1710–1770 MHz and 2110–2170 MHz Bands," July 22, 2002, downloaded from www.ntia.doc.gov. An excellent overview of the development of 3G U.S. services and key issues in regard to spectrum allocations.

Vitaliano, Franco. "How Work Becomes Remotely Possible." *BackOffice* (January 1996): 43–51. Portable technologies; includes a list of terms.

## GLOSSARY

*Bypass System:* A private/leased communications system that bypasses standard commercial and public systems.

*Cellular Telephone:* A personal communications tool based on frequency reuse and a monitoring design.

*Microwave and Laser:* Two wireless, line-of-sight communications systems.

*Personal Communications Services (PCSs):* A family of mobile services designed to meet the communication requirements of people on the move. Computer-to-computer and voice relays can be supported.

*Personal Satellite Communications:* A new generation of satellite can deliver personal communications services—you are not restricted by location. In one configuration, satellites are placed in low earth orbits.

*Third Generation (3G) Systems:* A newer generation of mobile communications systems that will support enhanced communications applications.

*Virtual Office:* Instead of working in a traditional office, you are equipped with a PC and mobile communications tools so you can work/communicate from the field. Sites have also been designed to serve this workforce when not in the field (for example, a place to plug in your PC).

*Wireless Local Area Network (WLAN):* As implied, you can tap a LAN's resources without a direct, physical connection.

**INFORMATION TRANSMISSION**

# III

## INFORMATION STORAGE

# 9 Information Storage: The Optical Disk and Holography

The optical disk emerged as an important information storage tool in the 1980s. A popular application is the compact disk (CD), a small, round disk that stores digital audio information in the form of microscopic pits. To retrieve this information, you place a CD in a player.[1] A laser subsequently reads back or recovers the information.

A laser's light scans the CD, and its beam is reflected to different degrees, in terms of its strength, when it passes over the pits and unpitted areas called *lands*. A light-sensitive detector picks up the reflected light, an optical representation of the stored information. After processing, the final output for a CD is an analog signal.

The CD and other optical disks are also constructed like sandwiches: These include the information layer with the code of pits and a reflective metallic layer. The latter enables the read or playback operation.

## OPTICAL DISK OVERVIEW

For our discussion, the growing optical disk family falls into two categories: nonrecordable and recordable media. CDs, conventional Compact Disk-Read Only Memory (CD-ROM) and Digital Versatile or Video disks (DVDs) are nonrecordable. Write Once, Read Many (WORM) and erasable systems, including members of the CD-ROM and DVD families, are recordable.

Both categories of disks share some characteristics:

1. Information can be stored in the form of pits, or in erasable systems, through other techniques. This information is also digital, with the major exception of the videodisk.
2. Optical disks are fairly rugged since the stored information is physically protected from fingerprints and scratches. The laser is also focused beneath a disk's surface at the information layer, so dust and other minor surface obstructions may not adversely affect a playback.
3. A disk is not subject to wear because the playback is conducted by a beam of light. The same disk can be played multiple times with no discernible loss of quality.
4. Disks are not indestructible, however. For example, deep scratches can affect a playback. Some older disks may also have a manufacturing defect—corrosion of a disk's metallic layer.[2]
5. Optical disks are high-capacity storage media. This is partly a reflection of a laser's capability to distinguish between tightly recorded information tracks.
6. The different systems incorporate sophisticated error-detection and checking

**Figure 9.1**
*We have the tools to produce professional quality recordings, which in this example, includes an audio CD. (Courtesy of Sonic Foundry; CD Architect.)*

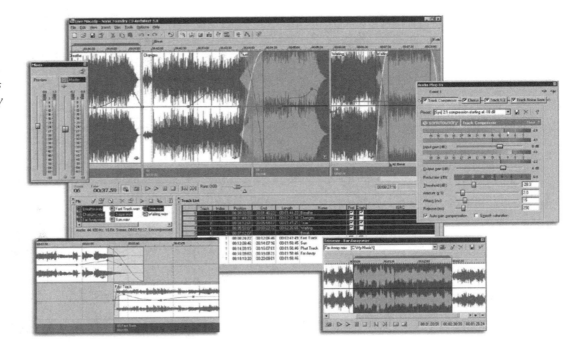

schemes to ensure the information's integrity. But the potential impact of errors has made this process more critical for a disk that stores data than for a CD.

7. Like a hard disk, an optical disk is a random access device. It provides almost immediate access to the stored information.

8. The optical storage field is expanding. Even though some formats may disappear, the list continues to grow.

## NONRECORDABLE MEDIA

### Compact Disks

A CD is a long-playing, high-fidelity audio storage medium, and its excellent sound reproduction qualities are a reflection of its digital and optical heritage. Interfering noise is reduced, and a CD player, equipped with a microprocessor, allows you to quickly access any of the disk's tracks and to select

a predefined playback order. These functions, in addition to a disk's small size, durability, and capacity have contributed to its popularity with consumers and radio stations.

As described in Chapter 2, the CD industry is also governed by an established set of standards. This factor played an instrumental role in the CD's widespread acceptance.

The CD has, however, been faced with potential competitors over the years, ranging from digital audiotape (DAT) to MP3 players. A DAT player uses digital tapes while an MP3 player, through compression, can store and play back music downloaded from your computer. Digital audio is covered in greater depth in a later chapter.

### Compact Disk–Read Only Memory (CD-ROM)

A CD-ROM is a high-capacity data storage medium. A CD-ROM, which looks like a

CD but can store 600+ megabytes of data, is preloaded with information and/or programs. A CD-ROM is also interfaced with a computer via a CD-ROM drive and special driver software.

An early CD-ROM release was the electronic text version of Grolier's *Academic American Encyclopedia*. This disk highlighted the CD-ROM's storage properties: An entire encyclopedia of some 30,000 articles was recorded on a single disk, with room to spare. It was also integrated in a PC environment, and word processing programs could retrieve information from the encyclopedia. More recent encyclopedias also incorporate sound, graphics, and animations.

Another interesting earlier disk, which spawned other releases, was the PC-SIG CD-ROM. The PC-SIG has been a source of public domain and user-supported software (shareware) written for IBM PCs. The programs have covered everything from computer languages to games, and the entire library could fill 1000 or more floppy disks.

This library was transferred to a CD-ROM. This type of application was and remains a valuable one for PC owners, since the CD-ROM is inexpensive when compared to an equivalent floppy disk library. This factor has made the CD-ROM an ideal distribution medium for computer software collections. Other applications include the following:

- Information pools can be compiled ranging from telephone number compilations to U.S. street maps.
- Companies have adopted CD-ROMs to complement their print lines. For software, a CD-ROM could hold tutorials and sample files referred to in the manual. Their size may have precluded their use when floppy disks were the primary distribution vehicle.
- Spacecraft images have been made available. Through the National Space Science Data Center and other sources, you could explore Venus and Mars from your armchair by viewing information generated from the Magellan and Viking Orbiter missions, respectively.[3]
- Complex software collections can be created on a single or multiple disks. This may include a desktop publishing program and an extensive clip art collection. Clip art is a library of drawings and pictures you can legally use. An earlier application included a bundle of several thousand images with Corel Systems Corporation's program Corel Draw, a high-end illustration program. The floppy disk equivalent was approximately 500+ disks.[4]
- The new generation of interactive games has a large storage appetite, especially if the games incorporate digital audio, video, or computer animations. CD-ROM media can, however, accommodate them.

CD-ROMs are also valuable for libraries, which typically face storage and budgetary problems. CD-ROMs are cost effective, can be used with a PC to search for specific information, and can save space.

These types of databases have, however, been replaced to a great extent by online systems. In this situation, a library contracts with a vendor to gain access to specific data pools, typically through an Internet connection.

## Compact Disk-Interactive

The compact disk-interactive (CD-I) was originally designed as a stand-alone unit equipped with an internal computer.[5] A goal was to make a CD-I system attractive to consumers since it was self-contained, simple to operate, and could be used with a television set. In one example, a Louis Armstrong disk could be used with a CD player, but when

**Figure 9.2**

*A CD-I disk. The art of the Czars: St. Petersburg and the Treasures of the Hermitage. An example of the rich pool of information electronic publishers have tapped. (Courtesy of the Philips Corp., photo credit Richard Foertsh.)*

played on a CD-I system, you could also view relevant graphics and retrieve information about Armstrong's work.[6]

### Photo Compact Disk

Kodak entered this industry with the Photo CD system. A disk could store up to 100 pictures that could subsequently be used in a compatible CD-ROM drive, or for consumers, in a player for viewing on a television set. The same system could also play audio CDs.

The capability to play different disks was an important one and an industry-wide development. Instead of buying multiple players and/or PC-based drives, you use one machine. This is particularly significant in view of the growing number of formats.[7]

### Digital Versatile (or Video) Disk

Designed for commercial and consumer applications, the DVD was initially headed for a format war, much like the earlier Beta and VHS scenario. But the companies and manufacturers that had a stake in this field reached a compromise in late 1995.

A DVD's storage capacity far exceeds a conventional CD-ROM or CD. This capability offers producers and consumers numerous advantages:

1. Unlike CD-ROMs, high quality audio-video cuts could be used in a game, educational title, or other multimedia production.[8]
2. It's a boon for computer data storage—a single disk can replace a collection of conventional CD-ROMs.
3. Consumers can view movies with support for wide screen television sets, advanced television systems, and other enhancements.

The DVD also replaced *videodisks*, the pioneer product of the optical disk family. The videodisk was produced in different formats, including a discontinued nonoptical version manufactured by RCA. Videodisks primarily supported two applications. In the first, the consumer category, high quality movies were distributed. It also supported a digital audio signal, even though the picture information was analog. In the second category, videodisks were interfaced with PCs to create a sophisticated interactive environment.

While a videodisk was a high quality distribution vehicle for movies, it never really caught on in the consumer market—the DVD has. Players range in sophistication, and you can buy a sophisticated Surround-Sound system, with multiple speakers, that will produce a high quality audio playback when viewing a movie or listening to a CD.

A DVD's data storage capability also enables producers to bundle additional information with a disk. This ranges from multiple language tracks to additional footage that might have been cut out of a movie's theatrical release. When these features are combined with a high quality playback that exceeds the typical videotape-based system, DVD players flooded the home and professional markets in the early 2000s.

## RECORDABLE MEDIA

### Compact Disk-Recordable and Compact Disk-Rewritable

A CD-Recordable (CD-R) enables you to store your own information on a disk. The CD-R is a write-once, permanent medium. After the information is recorded, it cannot be erased.

The CD-R has two important advantages. It is a cost-effective and high capacity storage medium. CD-Rs can also tap the enormous CD-ROM drive universe. CD-Rs you create can be used with conventional CD-ROM players. In one example, you can create or "burn" an audio disk for playback on a CD player.

In contrast, the CD-Rewritable (CD-RW) functions much like a high capacity floppy disk. You can reuse a disk multiple times, storing new and erasing old data.

The price for media and recorders has dropped over time, making them ubiquitous home and business fixtures. The software that drives the recording process has also improved. This has enhanced a recording's operational performance, in both speed and accuracy.

Magneto-optical (MO) configurations, which employ optical and magnetic principles to store and retrieve data, have also been manufactured. Like the CD-RW, it is a reusable medium.

### Digital Versatile (or Video) Disk

The DVD family, for its part, supports a recordable option in various flavors. They include the DVD-R, DVD+R, the DVD-RAM, and the DVD-RW.[9] Their primary advantage? A storage capacity of 2 or more gigabytes. This capability makes a DVD, which is the size of a CD, a highly portable mass storage medium. You can record movies, data, a library of audio-video clips, and complex animations. You can also create an interactive interface a viewer can use for various operations.

When designing a typical disk, options include the following:

- Selecting the audio-video clips (e.g., TV shows or a movie) you want to include on a disk.
- Creating on-screen menus that can help guide a user through the stored information.
- Creating on-screen buttons that are used to select and view, for example, a scene in a movie.
- Implementing a sophisticated audio design.
- Testing the DVD and its links. Using your computer, you can test different functions, through a simulated on-screen DVD player, before you spend time/money in creating an actual disk.
- Compressing, for instance, the video program that will be stored on the disk. As described in a later chapter, digital audio and video information have enormous storage appetites. Compression techniques help curb this appetite-in this case, the original information, which may have exceeded a disk's capacity, can now fit on the disk.

The software used for these tasks vary in capability, and programs have been released for the home and the professional/corporate markets.

In the early 2000s, however, DVD home design/recording could still present a challenge. It was, in a sense, a fairly recent

**Figure 9.3**

*A computer-based DVD simulation used to test a disk you're designing. DVD authoring programs can also be used to produce electronic slide shows in addition to storing other media-based information. (Courtesy of Sonic Foundry; DVD Architect.)*

INFORMATION STORAGE

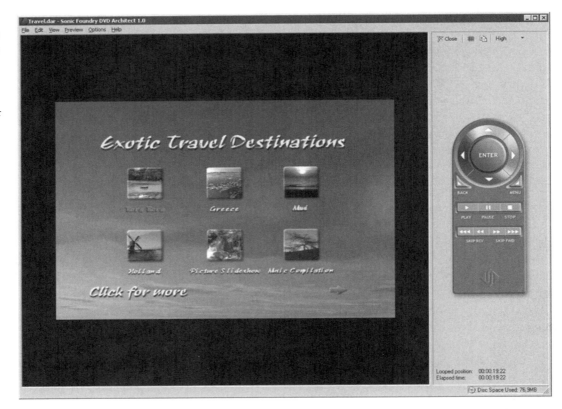

merger between new software, drives, compression techniques, and media.

In one case, the DVD you just created may not have played on a given DVD unit, based on a number of factors. This included the media itself, that is, the disks manufactured by one company versus another company.[10]

Another factor was the DVD drive. Initially, a drive may have only supported a limited number of DVD formats. A system may have been compatible with DVD-R but not with DVD+R media. Other drives that supported erasable DVD formats may have supported DVD-RAM but not DVD-RW media. The DVD-RAM has been used to back-up data since it has a high storage capacity. The DVD-RW, for its part, is analogous to the CD-RW, albeit with a much greater information storage capacity. But like many other fields, manufacturers worked to improve the compatibility situation.[11]

Stand-alone DVD recorders, which are not connected to a PC, have also be manufactured, with one goal of replacing VCRs.

As of this writing, the VCR is still a dominant player in the consumer market since it can play and record television shows and other video materials. Most DVD consumer players have been play-only devices. This may gradually change, though, as newer models are manufactured and media continue to improve and become more cost-effective.

Finally, it should be noted that another system, the write-once, read many (WORM) drive, has also been manufactured. An early permanent storage medium, which included a 14-inch optical disk configuration introduced by the Storage Technology Corporation in 1983, a disk could serve as an archival storage medium.[12]

An archival and permanent capability could also be important in certain applications, in contrast with an erasable disk. In one example, a financial institution could use an optical system to create a permanent record of transactions. This could be advantageous since the information could not be altered, and the disk could facilitate a future audit.

## Summary

Regardless of the optical drive or media, current and future systems provide us with the means to store large volumes of information. This is important: Data storage demands are increasing as the nature of information becomes more complex. In desktop video and multimedia applications, for instance, 24-bit graphics as well as digital audio and video clips can be used. While data compression can be applied, this information can still be data storage intensive.

The different classes of optical systems will also continue to coexist. At first glance, a conventional CD-ROM may appear obsolete in the face of some other configurations. But CD-ROMs are cost effective, have a large storage capacity, and a high market penetration.

Magnetic media, hard and floppy drives, will also continue to be used. Hard drives are generally faster than their recordable optical counterparts, making them a staple for PC-based video editing systems and other demanding data storage operations. In brief, as you store, edit, and retrieve the video through the editing program and the computer, the drive must be able to handle this high data flow.

Fixed storage hard drives can also be less expensive than some recordable optical drives, although this economic advantage decreases as the data storage requirement increases. Basically, when dealing with mass storage needs, it is cheaper to buy another optical disk than another hard drive.

But as introduced in Chapter 3, removable magnetic storage systems have added a new twist to this scenario. Iomega's widely adopted drives, for instance, have a range of data storage capacities. When you fill a disk with information, you simply buy a new disk rather than a new drive.

The late 1990s and early 2000s also witnessed lower prices for CD-Rs and CD-RWs—they fell below the $75 mark. Blank CD-R disks could also cost less than $1 each, in contrast with more expensive removable magnetic media. DVD drives, though more expensive, similarly benefited from price breaks. Many new PCs, including notebooks, were also outfitted with recordable CD-ROM and/or DVD drives.

In essence, each medium has its benefits and appropriate application areas. While you may use a CD-R to distribute audio cuts, a fixed hard drive may serve to store a video sequence you are digitizing. The media are complementary and, in the end, have resulted in a cost-effective and flexible storage bonanza.

## OTHER ISSUES

### Convergence
The rapid growth of the optical disk field highlights the growing convergence between different technologies and their respective applications. Products such as the CD-I married computer applications with standard television technology. A television became part of a computer-based entertainment and educational system.[13]

Products such as the Photo CD, for their part, cut across the traditional and silverless photographic fields. Pictures produced as standard prints or slides could also be stored on a disk and viewed on a television or computer. The same pictures could subsequently be manipulated on a computer.

The convergence factor has even bridged different application areas. For example, the mass storage capabilities of CD-ROMs were married to the instant update capability of online services. A CD-ROM stored data that would normally require an extended download time, that is, the period of time to relay the information from the company to you. The online service, for its part, would supply new and updated information.[14] The same scenario is playing out in certain DVD markets where DVD-Internet integration is an important goal and tool.

### Privacy
The storage capabilities of optical media have raised privacy concerns. In one case, a potential CD-ROM produced by the Lotus Development Corporation triggered a public outcry. Lotus planned to take advantage of a CD-ROM's storage capabilities and sell a disk loaded with demographic data about American consumers. The company received so many complaints, though, that the product was dropped from its line.

This example highlights a key issue of the information age: an individual's right to privacy. Although the new technologies can enhance our communications capabilities, they can also be invasive.[15] The available options to protect yourself include consumer pressure, employing other emerging technologies, and adopting privacy regulations, as was the case with the European Community (EC).

The EC had proposed a series of strict regulations governing the collection and dissemination of personal information. The rules recalled the World War II era when information from telephone records was used for political purposes. The proposed regulations were designed to help protect an individual's privacy.[16]

Although privacy is important, some people believed the regulations were too strict and would impede the flow of information between countries. A similar situation has prevailed in the United States. Some individuals, as well as government agencies, believe there has to be a balance between protecting an individual's privacy and the government's ability to retrieve specific types of information. Questions were also raised about privacy with regard to e-mail in a business environment.[17] This topic is covered in a later chapter.

## Information Retrieval

Through optical media, we can gain access to entire libraries of information stored on a series of small disks. Yet to tap this information effectively, suitable search methods must be devised.

In one example, a CD-ROM that serves as a database is equipped with software that functions as the information-retrieval mechanism. Depending on the package, it may also allow us to search through the information in different ways. Some systems employ keyword searches while others support more sophisticated mechanisms, including hypertext.

*Hypertext* is a sophisticated information management and retrieval mechanism that works in a nonlinear fashion. It cuts across magnetic and optical storage domains, the World Wide Web, and other information systems. Vannevar Bush, President Roosevelt's science advisor, originally conceived the concept for such a mechanism in the 1940s.

Hypertext operates in much the same way that humans think. You may, for example, use hypertext to conduct research about the new communications technologies. While exploring this topic, you see the term *laser diode*. Because it is important, the words are highlighted or underlined in our hypertext environment, indicating they are linked to other information. Next, move the on-screen cursor to a word and click a mouse button. The linked information is subsequently retrieved and displayed, either on a separate page or in a window. When

finished reading, you can return to the original page with another click of a button.

As you continue, the term *semiconductor* is similarly marked. You could then retrieve information, through a series of links, about semiconductors, the semiconductor industry, and related economic implications.

By following this pattern, you can retrieve information in a natural way, that is, the way people think. You follow a train of thought that enables you to make associations between diverse topics. In this context, hypertext is no longer just a search mechanism. It provides a new way to organize, link, and communicate bodies of information and knowledge. You can also reveal previously hidden connections between topics.

A hypertext system performs these tasks effortlessly, at least for the user. The various links can also be retraced so you can return to a given source.

This concept has been extended through hypermedia systems. The term *hypermedia* has been associated with *multimedia*, a subject discussed in a later chapter. Briefly, "multimedia is the integration of different media types into a single document. Multimedia productions can be composed of text, graphics, digitized sound . . . ," video, and other information. Hypermedia software, in turn, allows you to "form logical connections among the different media composing the document."[18] When you click on a highlighted term, you may see a picture or hear a sound, instead of simply a page of text.

With a hypermedia program, the links between the information, whether textual or graphic, can be represented in different ways. These include highlighted text, text placed between different symbols, and buttons. Buttons are visual markers, typically labeled, that can help create a more effective user interface.

This searching capability, regardless of the system, highlights the power of the PC when combined with a mass storage device.

INFORMATION STORAGE

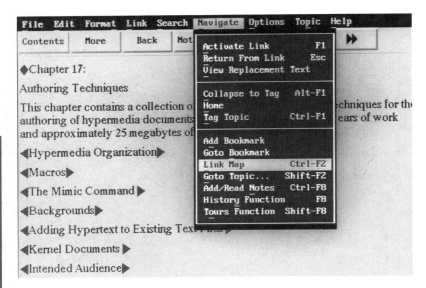

**Figure 9.4**
*Some of the navigational tools supported by hypermedia packages. (Software courtesy of Ntergaid, Inc.; HyperWriter!)*

Instead of looking through different books and magazines, you can let the PC do the work for you. If the data are stored on a disk, the retrieval process can be simplified and enhanced.

### Conclusion

Optical disk technology has played an important role in the communication revolution. As we generate more information, these disks serve as effective storage and distribution media. They are also cost effective and can accommodate a range of applications.

Some individuals believe high-speed communications channels for Internet connections, among other applications, may doom optical media to extinction. But it's important to remember that optical media are portable—they can go where you go and where there may not be a network connection. You can also deliver high quality video clips, and even movies, operations that may even tax high-speed data lines. In essence, both technologies and their respective applications will continue to coexist for a number of years.

Finally, as briefly covered in the next section, there is another form of informa-tion storage that taps a laser's capabilities—the hologram.

## HOLOGRAPHY

Holographic techniques were discovered in the 1940s by Dr. Dennis Gabor. A research engineer, he won a Nobel Prize in 1971 for his pioneering work as the father and creator of holography. But the field did not fully blossom until the 1960s. The develop-ment of a suitable light source, the laser, helped provide the key.

Even though it is beyond the scope of this book to explore holography's physical prop-erties, we can still present a working defin-ition. A *hologram*, a product of holography, can be considered a record of the optical information that composes a scene. It can store information about a three-dimensional object, and unlike a standard photograph, which

records light intensity . . . the hologram has the added information of phase . . . to show depth. When a person looks at a tree . . . he is using his eyes to capture light bouncing off the object and then processing the information to give it meaning. A hologram is just a convenient way to recreate the same light waves that would come from an object if it were actually there.[19]

This capability can be startling. Objects are quite lifelike, and they may appear to "jump out" of a scene. A standard photograph records a scene from a single perspective. A hologram breaks these boundaries.

Finally, many holograms are created through a special film and are chemically processed. But newer techniques have en-hanced this task.

### Applications

A statue can be recorded as a cylindrically shaped hologram.[20] You can see the recorded statue as you would the original object—from various views.

The advertising and security fields are also supported. Attention-grabbing ads can be produced, and small, inexpensive embossed holograms have been placed on credit cards for security purposes.[21]

Another application area is *holographic interferometry*. Unlike artistic endeavors, a conventional hologram is not created. Rather, in one variation, a double exposure of the same object is made. During one exposure, the object is stationary and at rest. In the second, the object is subject to stress through the introduction of the physical forces it may experience in actual use.[22]

When processed, the hologram will depict a series of fringes or lines across the object. They resemble the contour lines on a map. The fringes reveal the small differences between the exposures—the differences between the object while at rest and under stress.[23] Consequently, the hologram serves as a visual map of the areas that may be deformed or affected by the operation.

Holographic interferometry has been used to inspect products for manufacturing defects. Metals and other materials have also been subjected to stress to test for possible flaws.[24]

Holography has also provided us with the *holographic optical element* (HOE), a hologram that functions, for instance, as a lens or mirror.[25] HOEs are especially useful when a standard optical component may not fit or may be too heavy.

In one application, a head-up display, an image of an instrument panel could be displayed before a pilot's eyes. The pilot would not have to move his or her head to view the instruments.

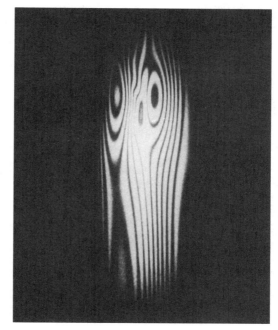

**Figure 9.5**

*Corrosion monitored: An example of holographic interferometry. Holography can, in effect, let you visualize the interior of a pipe from the outside. The circular fringe pattern, due to a small pressure change, indicates weakened regions of the pipe wall. (Courtesy of the Newport Corp.)*

The holographic field may also bring massive data storage capabilities to the desktop.[26] In one example, engineers developed "a CD-sized holographic storage disk that holds on the order of a terabit of information—a factor of thousands more than a CD. And . . . the disk is quickly searchable—a property that springs naturally from the holographic reading process."[27] Thus, its unique properties can make a holographic-based system a highly effective and efficient storage system.

The future may also witness 3-D television. Experimental work has already been conducted, and if combined with CD-quality sound, the system would support realistic computer games as well as entertainment and educational programs.

## REFERENCES/NOTES

1. Lasers initially store this information on a master disk. This disk plays a key role in producing copies.

2. Paul Freiberger, "CD Rot," *MPC World* (June/July 1992), 34.

3. Even in early 1992, the price was $20 for the first disk and only $6 per additional disk in a series.

4. Advertisement, "CorelDraw," *NewMedia*, April 1992, back cover.

5. A more sophisticated business model was later released.

6. CD-Interactive Information Bureau, "CD-I Launches Titles," *CD-Interactive News* (January 1992), 4.

7. These include the CD-ROM XA and other formats.

8. Frank Beacham, "Consortium Proposes New CD Format," *Computer Video* 2 (March/April 1995), 10.

9. There are also subgroupings in these general categories.

10. Ralph LaBarge, "DVD Compatibility Test," *DV* 10 (July 2002), 24.

11. Please see Ralph LaBarge, "The Burning Question," *DV* 11 (June 2003), 40–46.

12. Storage Technology Corporation, "7600 Optical Storage Subsystem," product description brochure. *Note*: Archival implies the disk and records will theoretically last past a certain number of years.

13. As discussed in a later chapter, the establishment of an enhanced and advanced television standard will only accelerate this trend.

14. Domenic Stansberry, "Going Hybrid: The Online/CD-ROM Connection," *NewMedia* 5 (June 1995), 37.

15. Caller ID triggered a similar response. See Bob Wallace, "Mich. Bell Offers Caller ID Service with Call Blocking," *Network World* 9 (March 2, 1992), 19, for a description of why blocking options were adopted.

16. Mary Martin, "Expectations on Ice," *Network World* 9 (September 7, 1992), 44.

17. See the e-mail chapter, Chapter 18, for specific information.

18. Ntergaid, Inc., "HyperWriter," flyer. HyperWriter (for IBMs) and HyperCard (for Macs) were two pioneer programs that you could use to create hypertext/hypermedia documents.

19. Brad Sharpe, "Hologram Views," *Advanced Imaging* 2 (August 1987), 28.

20. Thomas Cathey, *Optical Information Processing and Holography* (New York: John Wiley & Sons, 1974), 324.

21. These holograms would be difficult to duplicate and may have high expertise and physical plant requirements.

22. Edward Bush, "Industrial Holography Applications," *ITR&D* (n.d.), 30.

23. Cathey, *Optical Information Processing and Holography*, 324.

24. James D. Trolinger, "Outlook for Holography Strong as Applications Achieve Success," *Laser Focus/Electro-Optics* 22 (July 1986), 84.

25. Jose R. Magarinos and Daniel J. Coleman, "Holographic Mirrors," *Optical Engineering* 24 (September/October 1985), 769.

26. Tom Parish, "Crystal Clear Storage," *Byte* 15 (November 1990), 283.

27. "Holographic Disk Is Quickly Searchable," *Laser Focus World* 39 (February 2003), 13.

## SUGGESTED READINGS

Amadesi, S., et al. "Real-Time Holography for Microcrack Detection in Ancient Gold Paintings." *Optical Engineering* 22 (September/October 1983), 660–662; P. Carelli, et al. "Holographic Contouring Method: Application to Automatic Measurements of Surface Defects in Artwork." *Optical Engineering* 30 (September 1991), 1294–1298. Holography and the art/art restoration worlds.

Bell, Alan. "Next-Generation Compact Discs." *Scientific American* 275 (July 1996), 42–46; Philip Dodds. "Comparing Oranges and Mangoes: Another View of the Emerging Digital Videodisk." *SMPTE Journal* 105 (January 1996), 46–47; Dave Kapoor. "The Next 10 Years." *DV* 11 (January 2003), 28–34; Ralph LaBarge. "DVD Compatibility Test." *DV* 10 (July 2002), 20–29. DVD coverage; the Kapoor article covers various, future digital video projections, including the DVD; the LaBarge article is an excellent resource, as of this writing, for compatibility issues.

DeLancie, Philip. "Thinking Inside the Box Smart Set-Tops and Web-DVD Convergence." *EMedia Magazine* 14 (September 2001), 46–52; Philip De Lancie. "Untangling Web DVD Playback." *EMedia Magazine* 14 (February 2001), 40–45; Domenic Stansberry. "Going Hybrid: The Online/CD-ROM Connection. *NewMedia* 5 (June 1995), 34–40. The convergence of DVDs, CD-ROMs, and online services/the Internet.

Foskett, William H. "Reg-in-a-Box: A Hypertext Solution." *AI Expert* 5 (February 1990), 38–45. Traces the development of a hypertext system concerned with regulations about underground storage tanks.

Hand, Aaron, J. "Is Holographic Storage a Viable Alternative for Space?" *Photonics Spectra* 32 (June 1998), 120–124; "Holographic Storage Delivers High Data Density." *Laser Focus World* 36 (December 2000), 123–127. Holographic storage systems and technology.

Higgins, Thomas V. "Holography Takes Optics Beyond the Looking Glass." *Laser Focus World* 31 (May 1995), 131–142. Excellent tutorial about holography.

Ih, Charles S. "A Holographic Process for Color Motion-Picture Preservation." *SMPTE Journal* 87 (December 1978), 832–834. Using holography as a film preservation system.

Lambert, Steve, and Suzanne Ropiequet, eds. *CD-ROM: The New Papyrus*. Redmond, WA: Microsoft Press, 1986; Suzanne Ropiequet, ed., with John Einberger and Bill Zoellick. *CD ROM Optical Publishing: A Practical Approach to Developing CD-ROM Applications*, Vol. 2. Redmond, WA: Microsoft Press, 1987. Microsoft Press's two-volume CD-ROM book series. Although published in the 1980s, they remain excellent CD-ROM and electronic/optical publishing resources.

McCort, Kristinha. "Beyond Aesthetics." *Millimeter* 29 (March 2001), 37–41. Design, aesthetic, and material management issues in DVD authoring.

N.A. "Holography Drives Ford Car Design." *Laser Focus World* 35 (February 1999), 14–16; R.J. Parker and D.G. Jones. "Holography in an Industrial Environment." *Optical Engineering* 27 (January 1988), 55–66; David Rosenthal and Rudy Garza. "Holographic NDT and the Real World." *Photonics Spectra* 21 (December 1987), 105–106. Holography and industrial applications.

Nelson, Theodor. "Managing Immense Storage." *Byte* 13 (January 1988), 225–238. A description of the storage engine of the Xanadu project, a model for a new information storage and management system.

Pozo, Leo F. "Glossary of CD and DVD Technologies," downloaded. Excellent and comprehensive listing of terms and systems.

Waring, Becky, and Alexander Rosenberg. "New CD-ROM Hardware Swells the Consumer Market." *NewMedia* 2 (May 1992), 12–15. An earlier look at CD-ROM systems with sidebars on Kodak's Photo CD system and the role of game manufacturers in this market.

## GLOSSARY

*Compact Disk (CD):* A prerecorded optical disk that stores music. The CD player uses a laser to read the information.

*Compact Disk-Read Only Memory (CD-ROM):* A prerecorded optical disk that stores data. CD-ROM applications range from the distribution of computer software to electronic publishing.

*Compact Disk-Recordable (CD-R):* A permanent, recordable CD system.

*Compact Disk-Rewritable (CD-RW):* An erasable, CD recordable system.

*Digital Versatile (or Video) Disk (DVD):* A new, high-capacity optical disk developed in the 1990s. Designed for commercial and consumer applications, it could replace the home VCR.

*Digital Versatile (or Video) Disk-Recordable (DVD-R):* A permanent, recordable DVD system.

*Digital Versatile (or Video) Disk-Rewritable (DVD-RW):* An erasable, DVD recordable system.

*Erasable Optical Disks:* A class of optical disk where data can be stored and erased.

*Hologram:* A hologram is a record of the optical information that composes a scene. It can be used for applications ranging from advertising to security.

*Holographic Interferometry:* A holographic industrial application (e.g., for material testing).

*Holographic Optical Element (HOE):* An application where a hologram functions as a lens or other optical element.

*Hypertext:* A nonlinear system for information storage, management, and retrieval. Links between associative information can be created and activated. This concept has been extended to pictures and sounds (hypermedia).

*Optical Disk:* The umbrella term for optical storage systems.

# IV

## PRODUCTION TECHNOLOGIES

# 10　Desktop Publishing

This chapter examines desktop publishing (DTP), an application used to produce newsletters, brochures, books, and other documents. DTP tools include PCs, software, printers, and scanners. When combined, they create a publishing system that can literally fit on a desktop. DTP also provides us with an electronic composition tool in that a design is electronically composed.[1] A monitor's screen serves as a window in this process.

Various factors led to the proliferation of PC-based DTP systems in the late 1980s and early 1990s. More sophisticated PCs and complementary software flooded the market while the laser printer became an affordable option.

## HARDWARE—THE COMPUTER, MONITOR, AND PRINTER

### Computer

The hardware end of the typical DTP system consists of four major components: a monitor, printer, scanner, and this section's focus, the computer.

The Macintosh helped launch the PC-based DTP industry. It was more graphically oriented and easier to set up than comparable IBMs. A stock Mac, with its software library, was also better equipped to handle desktop publishing and graphics applications. When combined, these factors helped individuals who were not graphic artists to design their own projects. Graphic artists, on the other hand, could now experiment with different concepts.

IBM PCs, for their part, despite some initial hardware and software disadvantages, eventually emerged as another major force in this field. Their dominant position in the overall computer market, and the introduction of new equipment and programs, contributed to this development.

### The Monitor

When the DTP was taking off, many people still owned 13 to 15-inch monitors. While suitable for different tasks, they could not display a full page of readable text. In the late 1990s and early 2000s, 17 and 19-inch monitors became the norm as did LCD monitors. As indicated in Chapter 3, inch for inch, LCD monitors yield a larger working or viewing area than conventional monitors.[2] Consequently, users now have a larger electronic canvas to design and implement their publications. But despite this advantage, one practice still holds true— different *viewing modes* are used during the design process.

In the full-page mode, a page can be displayed in its entirety to reveal the placement and spatial relationship between graphic and textual elements. But the page outline is noticeably reduced in size. Depending on your set-up, most, if not all, the text may be replaced by small lines or bars (a process called *greeking*) since the characters are essentially too small to be reproduced on the screen.

Other modes provide magnified or enlarged views of specific page sections. The

PRODUCTION TECHNOLOGIES

text can be read, and fine details of the document's style can be checked. Special monitors have also been manufactured that have generated enhanced document views, especially in the full-page mode.

### Printer

A laser printer can produce high quality documents. Most printers support a 600+ dots per inch (dpi) resolution. In the context of this discussion, the term *resolution* refers to the apparent visual sharpness or clarity of the printed characters and graphics. This working definition is used throughout the chapter. There's a relationship between the dpi and a document's perceived quality. In general, a higher dpi figure could result in a higher quality document.

When the first reasonably priced laser printer appeared in the early 1980s, it created a stir in the computer industry. The printer could handle some of the printing jobs that had been reserved for traditional typesetting equipment. This trend, started by the Hewlett-Packard LaserJet and the Apple LaserWriter printers, has continued unabated.

As of this writing, most if not all printers designed for the general business and consumer markets share several broad characteristics.

1. A printer should be equipped with enough memory to take full advantage of its printing capabilities. The memory requirement increases for a color printer.
2. Various typefaces and fonts are available for the printers. The upshot? You can design a document that fits your publishing needs. A *typeface* is a unique print style. The different characters of a given typeface conform to a style, a set of physical attributes shared by all the characters. Two examples of common typefaces are Helvetica and Times Roman.[3] A *font* is a

typeface in a specific size. The size is measured in points, and as the point size increases, so too does the character's size. As a frame of reference, 72 points equal approximately one inch.

3. Although the typical printer cannot generate typeset-quality documents, it is satisfactory for creating newsletters, an organization's in-house magazine, and even a book on a tight production schedule and budget. High-quality line drawings, a series of black lines on a white background, can also be printed. These may range from a building to an interior view of an engine that's slated for a technical document.
4. The typical laser printer cannot support full-color output. Its graphics capabilities may also be limited for reproducing black-and-white photographs. Details may not be sharply defined or too few gray levels may be reproduced.

***Color Printing.*** Color printers became increasingly popular during the 1990s. They ranged from thermal wax to laser to dye sublimation units.[4] The latter two systems could support a high quality output, but had been comparatively expensive.[5]

Color laser printers also broke the $1000 price barrier. They are generally fast, and for volume printing, can be more cost-effective than other printer types.

However, the ink jet color printer has made the greatest impact in the general consumer/business markets. By incorporating various technological improvements, printing quality improved. For example, when a special paper that looks much like glossy photographic paper is used, ink smearing is reduced. When combined with other factors, an image that can look like a conventional photograph can be produced. Other developments range from inks with archival properties, so the image won't fade for a number of years, to specialized printers that can support large format prints.

Depending on your needs, you may even opt for commercial printing. While new hardware and software releases may make it easier to prepare color images for this task, it is still a complicated process.[6] Consequently, it is a good idea to discuss a project with a commercial printer to obtain the best results. You may also decide to leave the work to individuals who are well versed in this field.[7]

Finally, color can be an important DTP element. Color can catch a reader's eye and can help convey information more effectively. Think of a bar chart showing a radio station's ratings versus its competitors. Different colors may make it easier to differentiate the information. In another example, color may produce a more visually appealing ad.

## SCANNERS

A DTP project may include photographs, line drawings, and other artwork originally produced in hardcopy form. This operation is made possible by using a scanner, a piece of equipment interfaced with the computer. For example, a black-and-white photograph can be placed on the scanner, much like a piece of paper on a copy machine. The image is read or scanned, and the picture information is digitized and fed into the computer.

At this point, the image can be manipulated with graphics software. It can be edited, the contrast and brightness levels can be changed, and special filters can be applied. The now altered image can then be saved and imported by the DTP software.

If a hardcopy is produced with our laser printer, a digital halftone method is employed. Because a conventional printer cannot produce true shades of gray and only prints black dots on a page, the halftone method creates gray-level representations or simulated gray shades. The image is divided into small areas or cells. The picture's various gray shade representations are subsequently generated by turning dots in these areas either on or off. This varying density of black dots, and ultimately the cells, creates the various apparent shades of gray throughout the picture.

There is also a balance, especially when using a typical PC-based configuration, between the number of gray levels and the picture's resolution. As the number of levels increases, the apparent resolution drops.[8]

For desktop color work, the final print quality is affected by proper color registra-

# Different Point Sizes

18 points

24 points

## 48 points

# 72 points

**Figure 10.1**
*Different point sizes.*

PRODUCTION TECHNOLOGIES

tion, the quality of the printer, and other factors.

The typical scanner supports 256 levels of gray and 24+ bit color. Contemporary systems also capture an image in a single pass or scan. Three passes had been the norm.[9]

Scanners also come in different flavors. In one example, when desktop units were still expensive, manufacturers developed less expensive handheld scanners. You physically moved the scanner down the page to capture the information. While cost effective and suitable for various tasks, there were limitations.[10] Specialized scanners have also been created, including very high resolution units geared for commercial printing.

The second component of a scanning system is the software—the mechanism by which the computer controls the scanning process. Pioneered by Ofoto, the latest generation of software can automatically generate higher quality scans or images. This capability also makes the technology more accessible. You do not have to be an expert for general work. You supply the aesthetic framework, and the program can help you produce a better product. Manual adjustments and various image-editing functions are also supported.

In a related area, film/slide scanners are also popular. Newspapers have used such systems to send photographs between local and international locations. In a typical situation, a roll of film is developed, and a scanner digitizes select film images. They are then compressed, relayed, and processed at the home office.[11] This system saves time and has provided for a more error-free relay.

It should also be noted that scanners geared for the consumer/business markets may also support slides and/or negatives. However, they may not match the flexibility and capability of scanners dedicated to this operation.

## Optical Character Recognition

A scanner can also be used with optical character recognition (OCR) software. In a typical application, scan a printed document, and the software recognizes the text. The information can then be saved and used with a word processing program.

An OCR system can save time and labor but may have limitations. Only a certain number of typefaces may be recognized, and the characters must be legible and fairly dark. Some characters will also be incorrectly read and must be replaced during an editing session.

To overcome some of these problems, the program may support a learning mode. You teach the software/computer to recognize incompatible type. More recent packages also recognize a wider range of type and have reduced the number of read errors.

## DIGITAL/ELECTRONIC STILL CAMERAS

Images used for a range of DTP and video projects are also produced by digital still cameras. Through the emergence of electronic still photography (silverless photography), an image is electronically captured rather than using film.

**Figure 10.2**
*An OCR operation. The text is reproduced in the top of the screen (top window). The window in the left, bottom corner, shows the original scanned text. The learning mode is also activated so the system can be "taught" to identify letters, for example that were previously unrecognized. (Software courtesy of Image-In, Inc.; Image-In-Read.)*

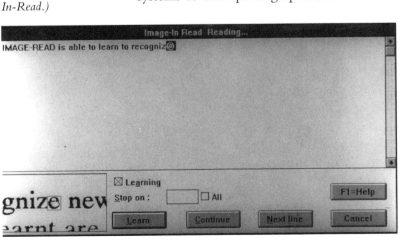

In this operation, film is replaced with a Charged Coupled Device (CCD), or other *electronic sensor*, and memory.[12] The images can be stored on a variety of media, typically a removable memory storage card. The images can subsequently be retrieved by a computer for processing and/or more permanent storage.

When first introduced, digital cameras geared for the consumer market could not match a film camera's cost or film's resolution. However, as newer models were introduced, various factors made digital cameras a hot commodity. These include the introduction of professional models that could actually rival and/or even surpass film's image quality under certain conditions.[13] Other factors are:

- price breakthroughs;
- enhanced sensors;
- immediacy—like a video camera, you can immediately view an image;
- the computer connection—you can readily manipulate and store the images in your PC;
- a new generation of cost-effective color printers;
- cost-effective—film and processing expenses are eliminated by the reusable storage card; and
- CCDs and other sensors are sensitive to low light, and their characteristics also make them conducive for specialized work, including astronomical photographs.[14]

On the flip side, digital cameras can still be comparatively expensive, particularly at the professional level. More pointedly, a professional film camera may be able to operate if its battery dies—you generally don't have the same option with an electronic camera. Film has other advantages as well, and like its electronic counterparts, is also improving with new enhancements.[15]

## SOFTWARE

The heart of any DTP system is the software. Our discussion focuses on two PC program categories, word processing and page composition programs.[16]

### Word Processing Software

Newer word processing programs can complete some of the jobs once reserved for DTP software. These include generating articles, business forms, and newsletters.

Word processing programs incorporate different functions to assist the writer. These include the following:

- spelling and grammar checkers;
- a macro capability—you complete a command with one or two rather than multiple keystrokes;
- a what-you-see-is-what-you-get (WYSIWYG) display on your monitor—as you compose a page, you visually see its margins, graphics, and other elements, you don't have to wait for the page to be printed;[17] and
- an ability to create tables, charts, and indexes.

A sophisticated word processing program can handle a range of applications, including those that do not demand the full power of a DTP system. A word processing program can also be easier to use, faster, and can produce a high-quality output.

### Page Composition Software

Page composition (DTP) programs also support a WYSIWYG display and an interactive interface. Like many word processing programs, as you move a graphic or column of text, the changes take place in real time. This visual feedback helps you determine if a page design is satisfactory, and it allows you to quickly experiment with different layouts.

DTP software also sports numerous enhancements. For example, text and graphics can be accurately placed via alignment aids, and there are extensive text and graphics manipulation modes. The discussion that follows provides a general overview of select operations.[18]

***Text Manipulation.*** When you create a document, the program displays an outline of a blank page. Next, you can open up one or more columns for text placement. Text and graphics can also be placed in resizable, movable blocks or frames. For example, you can create a newsletter that's formatted like a newspaper. The first page consists of columns of text, assorted illustrations, and the newsletter's masthead.

For the masthead, a mouse is used to create a long and narrow frame across the top of the page. An appropriate typeface can then be selected, as can stylistic elements. These include printing the text with a shadow effect.

A DTP program is also equipped with a word processing module. But you may continue to use your favorite software to type the text. It may be faster, and you may be more comfortable with its functions. The DTP program can subsequently import this file.[19] When retrieved, it can be placed in designated columns and spaces.

If the file is large and one column fills up, the text can be routed to another column. The routing can be automatic, or you can manually designate the next column the text should fill.

A program can also compensate for editing. As words are added or deleted, the text flows or snakes from column to column until the proper space adjustments are made.

This control over the text also extends to the physical spacing between individual characters and sentences. In kerning and leading, respectively, the space between specific pairs of letters and between individual lines can be altered. This capability can enhance a document's appearance and readability.

DTP programs can also import different file types. These include data from graphics, word processing, and spreadsheet programs. It may also be possible to rotate a line of text and to produce other special effects.

The trend appears to be toward the creation of more self-contained programs. As modules are added and refined, the program may be able to handle more tasks without tapping other software.[20]

***Graphics Manipulation.*** Many DTP programs can create simple graphics. But like the word processing function, you will probably continue to use a dedicated graphics package.

Once a graphic is created (for example, an artist's rendition of a mountain to illustrate an article in our newsletter) it is imported. At this point, the graphic can be resized. It can also be cropped, so only a portion of the entire image appears. You can then move the graphic to other positions, and if supported, wrap text around the image. Thus, instead of seeing separate and distinct blocks of text and graphics, they can be more integrated.

***Templates.*** A DTP program may be sold with templates. A template specifies the design of newsletters, books, and other documents.[21] Instead of spending hours to design a publication, you can use a premade template, which delineates the document's physical appearance. Custom templates can also be created, stored, and recalled when necessary.

Besides helping novices, a template can be useful to people who are in a hurry or who cannot create an effective design. For an organization, it can also bring a sense of order and uniformity to reports. A large company, for instance, can house several templates or more divisions can produce their own reports. If a format is not adopted, a report's structure could vary from division to division. This could hinder communication, especially if a document does not present the material in a clear and logical fashion.[22] It may also detract from the company's look or corporate identity, recognizable physical attributes that readily identify the company to internal and external parties.

## DESKTOP PUBLISHING GUIDELINES

Now that we've covered DTP basics, it's appropriate to discuss some general guidelines:

1. Institute a DTP training program. The software may be complicated and have a steep learning curve.

2. A DTP system will not turn everyone into an artist. It is simply a tool to present ideas and information more effectively. Paraphrasing Clint Eastwood in one of his *Dirty Harry* roles, you've got to know your own limitations (and strengths). Do what you do best. But if you need an artist or other professional, hire one. You will save time and money and are likely to get a superior final product.

3. Plan and effectively use a page's white space. It can provide visual relief for the reader and serves as a design tool. White space can highlight and focus attention on specific page elements. This concept also extends beyond the DTP industry. It is an established technical writing axiom.

4. Pay attention to the basics: grammar, typos, and spelling mistakes. Proofread the document after you are finished. Do not try to catch every kind of mistake in one reading.

5. When designing a document, keep it simple, if appropriate. Although a DTP program may support many typefaces and special printing effects, do not use them all on the same page. The document may be difficult to read, and the information may be lost in a maze of fonts and double-underlined text.

6. Read. Hundreds of DTP books and magazine articles have been written that cover everything from aesthetics to scanning to tips from professional DTP users. If you are learning how to use a DTP system, it is also important to practice your craft. Experiment with what you've learned and develop your own style.

7. Explore your software's other capabilities. For example, it may also support document generation for the Internet and/or other electronic publishing venues. Instead of printing on paper, you can print to an electronic medium, thus extending your publishing options via the same program.

As mentioned elsewhere, the same graphic may also be used for multiple distribution venues, ranging from a print-based brochure to a DVD to a web page. Thus, you may use compression and other techniques to optimize the graphic for each application and environment.

8. Use your imagination.

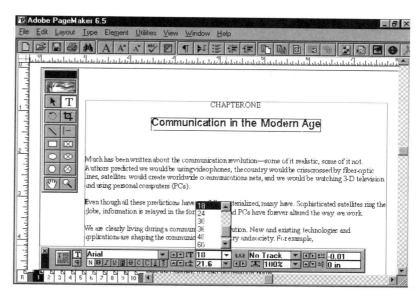

**Figure 10.3**

*The PageMaker's DTP program. Its Toolbox, is used for drawing, rotating text, and other options (upper left corner). The control panel, visible at the bottom, can be used to quickly manipulate text/graphics. (Software courtesy of Adobe Systems, Inc.; PageMaker.)*

PRODUCTION TECHNOLOGIES

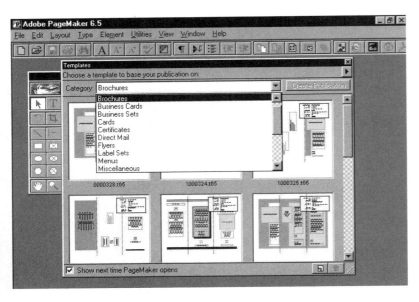

**Figure 10.4**

*DTP programs may also include templates or premade publication designs. In this figure some of PageMaker's brochure options are highlighted. (Software courtesy of Adobe Systems, Inc.; PageMaker.)*

## APPLICATIONS

### Personal Publishing

Prior to the DTP revolution, if you wanted to publish a book, you generally had two options. You signed a contract with either an established publishing house or a vanity press. For the latter, you paid a publisher for printing and possibly distributing your work. DTP systems provide authors with a third choice—to act as your own publisher.

As the author/publisher, you are in control of the entire process. You can make last-minute changes and updates, and a standard laser printer may suffice as the printing press. A higher dpi commercial unit can also be used for the final copy.

In a variation of this theme, you can initiate what Don Lancaster has called book-on-demand publishing. You do the work yourself, including the printing and binding. But instead of producing an initial run of 500 or more copies, you print a book only when someone orders it. This reduces the up-front costs for materials, and each copy could literally be an updated version of the original book.[23]

New authors/publishers do, however, face some constraints. These vary from individual to individual and may include your budget, experience, and DTP system's level of sophistication. They'll have an impact on your final product.

You must also face the dual problem of promoting and distributing your work. Although the structure for this market is evolving, it may still have to mature to provide a more established support mechanism. You also do not have access to the editorial and technical expertise afforded by a traditional publishing house.

### Other Applications

DTP systems have been employed in other applications. These include the publication of technical manuals, year-end reports, ads, information flyers, posters, and newsletters. They have also been adopted by traditional media organizations because DTP systems can save time and money.

The *New Yorker* slowly integrated DTP technology in its operation. The move was initiated to speed up certain tasks, and as stated, to save money.[24] In the newspaper industry, PCs equipped with the appropriate software have been used for photographic preparation and editing.[25]

Publishing companies have also used DTP technology. Since DTP systems are cost effective, an organization can potentially publish more books and take a chance with a manuscript geared toward a narrow audience. DTP may also be applicable for books that are regularly updated.

Individuals have also benefited. Besides book-on-demand publishing, you can use a DTP system to produce a document, such as a resume. Just as important, you can rapidly update the document and keep multiple versions—each one potentially targeted toward a different type of employer—on your computer.

Finally, in a related application, magazine publishers have created electronic versions of their products and have placed them on the Internet. Although the core of the information may remain unchanged, a conversion process is typically used to prep the information for this environment.

The Internet and related electronic media can offer a publisher additional capabilities through the use of interactive links, animations, and digital media cuts. A publisher can also update information in a more timely fashion and provide in-depth coverage. If space limitations force an article(s) to be cut from the print version, it can possibly be electronically housed. Back issues can also be made available, and this service could be used to attract new, potential subscribers.

## CONCLUSION

The DTP industry is still maturing. Besides the developments outlined in this chapter, the industry will benefit from the convergence of different technologies and applications. Entire font and clip art collections can, for instance, be stored on CD-ROMs. There is also an overlap in the area of graphics software. A computer-assisted design (CAD) program can be used to design a building. In a DTP project, the same program can create an illustration.

Advancements in one field can also have an impact on another field in the new technology universe. Desktop publishing is no exception. Newer, powerful PCs speed up various tasks, and for DTP, a project can be completed more rapidly. These same machines can also support sophisticated software that was once the domain of larger and more expensive computer systems. In one application, DTP users can correct and prepare large, high resolution images for publication.

In two other examples, progress in the overall laser market will have an impact on DTP systems (e.g., laser printer development), while the proliferation of networks could promote DTP-based operations. An electronic publishing environment would make it possible for more than one person to retrieve, review, and edit a document.

The concept of networking also cuts across national and international boundaries. High-speed digital lines can tie offices around the world in a global communications net. For DTP, you can gain access to the data stored on other networks, information can be rapidly exchanged, and ultimately important resources can be shared.[26]

The DTP field has also helped promote the growth of a *personal communications tool*. With a DTP system, an information consumer can now become an information producer. Information can also be tailored for a narrow rather than a mass audience, as may be the case with newsletters, pamphlets, and even book-on-demand publishing projects.

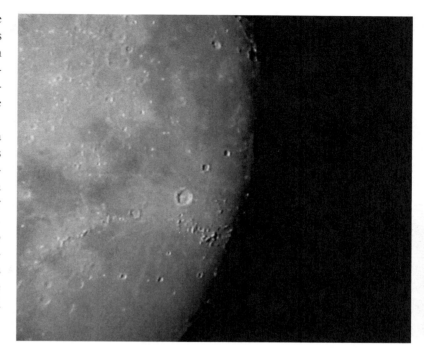

**Figure 10.5**
*Digital cameras have revolutionized picture taking, even in the astronomical field. In this case, amateur astronomers have used inexpensive digital cameras for imaging applications.*

## REFERENCES/NOTES

1. Please see the "Desktop Video and Multimedia Productions" chapter, Chapter 11, for a look at analogous systems for video production.

2. As covered in Chapter 3, a 17-inch LCD monitor has a larger viewing area than a 17-inch conventional monitor. For example, a conventional monitor's cover—the plastic supporting/around the tube—reduces the viewing area. For more information about the differences between monitor types, please see www.pc.ibm.com/us/infobrf/ibmon.html.

3. You can buy font libraries on CD-ROMs.

4. Please see Tom Thompson, "Color at a Reasonable Cost," *Byte* 17 (January 1992), 320, for a discussion of thermal wax units.

5. These printers also produce a continuous-tone output, in contrast to halftone-based systems, described in a later section of this chapter. Continuous-tone images look more like conventional photographs. For specific information, see Tom Thompson, "The Phaser II SD Prints Dazzling Dyes," *Byte* 17 (December 1992), 217.

6. John Gantz, "DTP Is Inching Toward Color, But Don't Hold Your Breath," *InfoWorld* 13 (June 10, 1991): 51. See also Janet Anderson, et al., *Aldus PageMaker Reference Manual* (Seattle, WA: Aldus Corporation, 1991), 76–83, for an excellent overview of color printing via a DTP system.

7. If you decide to go the commercial route, you can use process- or spot-color printing. For more information, see Eda Warren, "See Spot Color," *Aldus Magazine* 3 (January/February 1992): 45.

8. Image-In, Inc., Image-In (Minneapolis, MN: 1991), 178. *Note:* Line drawings are not affected by this factor.

9. *Note:* By capturing color and gray-level information, you can take full advantage of image editing software; scanning the same image multiple times, to produce a color output, has also been used in NASA's outer space probes (and missions).

10. Problems could include uneven scans.

11. Barbara Bourassa, "Mac Systems Speed Photo Transmissions," *PC Week* 9 (February 17, 1992): 25.

12. The sensor may include a complementary metal oxide semiconductor (CMOS) sensor.

13. This could include lighting and the camera's film speed-setting.

14. Digital cameras have, in one sense, helped advance amateur astrophotography when combined with new telescope/mount designs and computer software (for image processing).

15. For the general consumer market, film still holds the edge in resolution/contrast range.

16. Besides word processing and page composition software, another program category has supported page formatting/design options. A series of codes implements a page design. This type of program has been written for professional computer typesetting systems and PCs. A print preview mode may have also been supported, and this type of program could typically handle difficult formatting jobs.

17. Depending on the system, WYSIWYG may more aptly be called "what-you-see-is-probably-what-you'll-get" in the output. There may not be an exact one-to-one correspondence.

18. Note that the terms used can also vary from program to program even though the basic concepts hold true.

19. Most DTP programs can import from a broad range of word processing programs.

20. As described, word processing programs have also been enhanced. One price, though, for this increased sophistication, may be software that is almost too powerful for simple tasks. A program may also place a higher processing and data storage demand on the host PC.

21. Virginia Rose, *Templates Guide* (Seattle, WA: Aldus Corporation, 1990), 3.

22. When working with text, it's also possible to use a style sheet. In essence, a style sheet defines the attributes of a document's different

elements, such as a headline and body text. A headline may be centered and set in a specific typeface. To use this style, highlight the appropriate word(s) during editing, select the headline style, and the text will be automatically reformatted.

23. Don Lancaster, "Ask the Guru," *Computer Shopper* 9 (September 1989): 242.

24. James A. Martin, "There at the *New Yorker*," *Publish* 6 (November 1991): 53.

25. Jane Hundertmark, "Picture Success," *Publish* 7 (July 1992): 52.

26. Lon Poole, "Digital Data on Demand," *MacWorld* (February 1992): 227.

## SUGGESTED READINGS

*Aldus Magazine* 3 (January/February 1992). The following articles cover selecting a paper type, the history of paper, and a history of offset lithography: Mark Beach. "Paper in the Short Run." 33–36; Dirk J. Stratton. "Down the Paper Trail." 80; Nichole J. Vick. "Oil and Water." 19–22.

Bishop, Philip. "Crimes of the Art." *Personal Publishing* (May 1990), 19–25; Roger C. Parker. "Desktop Publishing Common Sense." *PC/Computing* (March 1989): 151–156; "Desktop Quality Circa 1992." *Business Publishing* 8 (January 1992): 23–29; "Publish Special Section; 101 Hot Tips." *Publish* 7 (July 1992): 63–88; Eda Warren. "See Spot Color." *Aldus Magazine* 3 (January/February 1992): 45–48. While the articles may be older, they offer an array of tips, guidelines, aesthetics, and effective design.

Bury, Scott. "Ready to Make the Digital Shift." *Electronic Publishing* 25 (February 2001): 24–28; Nancy A. Hitchcock. "Making a Splash." *Electronic Publishing* 25 (May 2001): 42–44; Matthew Klare. "Still Life in Pixels." *Interactivity* 4 (September 1998): 11–21; Michael J. McNamara. "Digital Dream What Makes This Camera Worth $9,000," downloaded from www.popularphotography.com/assets/download/3302003193951.pdf.; Michael D. Wheeler. "Information Processing: Law Enforcement Uses Digital Imaging and Storage to Track the Criminal." *Photonics Spectra* 32 (November 1998): 107–111. Digital /electronic cameras and imaging—overview and applications.

Dearmin, Thomas C. "Commercial Printing Drives Laser Development." *Laser Focus World* 37 (January 2001): 195–200; Alex Hamilton. "Pressroom of the Future." *Electronic Publishing* 25 (May 2001): 29–32. Printing technology developments.

Eggleston, Peter. "The Future of Color Printing: Beyond CMYK." *Advanced Imaging* 16 (April 2001): 28, 34; Noel Ward. "The Color of Print." *Electronic Publishing* 24 (July 2000): 31–38. Printing and color.

Gass, Linda, John Deubert, et al. *PostScript Language Tutorial and Cookbook*. Reading, MA: Addison-Wesley Publishing Company, 1985. Tutorial on PostScript.

Hitchcock, Nancy A. "How New Digital Papers Will Impact Designers." *Electronic Publishing* 22 (December 1998): 32–40; Bill Vaughn. "Are We Running Out of Trees." *Aldus Magazine* 4 (September/October 1994): 32–40. Two looks at paper: design and production issues (e.g., alternative sources/shortages).

Lodriguss, Jerry. "Scanning Deep-Sky Astrophotos." *Sky & Telescope* 105 (February 2003): 128–134. While the article focuses on astronomical photos, it is also an excellent overview of image scanning/correction in general.

Pennycock, Bruce. "Towards Advanced Optical Music Recognition." *Advanced Imaging* 5 (April 1990): 54–57; Noel Ward. "Digital Printing Goes Mainstream." *Electronic Publishing* 24 (February 2000): 32–36. Two desktop publishing applications.

## GLOSSARY

*Desktop Publishing (DTP):* A term that describes both the field and process whereby high quality documents can be produced with a PC, a laser printer, and software. DTP also implies that you have access to enhanced layout and printing options.

*Font:* A font is a typeface in a specific size.

*Optical Character Recognition (OCR):* Either a stand–alone unit or a software option for a scanner that makes it possible to directly input alphanumeric information from a printed page to a computer.

*Personal Publishing:* Desktop publishing systems make it possible for individuals to produce and potentially market their own work.

*Scanner:* An optical/mechanical device that is interfaced with, and subsequently inputs, graphics or text to a computer.

*Typeface:* A specific and unique print style.

# 11  Desktop Video and Multimedia Productions

The term *multimedia* can describe the integration of graphics, audio, and other media in a presentation or production. This chapter explores multimedia authoring software used to create such a production. Other topics include hardware, applications, and aesthetic considerations.

The chapter also covers desktop video (DT-V).[1] Analogous to desktop publishing, we can create our own video productions. PCs are an integral element in this process and are used for applications ranging from editing to creating graphics.

The DT-V and multimedia fields are also complementary. Desktop video tools can contribute to a multimedia production, and DT-V can be categorized as a component of the broader multimedia market.[2]

## MULTIMEDIA

Multimedia presentations are not new. Videodisks and accompanying software have served as a multimedia platform for years. But other products and factors, discussed in the sections that follow, have spurred the field's growth. Other multimedia resources that are not discussed in this chapter, but may also support multimedia applications, include conventional programming languages.

### Software Overview

In the context of our present discussion, the focus is on the dedicated multimedia authoring program. Many are user-friendly since you do not have to be a programmer to create a finished product. One software category uses a visual metaphor for this task. A series of icons, representing different functions, is linked to create the presentation. The icons are used as audio-visual and program control building blocks—one icon may play an animation while another may create a loop. When reached, a loop causes a series of events to repeat.

Other software categories include menu-based scripting interfaces. With such a program, commands that perform various functions are selected from pull-down menus. Depending on the software, it may also have a complementary programming language for more advanced functions.

As you create your presentation, you may also be able to use transitional effects. These could include a fade-to-black and a supporting audio effect.

*Impact.* Authoring software has opened up multimedia production to a broader user base, including the corporate and independent producer markets. If you are visually oriented, you can use visual tools to create a project, much like a desktop publishing document. You can import text and graphics and create on-screen buttons. These

**Figure 11.1**

*This shot highlights the integration between different software programs to enhance and speed-up certain operations. In this case, video editing and DVD authoring applications are integrated to streamline the production process. (Courtesy of Sonic Foundry; Vegas + DVD.)*

elements can subsequently be moved, re-sized, and linked to other information, or as described, to actions. From another stand-point, it is almost like using a PC-based Colorform set.

Nevertheless, although authoring software may be easier to use than conventional languages, you must still abide by a programming convention. To sustain an effective multimedia environment, the presentation must flow logically from event to event.

You should initially draft a set of criteria that drive the program's design. These should include the program's purpose and the best way to satisfy these goals with your software's tools. You must also adhere to technical and aesthetic guidelines.

***Other Capabilities.*** An authoring program can also create a highly interactive production. If you click on a specific area of a graphic, an event can be triggered. A digitized voice can be heard or a video clip played.

Authoring systems may also be more flexible than conventional programming tools for specific data handling tasks. The seamless integration of audio-visual elements in a multimedia production and hardware control are two examples. The key word in the last sentence is *seamless*. Unlike a conventional language that may require add-on software modules or special programming hooks, the capability is built-in and fully integrated in multimedia authoring packages.

A program may also extend your production capabilities to the Internet. Macromedia's Director, one of the leading multimedia production tools, has made it possible to play optimized versions of its projects in this environment. The same may hold true for presentation programs. In one example, a computer presentation slated for a group of investors could similarly be exported and reviewed via the Internet.

### Hardware Overview

Optical media have emerged as key multimedia distribution tools. This is a reflection of their mass storage capacities.

Another important tool is a PC's audio capability. Typical applications are playing

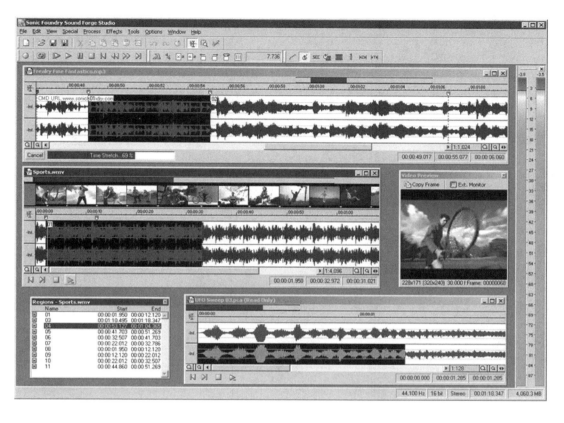

**Figure 11.2**

*PCs, when combined with software, can create an audio editing system. You can view an audio sequence and edit/manipulate specific sections. Video may also be accommodated to enhance the integration between, for instance, a sound track and a video sequence. (Courtesy of Sonic Foundry; Sound Forge Studio.)*

music and audio cuts, as well as editing your own pieces. Unlike a traditional setup where you may physically cut and splice tape, you work with a computer. You can digitize your voice, view it as a waveform, and subsequently edit and alter it. You can also select an audio effect, such as an echo or reverb, mix your voice with music, and save different variations of the same piece. In essence, the software provides you with a computer-based audio console.

Sound effects are also common and can serve as an audio cue. For one project, a bell, chime or other sound can provide a user with feedback confirming the selection of an on-screen button.

The video display is another consideration. The latest video cards can support life-like (photorealistic) images, and in conjunction with the PC and software drivers, may offer an accelerated operation.[3]

Beyond these devices, the primary hardware consideration is the computer itself. Multimedia authoring is generally hardware intensive, regardless of the platform. The PC should also be equipped with fast, high-capacity hard drives and as much RAM as you can afford.

At the dawn of PC-based multimedia production, Apple and Amiga computers were well ahead of the IBM PC world with respect to hardware compatibility. But the scale started to balance with later Microsoft Windows releases and, in fact, they helped fuel this market's growth.

Both computer families support a *plug and play* capability. The term means what it sounds like. You can theoretically plug in a common hardware peripheral and your computer will configure the setup for use. While this function works well in many cases, certain peripherals may require addi-

PRODUCTION TECHNOLOGIES

PRODUCTION TECHNOLOGIES

tional software drivers to operate. In other cases, you may have to scratch and shake your head a few times, go to an online help site, and drink several cups of coffee before you can get the system to work.[4]

## DESKTOP VIDEO

Desktop video is basically what it sounds like. You can set up a video production system on a desktop. PCs can control video equipment and are used to create video productions.

Some equipment may be geared for consumers, others target professionals, while a third category crosses both brackets (prosumer). The equipment's durability, speed, and sophistication can also vary, based on the intended application, budget, and buyer. Nevertheless, the new generation of hardware and software has blurred some of the distinctions between the professional and consumer worlds.

As discussed in a later section, this development also empowers people. Individuals and smaller organizations can now tap into the power of production tools that were once the domain of established media groups.

The following subsections cover the basic equipment and software used in desktop video applications. Although it is important to examine the capabilities of individual elements, it is also important to view them as a system.

A component may also have multiple applications: A PC may be used for creating graphics and for editing.

### Hardware Overview

*Video Capture Cards.* A capture system usually consists of a camera and an interface device. Much like a scanner, the system feeds

motion video and the accompanying audio to a computer.[5] This capability makes PC-based digital video production and editing systems a possibility.

A card's capabilities can vary. It may capture high quality, full-screen motion video. As covered in the next chapter, these types of cards are used in corporate and professional video production environments. Other cards may only capture information at a reduced size and frame rate. The technical quality of this information can also vary.

*Edit Controllers.* PC-based edit controllers are used to control editing VCRs. When you edit a production, selecting, organizing, and joining individual shots electronically create a story.

Desktop video systems can range from cuts-only to A/B roll to nonlinear configurations. In the first set-up, shots are linked together. In the second, you can make more sophisticated transitions, such as incorporating a dissolve, where one image is gradually replaced with the next. These transitions are more complex and require additional VCRs and other equipment.

In the third environment, you can edit a production using a visual metaphor—a timeline where video clips are arranged in sequential order. This topic is discussed in depth in the next chapter.

All in all, an editing system has also become a realistic production option for DT-V users. With the right video camera, you can produce near and, potentially, broadcast quality video. Combine this with a PC-based editing system, and you can create programs ranging from commercials to documentaries.

*Musical Instrument Digital Interface.* PCs can also take advantage of the Musical Instrument Digital Interface (MIDI). This standard made it possible for electronic musical instruments developed by different manufacturers to communicate with each other, and ultimately, with computers.

As part of a larger system, MIDI can also help create a powerful composing tool. In a typical application, a computer is linked with a synthesizer, an electronic musical instrument. This connection is made through the PC and synthesizer's MIDI ports, and a program can subsequently turn the computer into a sequencer. A sequencer essentially is a multitrack machine that can record and play back multitrack compositions.

You can manipulate the synthesizer's notes with the PC, since these notes, the events, are viewed by the computer as another form of data. The actual sounds are not recorded, but rather, information detailing the performance. The "speed at which the key was pressed" is one such piece of information.[6]

You can also create and edit musical compositions. A section of a track can be copied, the key can be transposed, and the tempo can be altered. The final piece can then be played back under the control of the PC.

The computer–MIDI marriage offers other advantages. A synthesizer can create a range of sounds that can vary from a harpsichord to a pulsating tone, a special effect suitable for a science fiction movie. The parameters that constitute a sound can also be saved on a disk, and it is possible to store, manipulate, and recall entire sound libraries. Collections of premade sounds are also available.

The MIDI revolution was brought about by the adoption of the standard in the early 1980s. Electronic equipment manufacturers agreed to follow this common standard to enable musicians to link different synthesizers, helping to fuel the growth of the electronic music industry. Previously incompatible and expensive instruments could now be interfaced, and musicians were handed a set of creative tools.

## Other Considerations

***Compression.*** The term *compression*, for our purposes, refers to data compression. Digitized video, audio, and image files can have enormous storage appetites. By using software and/or hardware schemes, data can be compressed and stored more efficiently. Compression is also used in other fields, especially the teleconferencing market, to support the relay of audio-video information over less expensive, lower capacity, communications channels.

Compression can be lossy or lossless.[7] *Lossy* means some of the data are lost through the compression scheme. For desktop publishing and certain other applications, this may not be a problem since the loss can be negligible.[8] *Lossless*, on the other hand, implies there is no loss of data. Lossless techniques are used in the medical field and other areas where data loss may be unacceptable.[9]

Popular compression options have included the following:

- Joint Photographic Experts Group (JPEG)
- Moving Picture Experts Group (MPEG-1), and by extension MPEG-2, MPEG Audio Layer-3 (MP3), and more recently, MPEG-4.

JPEG and MPEG-1 were originally and nominally geared for still and moving images, respectively. MPEG-2 has emerged as an important tool for video applications.[10] MP3 has been used for audio compression (music files). You can, for instance, download MP3 files from the Internet and store them on a CD. You can also purchase a standalone MP3 player, a small device that stores and subsequently plays back MP3 files.

PRODUCTION TECHNOLOGIES

MPEG-4, for its part, is "designed to be used to deploy complete new applications or to improve existing ones."[11] More versatile and "flexible" than its predecessors, MPEG-4 was introduced on the general market in the early 2000s.

Compression's importance also extends beyond the multimedia and DT-V fields. The development of efficient standards is critical for enhanced data relays through communications networks. Both developments, advanced communications systems and compression techniques, go hand-in-hand. Compression is similarly discussed in chapters covering high definition television, teleconferencing, and the Internet.

For the latter, this includes using, in a typical scenario, DT-V and multimedia tools to produce video clips slated for Internet distribution. In one example, the system you may use for editing a production may also support various Internet-based video delivery options. As stated at other points in the book, this type of production flexibility also equates to additional distribution venues and the need to convert this information to meet each venue's requirements.[12]

Finally, you may run across the term *codec* when working with audio-video systems. It stands for compressor/decompressor. In brief, a "codec is an algorithm, or specialized computer program, that reduces the number of bytes consumed by large files and programs."[13]

Codecs drive the compression/decompression processes and help make it possible to use less expensive and lower capacity communications channels for relays and/or to save storage space. Examples include MPEG-4 and those originally geared for the teleconferencing field.[14]

***QuickTime.*** The QuickTime architecture or standard, which taps compression's power, gave the multimedia and DT-V fields an enormous boost. Released by Apple in the early 1990s, it brought, in part, digital audio and video to the Macintosh at a reasonable cost. Designed to run on most models, QuickTime made it possible for Apple users to readily tap these resources. For example, playback did not require additional hardware, and digital video and synchronized audio became an integral component of the overall Mac system. Multimedia and DT-V support was now a built-in function, not an afterthought or a hardware and software kludge.

Since that time, enhanced QuickTime versions have been released. More important to the overall market, QuickTime supports Macs, IBMs, and Internet distribution. QuickTime serves as a representative example of how this type of standard can be used.

In a typical application, you can capture an audio-video clip via your Mac. Once saved, an editing program can manipulate the clip and accompanying audio.

One package, which illustrates this software's capabilities, is Premiere. The program can join clips using wipes, dissolves, and other transitions. Special creative and image correction filters can also be applied, and animations, still images, and multiple audio tracks, are supported.

Premiere's visual interface makes it an intuitive program. Clip sequences can be quickly rearranged, transitions can be added or deleted, and audio tracks, graphically represented as waveforms, can be manipulated.

Once you complete the project, it can be previewed. If it looks and sounds good, it can be made into a QuickTime movie, where its size and other output characteristics can be controlled.

The movie can be used by itself or with other QuickTime-aware or—compatible software. Instead of importing a still image, you import a movie. Place the movie in a document, click on it with a mouse, and the movie plays. It can turn a static document

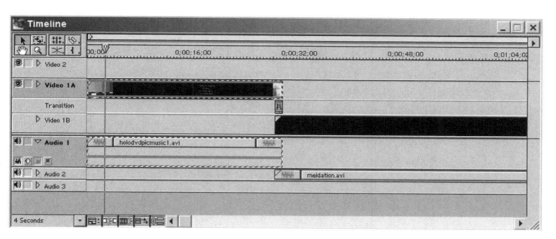

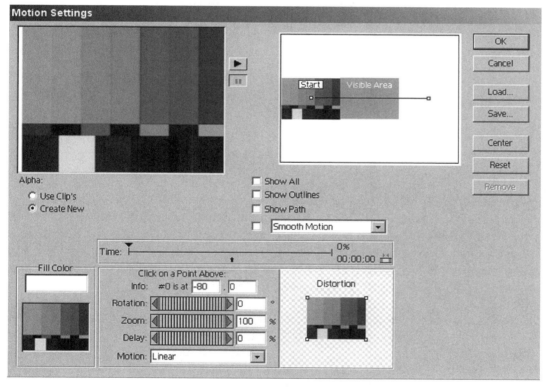

PRODUCTION TECHNOLOGIES

**Figure 11.3**
*Premiere's work environment. The construction window shows two clips that will be edited together. Premiere also supports a wide range of functions, including an option to animate (move) a video clip. (Software courtesy of Adobe Systems, Inc., Premiere.)*

into a moving one. These characteristics contribute to QuickTime's value as a communications tool. It can be used with numerous software packages, and as indicated, does not require special hardware for playback.

Hardware components do, however, have an impact on the system's performance. These range from the computer model to the graphics card to the use of a hardware-based versus a software-based compression system.[15]

It is also important to examine QuickTime and other standards from a systems approach. When first introduced, a QuickTime movie may have looked great to a computer user. It was cost effective and extended a PC's production capabilities. But a person working with video systems may have looked at the same movie and wondered what the fuss was all about.[16] Images were typically small, and as the movie's size was scaled up, clarity decreased. The frame acquisition and playback rates could also be low, which could make motions look jerky (not smooth).

The truth probably lies somewhere between both extremes. When first introduced, PC-based digital video generally did not match the quality of conventional high-end systems. But they were relatively inexpensive to implement and could be accommodated on a network.

Now, using a PC and the appropriate software, you can create a video production that can be published to the Internet or an optical disk. The quality can also vary, from a partial screen/lower frame rate playback, as was once the norm, to a high quality production.

It is also important to note that while QuickTime-type products have improved with age, DT-V producers may still work with some hardware/software limitations. Thus, the digital video you make may be the perfect complement for your presentation, but it may not be suitable for a television network.

There is another caveat in this discussion. FireWire-based editing systems proliferated in the late 1990s and early 2000s. This standard supports a high quality audio-video input and output and was matched by faster hard drives that could accommodate this data stream. What is the upshot of this development? Higher quality video editing, including the production of clips for multimedia projects, became available to a broader user base. But to fully take advantage of the standard, you still had to have a powerful computer fitted with the appropriate components.

## Convergence

What it boils down to is a sense of balance or perspective. Each production environment has its own relative merits and primary application areas. They should all be examined in the context of how they may fit in the overall communications system.

The convergence factor should also be considered. As indicated in different sections of the book, we continue to see a convergence of technologies and applications, including those taking place in the video editing market. In this case, the same software/hardware could be used to produce programming targeted for the Internet, an information kiosk, a DVD, or a television station.

## Videotape: It's Still Here

In this new world of digital video, A/V-ready hard drives, and optical storage, videotape is still here. It is a highly efficient and effective storage medium. For example, fast hard drives have supplemented and replaced tape in different broadcast applications. But these systems can still be expensive, and for the foreseeable future, may not match tape's high storage capacity/low cost ratio. Thus, both tape and nontape systems will coexist. Even when hard drives or other data storage systems become more cost effective, tape may still be used as an archival medium. This topic is covered in more detail in the next chapter.

## Graphics Software

Graphics programs contribute to DT-V and multimedia applications. Besides the typical software, other products are used. Computer assisted design (CAD) programs, for example, can serve as illustration tools. American Small Business Computers, creators of the DesignCAD software series, discovered that their programs were also used for business graphics, technical illustration, and desktop publishing.[17]

Other software may even be more specialized, and can include:

- programs that create realistic human figures;
- programs that generate landscapes ranging from Mt. Saint Helens to the planet Mars to surrealistic scenes created in your own mind; or
- character generator programs that create titles that can be overlaid on your video.[18]

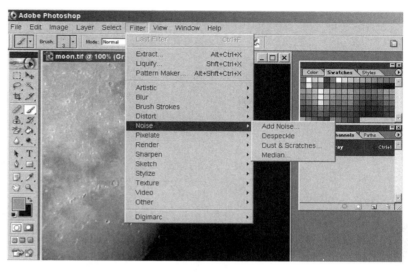

## Video Toaster

NewTek's Video Toaster is a product that spans the professional and nonprofessional markets. Introduced for the Amiga, the Toaster raised the level of DT-V production by a notch.

The Toaster brought professional video production capabilities to the desktop at a low cost. The system functions, in part, as a character generator, frame grabber, and a switcher. Switchers are used for image transitions. The Video Toaster also supports digital video effects, and LightWave 3-D, its 3-D graphics program, actually emerged as a cross-platform industry standard.

Depending on the production situation, you may also need a time base corrector (TBC) to take full advantage of these capabilities. A TBC "takes the unstable video from a VTR and acts as a shock absorber, outputting rock stable video that can be integrated with other video sources in a system to maintain good picture quality."[19]

TBCs were once out of the financial and technical reach of most DT-V users. The situation changed, though, when the Toaster's popularity and the growing DT-V market prompted the introduction of inexpensive TBCs, including internal models that fit inside a PC.

The newest version of the Video Toaster works with loaded IBM PCs. Loaded, in this context, refers to a fast processor(s), fast and high capacity hard drives, and as much RAM as you can afford. Like its predecessor, it supports multiple functions that are, with this generation, enhanced.

## APPLICATIONS AND IMPLICATIONS

Multimedia and desktop video systems have emerged as powerful information tools. In the business world, a multimedia production can make a speech more interesting and informative through the use of video and other media. This concept has been extended to teleconferencing and other fields.

## Education

The educational market is served by multimedia and DT-V products. In one application, a student can use a multimedia book that incorporates sounds and video clips. When combined with hypermedia links, the student can explore and experience this new

**Figure 11.4**

*Image editing and graphics programs play key roles in desktop video and multimedia production. In this example, an image can be altered though various tools, including filters that can be used for creative and image enhancement/correction applications. Other tools are visible on the left side of the screen. (Software courtesy of Adobe Systems, Inc.; PhotoShop.)*

PRODUCTION TECHNOLOGIES

PRODUCTION TECHNOLOGIES

world in a nonlinear fashion. The document's interactive nature can also help make learning active.

In the medical field, surgeons could use QuickTime, or another digital video product, to inexpensively document new techniques. Besides a written description, a surgical procedure could be annotated with video and voice. Digital video additionally offers a rapid turnaround time, a reasonable learning curve through programs such as Premiere, and the potential to exchange these data over the Internet and optical media.[20]

## Training, Sales, and Advertising

Training applications are well suited for DT-V and multimedia. A production could cover tasks ranging from car engine repairs to basic PC operations. Video clips of a real engine could be used, and the production could incorporate an interactive interface.

Store owners are also served. An electronic sales catalog can either replace or supplement a print version, and interactive kiosks where customers can get information about products have popped up in supermarkets and malls as well as other outlets.

The advertising and public relations industries have also benefited. Video clips and animations can be rapidly generated and used in a presentation. In a related area, a sophisticated multimedia system was developed to showcase the city of Atlanta, Georgia. The presentation was used to promote Atlanta as the site for the 1996 Summer Olympics.[21]

## Other Applications

The multimedia and DT-V fields have led to other applications and implications, including the following:

- Video artists adopting DT-V tools. Artists have used inexpensive video systems for years. Earlier projects included personal documentaries as well as video feedback, where different visual patterns could be created, controlled, and displayed on a television screen. Today, artists can tap sophisticated PCs and video equipment. In essence, video can be captured for either still or motion displays, images can be colored, and animations can be produced.

- The emergence of Canon XL video cameras, and other systems, as the tools of choice for independent documentary/movie producers. After shooting, the video would typically be edited with a PC-based system. This also includes Apple and IBM notebooks, generally configured with FireWire ports and external hard drives connected to the ports. In this set-up, the internal or system hard drive typically stores the software, and the external drive is used to store the audio-video files and the edited piece. An external drive offers another advantage: portability. To work on a different computer, simply disconnect the FireWire cable, move the drive to the other computer, and plug in that system's cable. This feature also enables you to share your files with a friend, colleague, or client.

- The use of DT-V and multimedia systems to extend the concept of the free flow of information. Like desktop publishing, production tools are now accessible to more people.[22] These tools have also contributed to development of a personal media. Instead of everyone receiving the same information, more personalized information can be created and received. For example, by using a video camera and a computer, we literally become producers and editors. Another implication is that we can now tap into more individualized information and entertainment pools.

**Figure 11.5**
*PCs, in combination with video equipment, have provided artists with a powerful set of tools. A still from "Godzilla Hey," by Megan Roberts and Raymond Ghirardo (1988). The work combines digital video imagery and sound with analog video synthesis. Produced at the Experimental Television Center with Amiga PCs and the FB-01 video frame buffer with proprietary software, designed by David Jones. (Courtesy of Megan Roberts and Raymond Ghirardo.)*

Various organizations, including MIT's Media Lab, have pushed both concepts beyond current boundaries. The Media Lab, as directed by Nicholas Negroponte, has been one of the world's premier research institutes. Besides exploring the convergence of different media, personalized interactive media have been investigated. Two examples are personal electronic newspapers and television.[23]

In this new world, information could be retrieved from different sources and subsequently delivered to you through the assistance of intelligent systems and human–machine interfaces. In one setting, a computer could scan a night's worth of programming and then summarize and possibly replay the portions that would be of personal interest to you.[24]

Other developments, including an interesting look at this institution, can be found in Stewart Brand's book *The Media Lab*.

Negroponte has also discussed this topic, and his view of our digital future, in his *Being Digital*.

## PRODUCTION CONSIDERATIONS

To wrap up this chapter, we should examine some basic production issues along with broader, aesthetic issues. For the latter, the growth of the multimedia and DT-V markets may have outstripped the development of a sound aesthetic base. An examination of traditional film and television frameworks can be valuable.

### Production Elements

Like desktop publishing, there are some basic conventions you should follow when creating a DT-V or multimedia project. These include the following:

1. Watch out for too many links. A reader can become lost in a document if there are too many consecutive links in an interactive multimedia production.[25] A prototype of the final project, like a storyboard for a television production, can help point out potential problems.

2. Know your audience. For example, the interface for a commercial catalog may be very different than one geared for the general public. The former may be designed to retrieve information as quickly as possible. The latter may have more visual effects and graphics to hold the audience's attention.

3. Remember the old adage "form follows function." This concept should be applied when you design your product.

4. As stated, examine your distribution venues. Are you going to use a CD-ROM, DVD, or the Internet? The distribution medium may, for instance, have an impact on the project's graphics. For the Internet, you may use compressed and/or smaller sized graphics to speed-up the information relay. Similarly, if you design a CD-ROM product, can it be readily adapted for the Internet? Does your software support this option? In one sense, you can now approach production from a systems approach in regard to program development and distribution.

5. Pay attention to other technical and nontechnical needs. Can your product be distributed on multiple computer platforms? Do you have to pay a fee for distribution rights? Are you using any copyrighted materials? Do you have permission to use these materials?

6. Check all your work on a conventional television or monitor while you are working. Colors may not be as rich and resolution can be lost. Some programs also have an option to help ensure the colors you use will technically conform to a television environment (e.g., NTSC-safe).

7. Do not place titles and other visual elements at the edge of the screen. They might be cut off when displayed on a television. Similarly, make sure the text is not too small to read. You should also avoid a fancy script typeface. Although a fancy typeface might be fine on the computer's display, it might be illegible on a standard television.

8. When recording, do not use old tape. Videotape that has been reused a number of times may not give you a clean recording.

9. *Be paranoid.* Always save your work and back it up—back it up—back it up.

## Aesthetic Elements

*Element Integration.* All the media elements that compose a production should be fully integrated, much like a film or television show. There should be motivation, a reason, for using any given element. Do not use an animation, for example, simply because you own animation software. Use one for a specific goal. It can range from demonstrating a new piece of equipment to serving as an attention grabbing device.

*Preproduction.* The issue of what elements to use can be worked out during a preproduction phase, the time before the program is created. The idea is to establish different criteria to help guide the program's design. The preproduction stage also serves a more practical purpose. It is less expensive to make changes at this stage than after the program's final assembly.

During the preproduction phase, different questions should be asked:

- What is the presentation's goal?
- Who is the potential audience?
- What are the budgetary limitations?
- What is the best way to satisfy the goal? For example, should video be used?

The last question is particularly important. Should you use conventional video or would a QuickTime clip suffice? This flexibility is one of the hallmarks of multimedia production. The key again, though, is ensuring the presentation's integrity. Use a tool only if you have a reason to use it.

***Prototyping.*** Besides a preproduction plan, you may want to prototype a small-scale version of the program. You can try out your ideas and plan any necessary changes.

A storyboard would also be helpful. A storyboard depicts, in sequential order, the major events in a production. It can be made of still pictures and may include audio. If the storyboard is computer based, you may be able to use some of its components in the final product. These can include graphics, animations, and QuickTime movies.

***Enhancement.*** Once you create your production, seek ways to improve your craft. Continue producing, watch related media products, including movies and television programs, and read.

If you are planning a DT-V project, you can learn how music can be an effective component by examining its role in certain movies. Music can heighten the tension in a scene or can serve as a counterpoint to what we're watching. This principle, observing for analytical purposes, also applies to lighting, scriptwriting, shooting, editing, and other production techniques.[26]

Finally, the goal of this process should be the development of your own style. Your production, whether it's video art, an electronic catalog, or a multimedia presentation, can have your own personal signature. With all the tools at your disposal, this should be an enjoyable prospect.

## REFERENCES/NOTES

1. The typical abbreviation for Desktop Video is DTV. However, as covered in the digital television chapter, DTV also represents digital television. To avoid confusion, DT-V will be used when discussing desktop video.

2. Tom Yager, "Practical Desktop Video," *Byte* 15 (April 1990), 108.

3. David A. Harvey, "Local Bus Video," *Computer Shopper* (July 1992), 181. *Note:* This concept can be extended to other computer peripherals to similarly speed up their performance.

4. In some cases, there may be hardware conflicts between the peripheral you are installing and the computer; there may also be software incompatibilities. In one example, an updated version of a video editing program may require an updated version of the operating system to run.

5. When using analog video, the information is also digitized.

6. David Miles Huber, *The MIDI Manual* (Carmel, IN: SAMS, 1991), 20. See Jeff Burger, "Getting Started with MIDI: A Guide for Beginners," *NewMedia* 1 (November/December 1991), 61, for additional information and for a chart that highlights MIDI's storage efficiency when compared with conventional digitized audio.

7. Bruce Fraser, "Scan Handlers," *Publish* (April 1992), 56.

8. The same principle applies to certain video production operations.

9. Chris Cavigioli, "Image Compression: Spelling Out the Options," *Advanced Imaging* 5 (October 1990), 64.

10. Please see *Media Cleaner Pro 4* (1999) software manual, by Terran Interactive, for an excellent discussion of compression schemes and their advantages/disadvantages (for example, pp. 31–35).

11. Richard Doherty, "The MPEG-4 Product Roll-Out: Digital Video Poised to Go the Distance," *Advanced Imaging* 16 (February 2001), 19.

12. This may include, for instance, using compression to make it possible to distribute a clip over the Internet.

PRODUCTION TECHNOLOGIES

13. Searchnetworking.com. Downloaded from http://searchnetworking.techtarget.com/sDefinition/0,sid7_gci211810,00.html.

14. When used in the telecommunications industry—to go from an analog to digital signal and back—codec generally stands for coder/decoder.

15. See Denise Salles and Judith Walthers von Alten, "Making, Playing, and Sequencing a Movie," Chapter 7 in *Adobe Premiere User Guide* (Mountain View, CA: Adobe Systems, 1992), for a comprehensive overview of the latter topic.

16. Ben Calica, "The Clash of the Video and Computer Worlds," *NewMedia* 1 (September/October 1991), 58.

17. American Small Business Computers, personal communication.

18. Character generators range from dedicated broadcast to PC-based units.

19. Tedd Jacoby, "Old Problems, New Answers," *Video Systems* (April 1988), 80. *Note:* TBCs are also used in high-end, conventional editing systems.

20. Steve Blank, "Video Image Manipulation with QuickTime and VideoSpigot," *Advanced Imaging* 7 (February 1992), 54.

21. Mike Sinclair, "Interactive Multimedia Pitches Atlanta Olympic Bid," *Advanced Imaging* 5 (March 1990), 38.

22. The produced electronic documents may present even more powerful messages than their paper counterparts.

23. Nicholas Negroponte, speech delivered at Ithaca College, Ithaca, NY, May 29, 1992.

24. Ibid.

25. The same principle applies to web site designs, among other operations.

26. Relevant television programs include *The Twilight Zone* and the PBS documentary *The Civil War.* In cinema, the list runs the gamut from older classics to more modern films: *Battleship Potemkin, Alexander Nevsky, Grand Illusion, The Adventures of Robin Hood, Citizen Kane, Casablanca, The Third Man, Psycho, The Godfather, Raging Bull, Glory, Braveheart, The Lord of the Rings*, and *Apocalypse Now.* As a group, the films serve as examples for production and aesthetic principles ranging from music to lighting to shot composition.

## SUGGESTED READINGS

Ashdown, Ian. "Lighting for Architects." *Computer Graphics World* (August 1996), 38–46. Architectural rendering, graphics, and the visual importance/impact of lighting (in an image).

Barnett, Peter. "Implementing Digital Compression: Picture Quality Issues for Television." *Advanced Imaging* 11 (April 1996), 30–33, 88; Richard Doherty. "The MPEG-4 Product Roll-Out: Digital Video Poised to Go the Distance." *Advanced Imaging* 16 (February 2001), 20–23; Lee J. Nelson "Video Compression." *Broadcast Engineering* 37 (October 1995), 42–46; Terran. *Cleaner 5 User Manual;* Ben Waggoner. *Compression for Great Digital Video.* Berkeley, CA: CMP Books, 2002. Compression, technical concerns, applications, and techniques. The Cleaner 5 manual is written for the Cleaner software. But it is also a rich information resource about compression and compression standards. The Waggoner book is very comprehensive and includes a companion CD.

Brain, Marshall. "How MP3 Files Work." *Verizon,* downloaded from wysiwyg://153/http://www22.verizon . . .r/Articles/article/?articleId=1023; Karl Heintz Brandenburg. "MP3 and AAC Explained." AES 17th International Conference on High Quality Audio Coding, downloaded from www.aes.org/publications/downloadDocument.cfm?accessID=14703162000122117. Articles about MP3 compression and applications. The Brandenburg article also covers other compressions standards.

Brand, Stewart. *The Media Lab.* New York:

Penguin Books, 1988. As stated in the text, the book provides an interesting look at MIT's Media Lab.

*Computer Graphics World*. The journal is an excellent resource for imaging, computer animation, and related topics. Three example articles are: Jenny Donelan. "Roamin' Ruins." *Computer Graphics World* 25 (August 2002), 33–34; Martin McEachern. "Double Headers." *Computer Graphics World* 25 (August 2002), 13–24; Karen Moltenbrey. "Preserving the Past." *Computer Graphics World* 24 (September 2001), 24–30.

De Leeuw, Ben. "Moving in Real-Time." *3D Design* (September 1998), 31–35; Craig Lyn. "Master Series, Part 9: Mapping." *DV* (January 1999), 68–69; Eni Oken. "Color. Color Everywhere." *3D Design* (September 1998), 53–63; Michael O'Rourke. *Principles of Three-Dimensional Computer Animation*. New York: W.W. Norton and Co., 1995; Barbara Robertson. "Staying Tooned." *Computer Graphics World* 24 (July 2001), 32–38. Graphics theory, techniques, and production; Oken's article is an excellent color primer.

Hall, Brandon. "Lessons in Corporate Training." *NewMedia* 6 (March 1996), 40–45. Corporate training and different tools and techniques.

Huber, David Miles. *The MIDI Manual*. Carmel, IN: SAMS, 1991. A detailed guide to MIDI systems and operations.

Kelsey, Logan and Jim Feeley. "Shooting Video for the Web." *DV* (February 2000), 54–62. Desktop video techniques for Internet-based projects.

Millerson, Gerald. *The Technique of Television Production*. Boston: Focal Press, 1990; Herbert Zettl. *Television Production Handbook*. Belmont, CA: Wadsworth Publishing Company, 1999. Video production texts. The topics include, depending on the book, field production, editing, shot composition, and aesthetics.

Negroponte, Nicholas. *Being Digital*. New York: Vintage Books. 1995. A fascinating look at the possibilities brought about by the communication and information revolution.

Ozer, Jan. "Building the Perfect Digital Video Studio." *EMedia Magazine* 15 (August 2002), 22–32; Ben Waggoner. "FireWire Gets Faster." *DV* 10 (October 2002), 28–32. Desktop video components and a faster FireWire interface (including FireWire versus USB for video production applications).

Wallace, Lou. "Amiga Video: Done to a T." *AmigaWorld* (October 1990), 21–26. An initial look at the Video Toaster.

**PRODUCTION TECHNOLOGIES**

## GLOSSARY

*Authoring Software:* Software that can simplify and enhance the creation of a multimedia presentation.

*Compression:* Compression refers to reducing the amount of space required to store information (e.g., video). Compression can also speed up information relays.

*Desktop Video:* Advancements in video technology have made it possible to assemble cost-effective yet powerful video production configurations. PCs play a major role in this environment.

*Musical Instrument Digital Interface (MIDI):* A MIDI interface makes it possible to link a variety of electronic musical instruments and computers. The MIDI standard also enables musicians to tap a computer's processing capabilities.

*QuickTime:* Apple Computer's PC-based digital media system.

# 12 The Production Environment: Personal Computers, Digital Technology, and Audio-Video Systems

Besides the desktop publishing and video revolutions, another revolution is sweeping the broadcast and nonbroadcast production industries. It is a revolution based on the adoption of computer and digital technologies. This chapter covers these developments and complementary topics. The latter range from convergence issues to digital recording.

## PRODUCTION EQUIPMENT AND APPLICATIONS

### Switchers and Cameras

The switcher has benefited from the integration of computer technology. In brief, a switcher is used to select the pictures produced by a video facility's multiple cameras. It also creates visual transitions, can be used in postproduction work, and serves other functions.

A computer-assisted switcher can help an operator in these tasks. In one application, a complex visual effect may be required. The actions to create the effect can be preprogrammed, stored, and recalled at the press of a button. This capability, to immediately execute a command in the middle of a production when time is always critical, is an important one.

Switcher configurations can also be stored and later retrieved. This option enables an operator to quickly reconfigure the switcher for different production situations (e.g., a news show versus a commercial).

The influence of computer technology also extends to video cameras and robotic camera configurations. In the former, various parameters can be set up with a computer-based control system, to help free an engineer's valuable time. In the latter, individual camera operators are replaced by a robotic system. Television news and other production situations generally call for set shots. A robotic system may prove acceptable in this environment.

A human operator can monitor and control a system's multiple cameras with various interfaces, including graphics tablets and joysticks. Stored camera shots and movements can also be recalled, which lends itself to repeatability.[1]

A robotic system may have some limitations. For example, it may not equal the speed or capabilities of individual camera operators. This is an important consideration for sporting events and other dynamic shoots. There may also be a cost factor, depending on a system's sophistication.

PRODUCTION TECHNOLOGIES

**Figure 12.1**
*One of the fallouts of the communication revolution: The ability to create sophisticated productions with PC-based products. This is a shot from Todd Rundgren's "Change Myself," produced by Rundgren using the Video Toaster. (Courtesy of NewTek, Inc.)*

## Digital Special Effects and Graphics Systems

A digital special effects system, also called a digital video effects (DVE) generator, manipulates a digitized picture or video sequence to create a special effect.[2] The final product may be recorded on videotape or used in real time.

A DVE generator can support 2-D and 3-D effects; an example of a manipulation is a compressed television picture. After the video signal is digitized and processed, the original picture can be reduced in size and repositioned on the screen. The original picture is still visible, but now it is physically smaller. Similarly, you can initiate an on-air zoom—a little "zooms in" to fill the screen. Other, more advanced effects include manipulation of an image so it appears to flip over, much like a page turning in a book.

You could use dedicated hardware or a PC with the appropriate software for this operation. Depending on the configuration, the PC could be less expensive, but you may pay a price in speed and flexibility.[3]

The same scenario applies to graphics systems. A PC can create graphics suitable for a production. This setup could also be cost effective, accommodate numerous applications, and allow you to tap into the free-lance market, which is the pool of people with PC graphics experience.[4]

Dedicated hardware, however, has its own advantages. A system can handle intricate 3-D images, digitized pictures, and large files with a faster turnaround. The latter may be critical for news and other time-sensitive productions.

## Nonlinear Video Editing

Computer and digital technologies have also influenced video editing. As described, editing can encompass the simple joining of scenes or more complex transitions. A newer form of this application is PC-based non-linear editing (NLE).

In brief, NLE allows you to retrieve stored audio-video sequences in random access fashion, much like conventional computer data. In a PC environment, the scenes are stored on a computer's hard drives for later use.

A linear system, in contrast, is in keeping with the more traditional editing method. You must search through a videotape to find the specified scenes. This process ultimately eats up more time.

When using an NLE system, the different audio-video elements are displayed on a monitor as you assemble the production. In one set-up, the video material can be represented by video frames, which serve as visual references for the scenes.

Some of the characteristics of working in this environment are as follows:

1. Audio-video is captured and, based on the system, you can select different options (e.g., audio quality level).

2. The different scenes are selected from an electronic pool or bin and are sequentially placed on a time line. The scenes can be quickly rearranged and/or deleted. For

many individuals, this visual metaphor has made the editing process more accessible. It can also lead to more creative freedom. You can rapidly experiment with edit points, different shot arrangements, and other production elements.

3. Transitions between the scenes are selected from an options list. At some point, if you want to delete a transition, it is a simple operation, much like deleting a phrase with a word processing program.

4. High-end systems may complete, and you may be able to view, dissolves and other transitions in real-time. Less sophisticated systems may require several or more seconds to complete or render a transition. Hardware acceleration, versus software-only systems, can be an important factor in this process.

5. You can generate titles and tap a PC's graphics capabilities. A graphic or animation, which you create, can be imported and used. You can also manipulate the audio tracks and add music.

6. You can print (record) the production to videotape; the quality is affected by various technical considerations.[5]

7. NLE systems flooded the market by the late 1990s and early 2000s. The final output improved while prices dropped. Both factors helped open the field to more users. Newer systems could also accommodate multiple video formats. These developments were fueled by the following:

- faster PCs
- new software/compression capabilities
- cheaper memory and hard drives
- the adoption of FireWire and other standards

FireWire facilitates the connection and linkage of, for our current discussion, different production equipment and PCs.[6] This standard also supports a fast data transfer rate and enables you to create a link with a single cable. Newer video cameras, including those based on the Mini-DV standard, are also

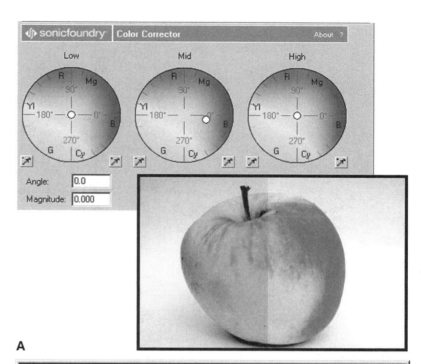

A

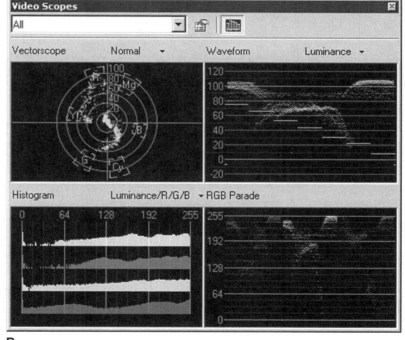

B

**Figure 12.2**
*The latest generation of NLE systems provide editors with powerful tools. In these screen shots are tools to analyze and correct video sequences. In (A) the color correction capability can be used to match the video shot with different cameras. (Courtesy of Sonic Foundry; Vegas 4.0.)*

equipped with a FireWire port. If you don't own a deck, you can use the camera itself to export recorded sequences to the PC. Once edited, you can use the camera to record your project.[7]

8. Although PC-based NLE systems are valuable, some factors should be taken into consideration:

- Install as much memory as you can afford.
- Be prepared for occasional lock-ups. System freezes can occur.
- Digital video has a high storage overhead, and to work efficiently and effectively, you may have to buy more drive space than you anticipated. Follow this simple rule of thumb: When you think you have enough storage, you'll probably need more.
- Although NLE systems can speed up certain tasks, such as rearranging scenes, others may be slower. Depending on the system, the latter include generating certain transitions and effects.
- Even though the visual interface can simplify the overall editing process, the software may have a high learning curve.
- Can the system handle the quality of video as well as the format required for your applications? Can the program convert a video clip, for instance, to formats compatible with DVD and Internet distribution or do you have to use another program?
- Does the company have a good technical support policy?
- Pay attention to aesthetics. Editing is still a craft that must be learned and practiced. The same concept applies to color correction. The editing system may support powerful color correction tools that can improve and enhance a video sequence's technical and visual qualities. But much like editing, proper

color correction techniques should be learned and practiced for their appropriate use.

## Audio Consoles and Editing

Audio equipment has also been influenced by computer and digital technologies. A computer-assisted audio console, for example, can help an operator to manipulate sound elements. As covered in the previous chapter, your voice or other audio piece can be digitized, edited, and manipulated. In this production environment, you can take advantage of random access editing and the visual representation of the audio sequence as a waveform on a monitor. This set-up provides you with aural and visual clues to help you quickly identify edit points.

For facilities with limited budgets, this application is supported by professional quality audio cards that work with PCs. Complementary software enables you to produce multitrack projects that can be recorded on tape, a hard drive, or other media. You can also create a sequence and use it in a video-editing project.

You can also buy a portable digital console for field and in-studio work. These small consoles may pack a powerful digital audio punch: "The new breed of console has just about everything needed to capture, edit, master, and burn audio masters, all condensed into one portable console."[8] In essence, you can literally carry an audio studio, which can perform a series of sophisticated functions, in your hands.

## DIGITAL RECORDING

### Why Digital?

Digital audio and video systems have certain advantages over their analog counterparts. Chapter 2 covered some of these characteristics, which include a more robust signal. A signal's quality is also preserved after multiple generations, unlike a typical analog system where the output suffers as you progressively "go down" a generation.

## Audio Recording and Playback

A digital audiotape (DAT) machine can record and play back digital tapes. The audio quality is equal to a CD, and the information can be stored on a compact cassette. DAT's recording and operational characteristics have made it particularly attractive to professionals. Studio and field machines have been designed, time code can be supported, and the output is excellent.

The consumer market, though, has been a different story. On face value, DAT systems should have been a success. Yet various factors combined to create a flat U.S. consumer response. For example, the recording industry claimed consumers would make CD-to-DAT copies, that is, digital-to-digital recordings, thus potentially reducing CD sales. Protection schemes were subsequently proposed, including the Copycode system. In practice, a DAT machine would shut down if it detected a special "notch" in prerecorded media. But this system was dropped because it did not work all the time, and some individuals indicated it affected a playback's quality.[9] When this factor was combined with the threat of litigation against manufacturers who sold unmodified machines, the result was the flat market.

But this picture has somewhat improved. DAT equipment has been embraced by elements of the audio industry and audiophiles.[10] A royalty and new protection scheme also defused the somewhat contentious atmosphere. Nevertheless, as of this writing, DAT has failed to achieve a broadbased U.S. consumer acceptance.[11]

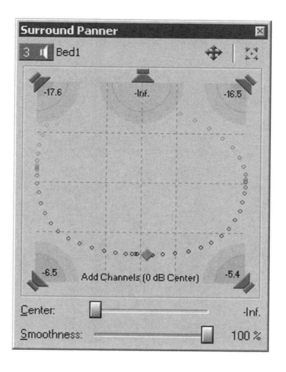

**Figure 12.3**
*Using software, you can create sophisticated audio compositions to complement a production—in this case, SurroundSound mixing tools. (Courtesy of Sonic Foundry; Vegas 4.0.)*

In the 1990s, Sony introduced a new digital audio format, the MiniDisc (MD). The equipment line has included a compact recording system with random access capability, digital quality sound, and compensation for physical jarring. Other models were later released for the professional market, including a unit that could replace the traditional cart machine, a common fixture in most professional facilities.[12]

Digital cart systems have also appeared on the market. Audio can be stored, much like computer data, for the immediate playback of different audio cuts. You can also tap your existing PC, if equipped with a high quality audio card, for this application.[13] Install the software, and you are greeted by virtual machines—on screen representations of this equipment. You can use your keyboard and mouse to subsequently control and playback the audio cuts.

What are the advantages? Random and immediate access to the different audio

PRODUCTION TECHNOLOGIES

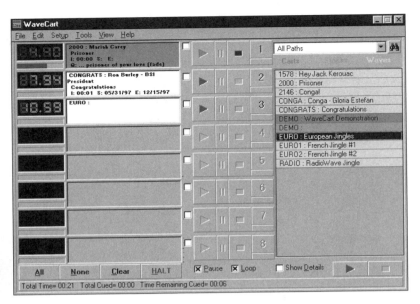

**Figure 12.4**
*A PC-based virtual cart system can replace multiple, traditional cart machines. The virtual carts are displayed on the left; the different music cuts, which are subsequently loaded in the carts, are displayed on the right. A cart can be started at the press of a key and can be played sequentially, automatically. (Software courtesy of Broadcast Software International; WaveCart.)*

tracks, computer control for playback, cost-savings for cart machines and media (carts), and the reliability afforded by this environment.[14] Your system may also be enhanced through a software rather than a hardware update (e.g., replacing expensive equipment), as would be the case with a conventional system. This last advantage may also hold true for other computer-based systems as well.

## Video Recording

As indicated at the beginning of this section, one of the advantages of a digital VTR is its high-quality, multigenerational capability. Two earlier formats include D-1 and D-2.

Other digital formats, and complementary equipment, have subsequently been produced. These include DVCAM and DVCPRO, two competing formats from Sony and Panasonic, respectively. The Mini-DV, another format, emerged as a prosumer standard supported by different manufacturers. Mini-DV systems also became popular professional tools. Cameras, for instance, are compact, produce a high quality output, and

sport FireWire ports and multiple shooting modes (e.g., wide screen).

## Tapeless Systems

Tapeless systems have also become popular. For our purpose, this term is an umbrella phrase. It describes equipment that does not use tape as its primary recording and/or playback medium. These include NLE stations equipped with fast hard drives and video servers. A *video server* is essentially a high speed and high capacity information processing/storage device analogous in function to a traditional server used in a LAN environment.

***Advantages***   In one application, a server can playback a television station's ads. Traditionally, a setup was expensive to maintain, prone to mechanical failure, and could be a complex mix of VTRs and a robotic control system. A server, which can require less maintenance and operator supervision, can replace this mechanical configuration. These devices can be more maintenance free and can run with less operator supervision.[15] The end result would be a more cost-efficient station.

A tapeless environment has also been extended to other areas. In one application, Avid helped introduce a disk–based camera for fieldwork. Instead of recording video on tape, it is recorded on a removable drive that can be married to a digital editing station. Thus, you can quickly go from a shoot to editing to on-air. In a similar application, portable hard drives allow you to record and then directly interface with an NLE system for postproduction work.[16] Other cameras support optical recording media instead of tape.

When viewed from a broad perspective, tapeless systems have other advantages:

1. Tapeless systems are less prone to physical breakdowns. There may be fewer

mechanical components and parts to wear out (e.g., a VTR's heads). A newer generation of equipment may also sport controls similar to those of traditional VTRs, thus, facilitating their acceptance and operation in a production environment.[17]

2. Tape is prone to wear and tear. When reused, defects, which can adversely affect a playback, may surface. A tapeless system's playback quality should remain unchanged.

3. The quantity of consumables, including replacements VTRs and tape, will be reduced.[18]

4. A tapeless system can simultaneously serve multiple users. In a typical scenario, the same video clip could be accessed by two editors to compose different stories.[19]

In sum, new systems are making their way to the market while the performance of existing systems have been enhanced. The variety of products and their capabilities also offer new production options. These range from the use of optical media and hard drives for recording video in the field to the creation of storage systems that may support one or multiple users. The latter may include using storage area networks (SANs) and other systems.

What is the bottom line? We now have access to a diversity of storage media for shooting and postproduction work that, when used appropriately, can *enhance the workflow across an organization*. It is also important to remember that the developments in one field may have an impact on another. In one example, as network technology is enhanced, be it through the communications line itself or other components/software, it will have an impact in the audio-video production field. Finally, please see the Suggested Readings section for detailed information about SAN-based technology/products and other storage options.

***Disadvantages*** Despite its advantages, there are some limiting factors as we migrate toward a tapeless environment:

1. Sectors of the broadcast industry are still heavily invested in a tape-based standard. Because an overnight switch would be too expensive, the changes will most likely be evolutionary. The technology base is also too new for some users—it has not met the test of time.

2. Technical limitations, including potential software problems, must be ironed out.[20] Basically, if there are software failures, whole systems could be rendered inoperable.

3. System redundancy must be considered. If a single VTR breaks in a traditional setup, another unit could easily replace it. In contrast, how effective are the server's backup and replacement capabilities?

4. Depending on the application, media costs may still be comparably high for tapeless systems. Videotape and other tape formats are cost effective, especially in view of their storage capacities. Hard drives, for instance, cannot as yet match this capacity-to-cost ratio.

Consequently, tape will still be used for an undetermined time. This is especially true for smaller stations with limited budgets.

We may also see tape/tapeless hybrids in the interim. In one example, servers used for video-on-demand applications may be supplemented by tape systems. The server could store frequently requested and timely information while other programming could be archived on tape. When requested, it would be transferred to the server and delivered. Data tape, rather than videotape, may also be used for archival purposes.[21] Ultimately, in this type of environment, we could draw on the relative strengths offered by both technologies and media.[22]

PRODUCTION TECHNOLOGIES

## INFORMATION MANAGEMENT AND OPERATIONS

Computers have been adopted for information management and control operations. These include scriptwriting, newsroom automation, and computerizing a station's traffic and sales departments. Even though these tasks may not be as visible as the production end of a facility, they are nonetheless vital.

### Scriptwriting, Budgets, and News

A critical job in any production facility is scriptwriting. The script is the heart of a program, and scriptwriting software has been released to support this function. The software can help free you from numbering scenes, formatting dialogue, and other time-consuming mechanical chores. The software completes these functions, and you can concentrate on writing.

A program typically supports a standard two-column television script and other formats. It may also link specific column sections. If changes are made, the corresponding audio or video section tags along.

The scriptwriting process can be further enhanced through supporting programs. In one instance, software can be used to create a storyboard, essentially a road map of a production. A storyboard can help you visualize the final product, and it may even be possible to use short digital video clips with other media.

Other specialized and general release programs are also widely used. For budgets, spreadsheets can track business expenses, crew fees, travel and postproduction costs, and equipment rentals. Another option includes the use of CAD programs for facility design.

For news, computer technology has helped transform the traditional news department into an *electronic newsroom*. By integrating hardware and software, different tasks are enhanced. For example, wire service stories can be fed directly to a computer. This information can be retrieved, printed, and saved.

A news department can also establish an electronic news morgue.[23] As part of a networked operation, reporters and editors can gain access to current and past stories. Digital video clips and other information could likewise be accommodated, particularly as compression techniques improve.

This digital vision, where information is networked and is readily accessible, could facilitate the overall production and news processes. Digital video workstations, for

**Figure 12.5**
*Multiple videotape formats have coexisted for years. This includes the S-VHS (right) and the newer, high quality Mini-DV formats (left). Note the size difference.*

PRODUCTION TECHNOLOGIES

PRODUCTION TECHNOLOGIES

**Figure 12.6**
*Computers play a key role in the broadcast industry. In this example, they are used to produce shots for weather forecasting. (Courtesy of AccuWeather Inc.)*

instance, could enable journalists, who may have traditionally worked in a text-based environment, to tap into video and related resources. For facilities with limited personnel, the same individual could write and edit the story, potentially all on the same desktop.[24]

Depending on the manufacturer and package, other applications could be supported. These include an interface to the station's production facilities where the newscast's text could be fed to a prompter and a closed captioning setup. A *prompter* is a device used by on-air talent to maintain good eye contact with a camera while reading news copy. *Closed captions* are the normally invisible subtitles for programming that can be displayed on a television set through a special decoder.

Consequently, a product can be used alone, as may be the case with a newswire service feed to a single PC. In another setting, an integrated system that links the newsroom with the production end can be created. In fact, this principle has been extended to tie various pieces of equipment in a central control network. It could potentially encompass the entire facility.[25]

## Operations and Information Management

Various media organizations have adopted computers for marketing research. Arbitron, for one, has served the broadcast industry for years. The company has tapped computer technology to speed up the delivery of information to client stations and to support different audience measuring techniques. The company even investigated an artificial intelligence-based passive system that could identify viewers without prompting.[26]

**Figure 12.7**

*An example of a virtual set—a computer generated set that could be used for a news program. Powerful computers and software have made it possible to create customized environments that may otherwise be impossible to build (e.g., money and time constraints). (Courtesy of Evans and Sutherland. Copyright 2000, Evans and Sutherland Computer Corporation.)*

Besides Arbitron, companies have supported telephone surveys with software. In brief, questionnaires could be designed and the data collected and analyzed to reveal the audience's characteristics. In one typical application, these new insights could be used to attract a specific demographic group.

### Programming, Traffic, and Sales

Software can also be used as an organizational and managerial tool. A package can generate a detailed list of a day's programming events and can handle other jobs. A radio station serves as an example.

In one application, a program can support a music library, essentially a sophisticated database. Depending on the system, songs can be cross-referenced, coded according to their tempo and intensity, and linked to age group and demographic appeal codes. These data are useful in attracting specific audiences. Similarly, the software may help a program director to devise the music rotation schedule—the list of songs played during the day.

Programs have also accommodated traffic duties. Log maintenance is an essential task, and computers have sped up this process.

A program could also automate the scheduling of commercials, and sales force performance could be evaluated in different categories. An accounting package may then be used to complete a comprehensive system that could cover everything from music rotation to commissions. Accounting software performs a multitude of billing, payroll, and projection functions.

Other packages and hardware support station automation—running a station with a minimum of human intervention. Systems can range from sophisticated to PC-based operations that can automatically play a series of prerecorded songs and commercials for a local radio station. In one application, once programmed, an educational station with a limited student staff could remain on the air even during the summer and holidays.

## CONCLUSION

The broadcast and nonbroadcast worlds have been influenced by computer and digital technologies. Some of these effects have included the following:

- the introduction of new audio systems
- the widespread adoption of digital and tapeless recording systems
- the birth of the electronic newsroom
- the ability to create amazing visual effects
- the impetus for an all-digital production facility

The last item can be considered an evolutionary process. Digital equipment initially had to operate in what was an analog sea, and an all-digital plant would reverse the situation—for example, a video camera's signal would be immediately digitized. This process would streamline certain operations and could help preserve a signal's integrity.

In view of its superiority, engineers are designing such digital facilities. This development also complements the creation of a fully integrated facility that can tie different equipment and systems in a centralized communications and control network. Although stand-alone systems will continue to be used, the trend is to unite these elements, as may be the case with an automated station.

But even though these tools are powerful, they still require human input. For a graphics system, this may be the creative ability to visualize a graphic and the skill and aesthetic judgment to execute the final product.

All of these developments have "pushed the envelope." Only this time, it is a creative envelope, and one whose potential may be unlimited.

## REFERENCES/NOTES

1. Robert Saltarelli, "Robotic Camera Control: A News Director's Tool," *SMPTE Journal* 100 (January 1991), 23.

2. DVE is a registered trademark of NEC America, Inc.

3. You also need an interface to transfer the digital information from the computer to, for instance, a VCR.

4. Linda Jacobson, "Mac Looks Good in Video Graphics," sidebar in "Macs Aid Corporate Video Production," *Macweek* (December 3, 1991), 40.

5. These include the system's capabilities, the quality of the video, and the compression ratio (if relevant) when the video is captured.

6. Frank Beacham, "Camcorders Take Another Great Leap," *TV Technology* 13 (November 1995), 1. *Note:* The standard has also been enhanced and has moved to a multigenerational status, as has the USB connection. But unlike the current USB standard, FireWire is designed to accommodate a high and sustained data rate and flow—as is the case with a video production environment.

7. Using the camera for these tasks can, however, have an impact on its operational lifetime—the extra wear and tear.

8. Andre Rocke. "Reduce Your Drag." *Videography* 28 (January 2003), 22.

9. Brian C. Fenton, "Digital Audio Tape," *Radio-Electronics* 58 (October 1987), 78.

10. An audiophile, like a videophile, demands the best performance from equipment—in this case, audio equipment.

11. Part of the problem is the lack of software, prerecorded music tapes. For details, see Jenna Dela Cruz, "Digital Audiotape," in August

PRODUCTION TECHNOLOGIES

E. Grant, ed., *Communication Technology Update* (Boston: Focal Press, 1994), 259.

12. Ken C. Pohlmann, "Digital I/Os Added to Sony's Latest MD," *Radio World* 19 (March 22, 1995), 4.

13. There are also dedicated audio systems designed for this application.

14. These include the elimination of tape—via the cart—breakage and media replacement.

15. Claire Tristram, "Stream On: Video Servers in the Real World," *NewMedia* 5 (April 1995), 49.

16. This has included the DTE Technology line of drives that were compatible with a variety of NLE systems.

17. Please see Mark J. Foley, "VTR Killer," *Videography* 28 (April 2003), 22 for an example of such a system.

18. Chris McConnell, "Curt Rawley: Avid Advocate for a Disk-Based Future," *Broadcasting & Cable* (April 3, 1995), 64.

19. Avid, "Server Control Production," *Across Avid* 2 (Issue 1), 6.

20. Peter Adamiak et al., "Digital Servers," *Broadcasting & Cable* (April 3, 1995), insert, S-3.

21. Mark Ostlund, "Multichannel Video Server Applications in TV Broadcasting and Post-Production," *SMPTE Journal* 105 (January 1996), 11.

22. Claire Tristram, "Bottleneck Busters," *NewMedia* 5 (April 1995), 53.

23. The idea of security is an important one since a newsroom computer, like any other computer, may be vulnerable. The electronic newsroom should also be equipped with backup systems to maintain at least a basic level of operation in times of emergency, when the automated system may be rendered inoperable. For information, see William A. Owens, "Newsroom Computers . . . Another View," sidebar in James McBride, "Newsroom Computers," *Television Engineering* (May 1990), 31.

24. McConnell, "Curt Rawley: Avid Advocate for a Disk-Based Future," 65.

25. Conversation with Basys Automation Systems, October 1992. *Note:* A still store "stores" images for later retrieval and display.

26. The Arbitron Company, "At Arbitron, These Technologies Aren't Just a Vision, They're Reality," brochure.

## SUGGESTED READINGS

Braverman, Barry. "Understanding Your Camera's CCD." *DV* 11 (March 2003), 22–30; Ken Kerschbaumer. "SONY Pushes Tapeless ENG." *Broadcasting & Cable* (February 10, 2003), 1, 38. The first article is a comprehensive overview of CCD-based cameras and production techniques.

*Broadcast Engineering* is an important publication for the broadcast industry. Sample articles that covered various issues include: Steve Epstein. "The Advantages of Tape." 39 (February 1997), 100–102; David Hopkins. "Digital Effects Systems: Moving to Software-Based Open Systems." 39 (February 1997), 88–92; Robert Streeter and Thomas Drewke. "NBC's Newsroom Communication System." 44 (August 2002), 74–76; Tom Tucker. "Audio-to-Video Delay Systems for DTV." 40 (November 1998), 82–85.

Chan, Curtis. "Advances in DAT Recorders." *Broadcast Engineering* 36 (August 1994), 34–40; John G. Garrett. "Really Remote Audio." *DV* 10 (July 2002), 42–50; Dave Hansen. "Broadcast and Production Audio Consoles." *Broadcast Engineering* 44 (September 2002), 58–66; A look at DAT—early to more recent systems—and other audio concerns.

Christiansen, Mark. "Creating Realtime 3D Graphics for Broadcast." *DV* 8 (May 2000), 46–54; Francis Hamit. "Imaging and Effects at *Star Trek*: New Universes to Conquer." *Advanced Imaging* 10 (April 1995), 22–24; Barbara Robertson. "The Grand Illusion." *Computer Graphics World* 21 (January 1998), 23–34; Barbara Robertson. "Reality Check." *Computer Graphics World* 24 (August 2001), 24–32; Roger Thorton. "Paintbox Re-Born for the Modern Age." *Videography*. 27 (December 2002), 52–56. Computer graphics and special effects.

Christiansen, Mark. "The Theory Behind Color." *DV* 9 (August 2001), 36–44; Jim Farmer. "Observing China's Cable TV Market." *CTI* 12 (March 2001), 24–31; Steve Hullfish and Jaime Fowler. *Color Correction for Digital Video.* Berkeley, CA: CMP Books, 2003; Oliver Peters. "And You Thought It Was a Simple Decision!" *Videography* 28 (January 2003), 40–44. Miscellaneous topics; the color correction book provides a comprehensive and practical view of this field and complementary techniques. It includes a companion CD.

Davidoff, Frank. "The All Digital Studio." *SMPTE Journal* 89 (June 1980), 445–449; Arielle Emmett. "Better Calling by Design." *CustomerInterface* 14 (November 2001), 21–23; Christian Mitchell. "Planning New Facilities." *Broadcast Engineering* 43 (May 2001), 68–73; D. Nasse et al. "An Experimental All-Digital Television Center." *SMPTE Journal* 95 (January 1986), 13–19; Michel Prouix and Randy Conrod. *SMPTE Journal* 104 (September 1995), 582–587. Early to more recent (including digital) studio design; the Emmett article covers call center design—but certain features are also applicable for production facilities (e.g., ergonomic furniture).

De Lancie, Philip. "Storage Scenarios." *DV* 11 (April 2003), 38–43; Ken Paulsen. "Prospective for Global Storage Networks." *TV Technology* 19 (June 13, 2001), 34–36; Greg Reitman. "Streaming Video with Storage Area Networks." *SMPTE Journal* 110 (August 2001), 517–522; Storage Network Industry Association (www.snia.org). Information about SANs and other storage systems/production applications; the SNIA web site also features white papers and other information about SANs systems.

Douglas, Peter. "Automating Master Control for Multichannel." *Broadcast Engineering* 40 (April 1998), 106–111; Ken Kerschbaumer. "Reshaping Broadcasting." *Broadcasting & Cable* (October 14, 2002), 30–36. Automation systems.

Houston, Brant. *Computer-Assisted Reporting.* New York: St. Martin's Press, 1996; Stevan Vigneaux. "Digital News Gathering on a Desktop." *Broadcast Engineering* 35 (September 1993), 50–56. Practical guide for learning computer-based tools for journalists and nonlinear editing, production, and news operations.

Ostlund, Mark. "Multichannel Video Server Applications in TV Broadcasting and Post-Production." *SMPTE Journal* 105 (January 1996), 8–12; Karl Paulsen. "Servers Found on Video Menu." *TV Technology* 14 (February 9, 1996), 44–45, 65; Todd Roth. "Video Servers: Shared 'Storage' for Cost-Effective Realtime Access." *SMPTE Journal* 107 (January 1998), 54–57; Claire Tristram. "Stream On: Video Servers in the Real World." *NewMedia* 5 (April 1995), 46–51. Video server applications and operations.

Reed, Kim. "FireWire's Future." *DV* (January 1999), 35–44. Excellent look at FireWire.

Yamanaka, Noritada et al. "An Intelligent Robotic Camera System." *SMPTE Journal* 104 (January 1995), 23–25. Intelligent robotic camera systems enhance operations.

## GLOSSARY

*Audio Console:* The component that controls microphones, CD players, and other audio equipment.

*Computer-Assisted Editing:* The process of using a computer to help streamline and enhance the editing process.

*Digital Audiotape (DAT):* A recordable digital tape system that can match a CD's audio quality. DATs have also been used for computer data backups.

*Digital Effects Generator:* A production component that digitizes and manipulates images to produce visual special effects.

*Electronic Newsroom:* A newsroom equipped with computers for newswriting, creating databases, and other functions.

PRODUCTION TECHNOLOGIES

*Graphics Generator:* The component that is used to create computer graphics. The graphics can be produced either with special dedicated stations or PCs.

*MiniDisc (MD):* A newer digital audio system.

*Nonlinear Video Editing (NLE):* Allows you to retrieve stored video sequences in random access fashion, much like conventional computer data. In a PC environment, the scenes are digitized, and a visual interface is used for editing.

*Switcher:* The video component that is used to select a production's camera and videotape sources. It can also be used in postproduction work and may incorporate computer technology.

*Tapeless Systems:* An umbrella phrase describing equipment that doesn't use tape as the primary recording and/or playback medium.

# 13 Digital Television and Digital Audio Broadcasting

This chapter focuses on digital television (DTV), an application designed to produce an enhanced television display. High definition television (HDTV), which can fall under the DTV designation, has been a goal of companies and countries for years. The chapter concludes with an overview of an analogous system geared for radio, digital audio broadcasting (DAB).

## HIGH DEFINITION TELEVISION

High definition television was touted as the new premier television standard. Its display would visually approach the quality of film, and the system would support an enhanced audio signal.[1]

An HDTV set would also have a 16:9 aspect ratio, that is, the ratio between a screen's width and height. In contrast to a conventional 4:3 configuration, an HDTV screen would present a more powerful image to viewers.

The push for HDTV, which was included under the Advanced Television (ATV) designation, stemmed from a natural process: the improvement of technology and the growth that takes place in any industry. The trend toward larger television screens accelerated this research. In general, as the screen size increases, the picture quality decreases. But HDTV and other digital systems can provide enhanced displays.

### Background

***Japan and CBS.*** The Japan Broadcasting Corporation, NHK, has been one of the world leaders in the HDTV field, and its personnel conducted breakthrough research. This included experiments to test technical and nontechnical parameters, such as the relationship between the number of lines in a picture, the optimum viewing distance from a screen, and the screen's size. For example, viewers judged the quality of television pictures composed of different lines of resolution from various distances.[2] These considerations were then weighed against bandwidth limitations and other factors.

NHK also developed HDTV equipment and held trial HDTV transmissions in its own country and the United States.

In the United States, CBS was an early HDTV supporter. The company promoted a two-channel, 1050-line component scheme, in contrast to NHK's 1125-line configuration. The CBS operation was also designed for a DBS relay. A 525-line signal and enhanced picture information would have been carried in two separate channels. A receiver would have combined both to produce a 5:3 HDTV display.[3] For viewers without HDTV receivers, the single 525-line channel would have been converted for viewing on conventional sets.

The work conducted by NHK, CBS, and others helped spur the HDTV field's

**Figure 13.1**
*Engineers set up the encoding room at the foot of the transmission tower at WRC-TV/Washington in preparation for the first over-the-air digital HDTV simulcast, in September 1992. (Courtesy of the Advanced Television Research Consortium; AD-HDTV.)*

growth. This development has also been an evolutionary one, with a minor revolution thrown in to boot. The evolutionary phase encompassed an emerging infrastructure built to support HDTV productions and relays. The revolution was the birth of an all-digital system.

### Global Issues

*"He who controls the spice, controls the universe."*
    —From the movie version of *Dune.*

During the 1980s, attempts were made to create an international standard. There was some progress, but technical, political, and economic issues stood in the way. One problem was a concern that Japan would gain an edge if its HDTV system dominated the industry. By extension, the situation could have affected consumer electronics sales.

For Europe, manufacturers believed their home markets could have been swamped by Japanese-manufactured HDTV goods.[4]

Other issues stemmed from the European Community's overall goal to promote European technologies and programming.

In the United States, official support was initially given for a standard based on Japan's system. But this support faded as the prospect for a global standard faded.[5]

Another factor was the American semiconductor industry. To some, HDTV represented this industry's future, since HDTV systems would be heavily tied to semiconductor technology.[6] The semiconductor industry claimed that if the United States was not a leading HDTV manufacturer, the country's overall semiconductor industry could suffer and, by extension, PC and dependent markets.

Thus, an international consensus could not be reached. It was feared the country that dominated this industry would have a hammerlock on a multibillion-dollar business and related industries.

This concern and perception is best summed up by the line that opened this section, from *Dune*, Frank Herbert's classic science fiction work: "He who controls the spice, controls the universe." Only in this case, the spice was HDTV.[7]

## United States

*History.*    Much of the focus and regulatory maneuvering in the United States has been on the terrestrial system, over the air broadcasts. This is a reflection of the system's unique status. Unlike cable or another optional service, terrestrial broadcasts are free. Even though the industry is supported by commercials, we do not pay a set fee to view the programming.

The FCC accelerated the development of HDTV operations in the late 1980s and early 1990s. In a 1988 decision, for example, the agency supported, at least for terrestrial transmissions, a configuration that would conform to existing channel allocations and would be backward compatible.

An HDTV relay also packs in more information than a standard relay. For the United States, it would exceed, under normal circumstances, current 6-MHz channel allotments. Thus, various organizations attempted to develop a system that operated within these constraints.

In one example, the David Sarnoff Research Center supported its upgradable configuration, the Advanced Compatible Television (ACTV) system. Under ACTV-I, an enhanced picture, but not one equal to the 1125-line standard, would have been relayed over existing channels. At some future date, an augmentation channel, a second channel with enhanced information, would have been integrated in the system (ACTV-II) to produce a higher quality picture.

Related issues have included the differences between the broadcast, cable, and satellite industries. Limited spectrum alloca-tions translated into tight channel requirements for broadcasters. Cable and satellite operators had more flexibility and could launch, as another option, HDTV pay services.

*Evolution and Revolution.*    The FCC refined its decision in the early 1990s by announcing its support for simulcasting. In this operation, a station would continue to relay a standard signal. The station would then be assigned a second channel for an HDTV relay. This approach was viewed by some as having an important advantage over using an augmentation channel. With simulcasting, the new system would not be tied to the NTSC standard. Thus, a superior, noncompatible standard could be developed.[8]

The final piece to the puzzle was added when the FCC, led by Chairman Alfred Sikes, encouraged the development of a digital system. If successful, this configuration would have advantages over analog operations.

Consideration was also given to a flexible standard that could handle future growth. Three terms were associated with this philosophy: scalability, extensibility, and interoperability.

In a scalable video system, the aspect ratio, number of frames per second and the number of scan lines can be adjusted . . . to the requirements of the individual picture . . . or viewer's choice. Extensibility means that a new television system must be able to operate on diverse display technologies . . . and be adaptable for use with new, higher resolution displays developed in the future. Interoperability means a television system can function at any frame rate on a variety of display devices.[9]

Basically, part of the deliberation process was the consideration of a standard that could accommodate different configurations and enhancements.

PRODUCTION TECHNOLOGIES

PRODUCTION TECHNOLOGIES

In preparation for this development, several proposed terrestrial systems were scheduled for testing by the Advanced Television Test Center. Following the tests, the FCC was slated to select a standard in early 1993, which was subsequently moved to a later date.

The FCC was also working on a terrestrial broadcast timetable. This covered items such as the length of time conventional broadcasts would be supported.

But many broadcasters were not happy with these decisions. Since DTV would be a new ball game, both producers and consumers would have to buy new equipment to play HDTV.[10] The cost would be particularly painful for small market stations, even with eventual price reductions as equipment became widely available.

Other broadcasters, at this time, were dissatisfied with the FCC's timetable and mandate. It was proposed the second allocation could be used for digital multichannel television relays, or possibly, for interactive and/or data services. The goal was to provide broadcasters with additional programming options so they could better compete with cable and satellite services. Eventually, HDTV relays could be phased in.

### The Mid-1990s: A New Name and Standard.
After this initial work, a standard was finally hammered out, which now fell under the DTV designation. It was recommended to the FCC as the U.S. digital television standard.[11] As summed up in an FCC document,

The Commission proposed adopting, as the technology for terrestrial broadcast in the United States, the Advanced Television Committee (ATSC) DTV standard. . . . The proposed standard is the culmination of over eight years of work by the federal Advisory Committee on Advanced Television Services (ACATS), the ATSC, the Advanced Television Test Center (ATTC), and the

members of the Grand Alliance [an "alliance" of initial HDTV competitors that joined forces] . . . . The technology provides a variety of formats that will allow broadcasters to select the one appropriate for their program material, from very high resolution providing the best possible picture quality to multiple programs of lower resolutions, which could result in more choices for viewers. Even at the lower resolutions, the recommended system represents a clear improvement over the current NTSC standard. The recommended system also permits transmission of text and data.[12]

The proposed standard's video compression was based on MPEG-2. A relay would also sport an enhanced audio output.

For the broadcast industry, a station would not be locked into a single format. There was an option for "multiple simultaneous" standard definition or digital television (SDTV) relays. While not true HDTV, a station could accommodate multiple feeds.[13]

### A Funny Thing Happened on the Way to High Definition Television.
Zero Mostel and a supporting cast helped make *A Funny Thing Happened on the Way to the Forum* a hit play and movie. In 1996, the computer and film industries, among others, helped bring the standards adoption process to a crashing halt. The proposed standard ran into a technical roadblock.

As stated by FCC Chairman Reed Hundt, who also expressed his own concerns,

. . . the dedicated and hard-working members of the Advisory Committee tried in good faith to produce a consensus standard. Unfortunately, they did not succeed. Important players in two huge American industries—Silicon Valley and Hollywood—object strongly to some elements of the standard. These groups support much of the Grand Alliance's work . . .

But the failure to reach consensus over the *interlaced format* and the *aspect ratio* [my emphasis] has led to a time-consuming and important debate in which all advocates are making serious points.[14]

It was a contentious atmosphere. Opponents indicated, for instance, that scanning and aspect ratio incompatibilities would hamper the marriage between computer and television products and would have an impact on the film industry. Supporters countered that the other industries were included in the standards deliberation process. More pointedly, this was essentially a television broadcast standard—the broadcast industry should take the lead.[15] Other contributing elements included the *Dune* factor, convergence, and the government.

- *Dune* factor: Which industry was going to control the flow of "digital media coming into the home," basically, who was going to control the spice?[16] This had broad economic implications for broadcasters and computer manufacturers, among other players.
- Convergence: What were once separate entities, such as the broadcast and computer industries, were now competing as well as collaborating with each other. Because multiple industries had an interest in DTV, it became the game ball to toss about.
- The government: The U.S. government was seemingly unwilling to grapple with this "hot potato." The FCC was divided, and since it was an election year, it was suggested the Clinton administration did not want to anger any particular group by taking sides.[17]

In late 1996, however, a consensus was reached. As described in a *Broadcasting & Cable* article, the "deal called for the adoption of the Grand Alliance standard minus the controversial picture formats that had divided the industries."[18] Basically, the core of the original proposal was retained. But

now, the broadcast and computer industries, among others, could support different display formats (e.g., different types of scanning). The deal helped clear the road for digital television.

**Figure 13.2**
*An HDTV display. Note the size and shape of the screen compared with the typical TV in current use. (Courtesy of the Advanced Television Research Consortium; AD-HDTV.)*

### Other Considerations

Other key issues have shaped the DTV landscape. Examples include the following:

1. Regular HDTV broadcasts began in the United States on April 1999, on *The Tonight Show*.[19] Only special programming, including coverage of John Glenn's launch on a space shuttle mission, were broadcast prior to this time in HDTV.[20]

2. The number of hours of DTV programming is increasing with each passing year. According to the National Association of Broadcasters (NAB), by early 2003, 97% of U.S. TV households could receive one or more signals.[21]

3. Although much of the media focus has been on over the air broadcasting, other media outlets can deliver digital programming, potentially on an optional basis and on a more flexible timetable. DirecTV, for one, helped pioneer the delivery of digital signals to consumers.

PRODUCTION TECHNOLOGIES

4. A 2002 GAO report indicated that many Americans were vaguely, if at all aware, of the DTV transition.[22] Consequently, as broadcasters hurled toward this digital realm, many Americans were not coming along for the ride—if they knew that it even existed. This did not bode well for a quick and widespread adoption, at least during the early years of the DTV rollout. This situation was exacerbated by consumer confusion over "true" HDTV and DTV units and programming.[23]

5. Stations would stop their analog transmissions at a designated date.[24] A station would subsequently surrender its second channel for release for other services. A station could receive an extension, however, based on market conditions.[25]

Money is also a key player in this scenario. As envisioned, portions of the recovered spectrum space were slated for the auction block.

6. The FCC adopted different measures to help speed-up DTV integration. For example, the Commission proposed that DTV tuners, which would receive over the air broadcasts, would be required for 13-inch and larger TV sets. A timetable was established for the plan's gradual implementation.[26]

But various consumer groups and manufacturers were against this measure. The Consumer Electronics Association, for example, indicated that many Americans received their programming through cable or satellite—a tuner would not be necessary in these cases. A tuner would also make a set more expensive. Even though costs would drop over time, the added expense could be a financial burden.

7. One or more obstacles may have delayed a station's DTV implementation timetable. They ranged from monetary—to pay for the conversion—to problems with the broadcast tower. The latter included bad weather, which would interfere with working on a tower, be it a new or an existing structure.[27]

8. An DTV system could have numerous nontelevision applications. A display could be attractive for videoconferencing activities, as described in another chapter. A satellite could also deliver signals to special movie theaters equipped with large, high-resolution screens. The programming could include movies, concerts, and sporting events.

Stations could also support datacasting—using their channels to handle "everything from broadcasting stock quotes to downloading an electronic catalog."[28] As demonstrated in early 2000, stations could "create web sites with multiple streams of video and broadcast them to PCs over their DTV channels."[29] Stations could also support, as described, multiple SDTV relays of more conventional programming.

9. The United States was not alone in DTV developments. In Europe, for instance, a new satellite-based service, Euro1080, was proposed in the early 2000s. One application called for delivering programs to suitably equipped theaters or electronic cinemas.[30] But as has been the case with the United States, progress for global terrestrial DTV operations was mixed.

10. Ultimately, the central question as to DTV's success lies in consumer hands. Does the average consumer want DTV? Is the average consumer willing, for instance, to pay a substantially higher price for true HDTV sets?

Japan could serve as an example. When first introduced, only a limited number of true and expensive HDTV sets were sold. In contrast, an enhanced definition widescreen system was more successful. Consequently, is HDTV that critical?[31] Or is an improved picture, on a wider screen, a more important factor for consumer acceptance?

This scenario has been supported in some U.S. studies. In one case, it is believed that some consumers could opt for a less expensive alternative to buying an HDTV set. This could include using a set-top converter boxes rather than an HDTV set.[32]

11. In a related area, will PCs be used for entertainment and information programming? Will the PC, or television set for that matter, emerge as the core of an advanced home entertainment and information center?

## Conclusion

As of the early- to mid-2000s, the DTV field was in a fluid state. While there were major advances, the widespread adoption of HDTV sets, for instance, remained somewhat unfulfilled. The problems ranged from broadcast station implementation delays to expensive HDTV sets to lackluster consumer demand. Other issues include:

- Copyright concerns: Broadcasters, for one, supported an anti-copying or piracy system (broadcast flag) to prevent the distribution of digital programming via the Internet and other conduits. In contrast, opponents believed such a system would infringe on consumer rights.
- Cable companies carrying DTV programming.[33]
- The charge that most stations were operating low rather than high power "DTV facilities."[34] This had technical (e.g., reception problems) and other implications.
- Even more acronyms to remember. When reading further about DTV, you may run across the term DTT—it simply means digital terrestrial television.

## DIGITAL AUDIO BROADCASTING

### History

Much like DTV, companies and governments worked to improve radio and related audio systems. What is the bottom line? To enhance radio program delivery/quality and, if possible, to bring it to the digital age.

Pioneered in Canada and Europe, digital audio broadcasting (DAB) can deliver a CD-quality audio relay. This was extended to a satellite-based operation, a satellite digital audio radio service (SDARS), which would serve car owners, among others, through a roof-mounted antenna.[35]

A frequency allocation (L-band), granted during a 1992 World Administrative Radio Conference, was initially supported. But the United States backed another plan since this allocation was already used for other services.

The National Association of Broadcasters (NAB), for its part, was cautious in its approach toward the technology. It "recommended that any domestic inauguration of DAB be on a terrestrial only basis, with existing broadcasters given first opportunity to employ the technology."[36] Broadcasters should be given the first crack at this technology, not potentially competing services.

This recommendation stemmed from two related concerns. The first was that a satellite-delivered service could adversely affect the broadcast industry. The second concern was localism—an SDARS operation could not match a conventional radio station's public interest commitment to the local community.[37]

In contrast, companies that proposed satellite-delivered services indicated there were numerous benefits. National programming could supplement local radio broadcasts. Listeners with limited radio choices

**Figure 13.3**
*Side-by-side comparisons of standard and HDTV sets. Note the aspect ratio and size differences. (Courtesy of the Advanced Television Research Consortium; AD-HDTV.)*

PRODUCTION TECHNOLOGIES

could also have a broader selection, while pay services and narrowcasting could be supported. A complementary terrestrial and satellite system could likewise develop.

Other factors also played a role in this process. This included work on in-band on-channel (IBOC) systems. Briefly, special techniques could make it possible to relay digital programming using existing allocations. This configuration could accommodate, at least in terms of spectrum space, the current radio broadcast industry. Much like DTV, it could also provide stations with an upgrade path to digital relays.[38]

By the mid-1990s, little had changed on the domestic front. U.S. efforts were still unfocused even though the international community had, with some reservations, adopted the Eureka-147 DAB standard.[39] The debate between radio broadcasters and satellite operators also continued, especially in light of the FCC's decision to allocate spectrum "for a satellite-based industry."[40] But the FCC also repeated its support for local broadcasters.

## Update

1. In late 1999, the FCC began a rule-making procedure to explore the possible ways to initiate digital audio broadcasting. The goal? To create a digital broadcast path that terrestrial station owners could follow. The criteria would include the capability to launch enhanced audio services without disrupting the current infrastructure.

2. The FCC investigated and approved a hybrid IBOC system in the early 2000s.[41] Developed by the iBiquity Digital Corporation, its HD radio technology provided broadcasters with a digital option while maintaining their current infrastructure. More pointedly, it provided for an efficient spectrum use plan that could deliver higher quality broadcasts. As stated by the FCC,

The iBiquity IBOC system is spectrum-efficient in that it can accommodate digital operations for all existing AM and FM radio stations with no additional allocation of spectrum. The NRSC tests show that both AM and FM IBOC systems offer enhanced audio fidelity and increased robustness to interference and other signal impairments. Coverage for both systems would be at least comparable to analog coverage. Considering that iBiquity's IBOC systems achieve these objectives in the hybrid mode, in which the relatively low-powered digital signal must coexist with more powerful analog signals, we expect that audio fidelity and robustness will improve greatly with all-digital operation.[42]

Much like DTV, analog and digital broadcasts would coexist. New receivers equipped with the proper components would subsequently receive the digital signal. Potentially, this path could lead to the implementation of an all-digital radio broadcasting system. Analog relays would be phased-out and additional services, which would tap this digital capability, could be introduced.[43]

3. Satellite-based systems also made a big move in the early 2000s. XM Satellite Radio Incorporated, for instance, offered an array of digital channels through two geostationary satellites and a supplemental repeater network.[44] You could listen to music ranging from rock to classical and jazz by using a small receiver and antenna. As an option, you could purchase a portable system that could be transferred between cars and likewise be used in the home. You could also order a new car with satellite radio capability.

Unlike a broadcast station, the XM service was more like cable or satellite television—you paid a subscriber fee for the service. When first introduced, it was under $10 a month.

## Conclusion

The DAB arena was volatile. Different proposals were floated to offer broadcasters a digital option, much like their television counterparts. For some, a key factor was providing broadcasters with a tool to compete with their satellite-based competitors. The latter, for their part, offered consumers a new, high quality service, which could provide national coverage with a broad range of music programming options. As was the case with DTV, however, the question was one of consumer acceptance—would enough individuals buy a system and/or become a subscriber?

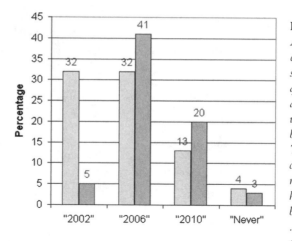

**Figure 13.4**
*A view of DTV developments. A GAO survey asked the question: If there wasn't a government mandate, when would your station begin DTV broadcasts? The answer: Many current DTV stations reported they would have "broadcast digitally by the end of 2002 . . . most transitioning stations . . . would have begun broadcasting digitally much later . . ."; some stations would not initiate DTV broadcasts. Current DTV stations are represented by the first bar for a given year; transitioning stations by the second bar. (Source: GAO Reports, "Many Broadcasters Will Not Meet May 2002 Digital Television Deadline," GAO-02-466, April 23, 2002.)*

PRODUCTION TECHNOLOGIES

## REFERENCES/NOTES

1. Technically speaking, it may not equal film's resolution.

2. Tetsuo Mitsuhashi, "Scanning Specifications and Picture Quality," *NHK Technical Monograph* 32 (June 1982), 24.

3. CBS/Broadcast Group, "CBS Announces a Two Channel Compatible Broadcast System for High-Definition Television," press release, September 22, 1983.

4. Elizabeth Corcoran, *Scientific American* 266 (February 1992), 96.

5. Even though there may not have been an official agreement, elements of the international production community did embrace the production standard.

6. "EIA Sets Itself Apart on HDTV," *Broadcasting* (January 16, 1989), 100.

7. Spice, also known as melange, was a rare and valuable substance.

8. "FCC to Take Simulcast Route to HDTV," *Broadcasting* (March 26, 1990), 39.

9. Frank Beacham, "Sikes: A New Ballgame for HDTV," *TV Technology* 10 (May 1992), 3.

10. "HDTV: A Game of Take and Give," *Broadcasting* (April 20, 1992), 6.

11. Chris McConnell, "Broadcasters Arm for ATV Fight," *Broadcasting & Cable* (October 21, 1996), 6.

12. FCC, News Report No. DC 96-42; "Commission Proposes Adoption of Digital Television Broadcast Standard," MM Docket No. 87-268, May 9, 1996.

13. Richard Ducey, "An Overview of the American Digital Television Service Program," Paper, downloaded from www.nab.org.

14. Reed Hundt, Chairman, FCC, "A New Paradigm for Digital Television," September 30, 1996, downloaded from www.fcc.gov. *Note:* TV uses an interlaced format; computers use a progressive scanning process.

15. Frank Beacham, "Pressure Builds for DTV Compromise," *TV Technology* 14 (October 25, 1996), 27.

16. Ibid.

17. McConnell, "Broadcasters Arm for ATV Fight," 7.

18. "DTV Standard: It's Official," *Broadcasting & Cable* (December 30, 1996), 4.

19. Kristine Garcia, "Heeere's HDTV," *Digital Television* 2 (May 1999), 1.

20. Paige Albiniak, "HDTV: Launched and Counting," *Broadcasting & Cable* (November 2, 1998), 6.

PRODUCTION TECHNOLOGIES

21. NAB, "11 New Stations on Air with DTV," Press Release, March 3, 2003, downloaded from www.nab.org/newsroom/pressr1/1303.htm.

22. Phillip Swan, "Give Us Some Credit," *Electronic Media Online,* downloaded from wysiwyg://4/http://www.emonline.com/technology/121602give.html.

23. Jennifer Davies, "The HDTV Revolution Will Be Televised; But Will Anyone Be Watching,?" *The San Diego Union-Tribune* (December 2, 2001, Sunday), downloaded from Nexis.

24. The year 2006 as of this writing.

25. Please see the following for details: FCC. "FCC Initiates Second Review of DTV Transition," Press Release, January 27, 2003, downloaded from http://hraunfoss.fcc.gov/edocs_public/attachmatch/DOC-230562A1.doc.

26. FCC. "FCC Introduces Phase-in Plan for DTV Tuners," Press Release, August 8, 2002, downloaded from http://hraunfoss.fcc.gov/edocs_pubic/attachmatch/DOC-225221A1.pdf?date=020208.

27. GAO Reports, "Many Broadcasters Will Not Meet May 2002 Digital Television Deadline," GAO-02-466, April 23, 2002, downloaded from Nexis.

28. Glen Dickson, "Getting Together Over Data," *Broadcasting & Cable* (March 27, 2000), 9.

29. Harry A. Jessell, "Broadcasting's Killer App?," *Broadcasting & Cable* (March 27, 2000), 10.

30. "Is HDTV Finally Coming to Europe," *TV Technology* (January 30, 2003), downloaded from www.tvtechnology.com/dailynews/one.php?id=837.

31. Mario Orazio, "HDTV: Hot Dang-Total Video," *TV Technology* 13 (December 1995), 25.

32. Marcia L. De Sonne, "HDTVs-It's Where the Buyers are Consumer Interest, Set Availability, Set-Top Converters," downloaded from www.nab.org. The National Association of Broadcasters web site is a rich resource for broadcasting topics, including DTV.

33. Edward O. Fritts, "Broadcasters' Breakthrough Year," Speech to the NAB ATSC Annual Membership Meeting, March 11, 2003, downloaded from www.nab.org/Newsroom/PressRel/speeches/031103.htm.

34. Broadcast Engineering, "Broadcasters and FCC Mismanaged DTV Transition, Consumer Federation Says," downloaded from http://editorial1.industryclick.com/microsites/index.asp?srid=11266&pageid=6952&siteid=15&magazineid=158&srtype=1#dtv.

35. Carolyn Horwitz, "DAB: Coming to a Car Near You?," *Satellite Communications* (October 1994), 38.

36. NAB. "Digital Audio Broadcasting," *Broadcast Regulation* 1992; A mid-year Report, 140.

37. Ibid., 142.

38. Conversation with FCC, October 1992.

39. "U.S. DAB on the Slow Track," *Radio World* 19 (December 27, 1995), 37.

40. Alan Huber, "FCC Lays Groundwork for Satellite Radio," *Radio World* 19 (February 8, 1995), 1.

41. Downloaded from www.fcc.gov/mb/audio/digital/index.html.

42. FCC, Digital Audio Broadcasting Systems and Their Impact on the Terrestrial Radio Broadcast Service, MM Docket No. 99–325, October 10, 2002, downloaded from http://hraunfoss.fcc.gov/edocs_public/attachmatch/FCC-02-286A1.doc.

43. This may include SurroundSound. Please see the iBiquity web site for additional information; www.ibiquity.com.

44. XM Satellite Radio Inc., "General FAQ," downloaded from www.xmradio.com.

## SUGGESTED READINGS

ATSC. Like the FCC, this is a rich resource for the digital television standard. (www.atsc.org) Example documents have included: "A/53 ATSC Digital Television Standard" (September 16, 1995) and "A/54 Guide to the Use of the ATSC Digital Television Standard" (October 4, 1995).

Beacham, Frank. "Digital TV Airs 'Grand' Soap Opera." *TV Technology* 14 (August 23, 1996), 1, 8; Frank Beacham. "Pressure Builds for DTV Compromise." *TV Technology* 14 (October 25, 1996), 1, 27; Chris McConnell. "Broadcasters Arm for ATV Fight." *Broadcasting & Cable* (October 21, 1996), 6–7, 12. The digital television standard controversy.

Boyer, William H. "Don't Touch that Dial." *Satellite Communication* 22 (May 1998), 44–47; Peter J. Brown. "Satellite Radio." *Via Satellite* 16 (November 2001), 18–24; Melanie Reynolds. "DAB Firms Tune in for Product Launch." *Electronic Weekly,* downloaded from Nexis; Lucent Digital Radio. "Submission to the National Radio Systems Committee." January 24, 2000, downloaded from www.nab.org.; Lynn Meadows. "AT&T Pulls IBOC from DAB Tests." *Radio World* 20 (October 16, 1996), 1, 11; XM Radio. www.xmradio.com. DAB and SDAR developments and the XM Radio web site has a range of documents outlining its satellite-based service.

Brown, Peter E. "PC OEMs Look to DTV as Panacea for Home Market Ills." *Digital Television* 2 (November 1999), 16; John Rice. "What Flavor Is Your DTV?" *TV Technology* 16 (November 30, 1998), 28. PCs and DTV.

Davies, Jennifer. "Unclear Signal; HDTV's Sizzle May Fizzle as Cost and Marketability Inhibit Growth." *The San Diego Union-Tribune* (December 2, 2001), downloaded from Nexis; Jimmy Schaeffler. "HDTV's Coming But Obstacles Remain." *Satellite News* (October 21, 2002), downloaded; Tanjay Talwani. "FCC: This Time We Mean It." *TV Technology* 20 (July 10, 2002), 1, 8; Sanjay Talwani. "FCC Mandates DTV Tuners." *TV Technology* 20 (September 4, 2002), 1, 12. DTV developments.

deCarmo, Linden. "Checkered Flag." *eMedia* 16 (May 2003), 34–41. Comprehensive review of the broadcast flag (copy protection) scenario; includes web site resources.

Ducey, Richard V. "An Overview of the American Digital Television Service Program." A paper written by an NAB Senior Vice President. An excellent look at the development of the U.S. DTV system.

FCC. The FCC is a rich resource for HDTV, DAB, SDAR, and related documents (www.fcc.gov). Example documents have included: Commission Begins Final Step in the Implementation of Digital Television (DTV). MM Docket No. 87-268, July 15, 1996; and IBOC Digital Radio Broadcasting for AM and FR Radio Broadcast Stations.

Freeman, John. "A Cross-Referenced, Comprehensive Bibliography on High-Definition and Advanced Television Systems, 1971–1988." *SMPTE Journal* 101 (November 1990), 909–933; David Strachan. "HDTV in North America." *SMPTE Journal* 105 (March 1996), 125–129; U.S. Congress, Office of Technology Assessment. *The Big Picture: HDTV and High Resolution Systems.* OTA-BP-CIT-64 Washington, DC: U.S. Government Printing Office, June 1990. Comprehensive coverage of HDTV developments, particularly the earlier years.

Hallinger, Mark. "Globally, DTV Struggles to Break Through." *TV Technology* 21 (March 19, 2003), 12, 16; "TV Transitions from a Global Perspective." *Broadcasting* (October 15, 1990), 50–52. DTV and a global perspective, earlier to more recent years.

*TV Technology.* "Mario Orazio" writes an ongoing column that covers a wide range of topics, including those related to HDTV.

Whitaker, Jerry. *DTV The Revolution in Digital Video.* New York: McGraw-Hill, 1999. A comprehensive look at DTV history and developments; includes an overview of imaging system principles.

Zou, William Y. "SDTV and Multiple Service in a DTV Environment." *SMPTE Journal* 107 (October 1998), 870–878. Review of SDTV elements.

**PRODUCTION TECHNOLOGIES**

## GLOSSARY

*Advanced Television (ATV) Systems:* A generic name for higher definition television configurations.

*Digital Audio Broadcasting (DAB):* A CD-quality audio signal that could be delivered to subscribers by satellite or terrestrial means.

*Digital Television (DTV):* A generic designation for digital television systems.

*High Definition Television (HDTV):* The field concerned with the development of a new and improved television standard. Superior pictures will appear on a wider and larger screen; both digital and analog configurations have been developed.

*Satellite Digital Audio Radio Service (SDARS):* A radio programming service delivered by satellite to, for instance, cars equipped with a special receiver and small antenna.

# 14 The Production Environment: Colorization and Other Technology Issues

We have explored production technologies that have influenced industry, and through a rippling effect, society. This chapter examines representative examples of such technologies as well as important applications and implications.

## COLORIZATION

Colorization is the process in which a black-and-white movie is manipulated by computer to produce a colorized version of the original film. Black-and-white television programs have also been affected.

In brief, when NASA started the exploration of the Solar System, thousands of photographs of other worlds were transmitted to Earth. They were fed to computers and subsequently underwent image processing. This operation had two primary goals: image correction and enhancement. If a picture was marred by noise and other defects, they could be eliminated or minimized. Enhancement techniques helped an individual to better interpret the photographic data. In one setup, an image's contrast could be altered to highlight a characteristic of a region of Mars.

Cost-effective and powerful computers extended image processing to a broader user base. Physicians processed X rays to reveal previously invisible details. Astronomers adopted a technique called *pseudocoloring*.

In a black-and-white image, different regions of a galaxy or other celestial object may be reproduced as almost identical gray shades. This may make it impossible to visually discern its physical characteristics.[1] This is where computer processing steps in. We assign colors to the different gray shades, which are then reproduced in the appropriate colors. This image will highlight the object's physical features because they are now represented by different and contrasting colors.[2]

Image processing has also influenced the communications industry. The product may change, but the intent is essentially the same: to manipulate a picture for a specific purpose. In this case, for a desktop publishing project or for colorization.

### Yankee Doodle Dandy
Starting in the 1980s, the colorization technique was applied to black-and-white films. Movies such as *It's a Wonderful Life* and *Yankee Doodle Dandy* were colorized to the cheers and jeers of supporters and opponents alike.[3]

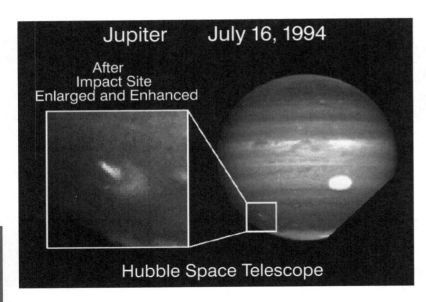

**Figure 14.1**
*The Shoemaker-Levy/Jupiter collision. Computers play a central and critical role in producing images of celestial events that we can subsequently view and study. (Courtesy of NSSDC/HST Science Team.)*

Colorization is a multifaceted task. A videotape copy of the film is created, and in one step, colors are assigned to appropriate objects in a frame. If possible, the colors correspond to those used when the movie was shot. A computer eventually processes the film, frame by frame, and a colorized version of the original is generated.[4] A former gray hat and jacket may emerge as yellow and blue.

Colorization proponents indicated the process would introduce older, and possibly classic movies, to new generations of viewers. If you grew up with color films and television, would you be willing to watch black-and-white products? Colorization also did not harm the original film, and you could still watch the black-and-white version.

Opponents countered that colorization should be halted for aesthetic and ethical reasons. Although a company may have the legal right to colorize a film, a distorted version of the original may be created. Colors may bleed, the lighting and makeup, which were originally designed for a black and white medium, are altered, and physical details may be lost.[5]

But a more salient point is the ethical issue of altering another person's work. According to many artists, including Frank Capra who directed *It's a Wonderful Life*, no one has the right to change a film or another piece of art. The artist's original vision and the work's integrity are destroyed.

Finally, ask yourself this question—even if colorization is flawless, would this make the process acceptable?

## IMAGE MANIPULATION

Computers have also been used to manipulate advertising and news images. The problem with the second scenario is the public's perception. People generally believe a news photograph portrays reality, an event as it occurred. By altering the image, the public's trust could be broken.

But image editing and manipulations are not new phenomena. The photographic process itself, in which you select a specific lens and compose a shot, is a form of editing. Image manipulations are also established darkroom fare.

Nevertheless, the new electronic systems raise these processes by a notch. It is now easier to manipulate an image, and the prerequisite tools, such as a PC and graphics packages, are readily available.

In one example, you can take a picture of a site and electronically add a proposed building. In another example, you can take a picture of a swamp and electronically alter it to make it look like prime real estate. Although each manipulation produces a new image, there are differences in the context of our discussion. If the first picture is created as part of an environmental impact study, and is clearly labeled as a simulation, not many people would have a problem with it. In fact, it can actually provide a service. But if you do not label the swamp for what it is, that's when problems start cropping up.

Newspapers and magazines with ready access to even more sophisticated systems have been guilty, at times, of essentially turning swamps into valuable land. In two examples, *National Geographic* and the *St. Louis Post-Dispatch*, respectively, moved a pyramid in one scene and eliminated and replaced a can of soda in another.[6] On face value, a can of soda may not be that important. Yet in the context of this photograph, it had a specific meaning in regard to the individual pictured in the photograph.

It also raised other questions. What else can you remove, add, or replace in other images? Will a government manipulate a picture to portray an event not as it occurred, but as it wants it to appear? What about abuses with legal evidence and the potential manipulation of video imagery? Do digital still cameras exacerbate the situation since their output is already in a more malleable form than a traditional film camera?[7] In essence, because a digital manipulation could be very hard to detect, if at all, an individual's ethical standards may be one of the only safeguards you have and can trust.[8]

## MULTIMEDIA LEGAL AND BROADER IMPLICATIONS

### Legal Implications

Multimedia and desktop video productions have legal implications. In a patent infringement lawsuit over hip prostheses, for instance, a firm created an interactive presentation that depicted how the devices worked. The goal was to show the devices in action and to make the information accessible and interesting to the jury.[9] The actual production process included using scanned images, image editing software, and animations.

Photo of Existing Conditions

Simulation of Action - Proposed Stack and Plume

New York State Electric & Gas - Clean Coal Technology Project at Milliken Station
Prepared by: Young Associates - Landscape Architecture

**Figure 14.2**
*Computer manipulation, in this instance, is performing a valuable service: revealing the potential impact of a proposed physical plant alteration. Note the top photo of the existing condition and the bottom photo that depicts the proposed changes. (Courtesy of David C. Young/Young Associates-Landscape Architecture.)*

Animations and computer graphics have also gained a wider acceptance in the courtroom. In one example, a court indicated that "animations were no different than drawings that were used by experts to illustrate their testimony. As long as the animation is a 'reasonable representation,' there is no barrier to using such techniques in court."[10] In essence, computer graphics and other computer-generated materials are being viewed much like traditional illustrative tools used by expert witnesses.

Computer graphics and animations have also been used to recreate real-life events. In one case, an animation of a roller coaster depicted the G-forces an individual would be subjected to during a ride. The data were gathered by instruments and incorporated in the animation to show how these forces caused a rider to suffer a stroke through a ruptured blood vessel.[11] Other cases have ranged from pinpointing the causes of fires to depicting how automobile and aircraft accidents could have taken place.

PRODUCTION TECHNOLOGIES

**Figure 14.3**
*Computer graphics, animations, and the law: from a case involving an aircraft accident. (Courtesy of FTI Corporation, Annapolis, MD.)*

### Exploring the World

As you know, the world is an intricate and rapidly changing place. Volumes of information are generated with each passing hour, and we must cope with this continuous information stream. To make matters worse, much of this information is difficult to comprehend.

Graphics, animations, and audio-video clips can help solve this problem by distilling a mountain of information into a more accessible form. For a report that examines the growth rate of urban centers, a series of graphics could potentially replace pages of census material.

Multimedia and desktop video productions can also be used to explore complex events. In one example, a series of graphics and animations has depicted the Strategic Defense Initiative (SDI), also known as "Star Wars." This satellite-based plan, proposed during the Reagan administration, was the core of a multilayered defense system. The missiles, satellites, and other components, which all played a role in the SDI plan, were brought to life via the computer.

Animations made it possible for viewers to at least grasp how the system would theoretically function. The interactions between the different SDI elements were also depicted in a dynamic fashion. With only a text-based description, this information would not have been conveyed as effectively. The subject may have been too difficult to comprehend.

But despite this positive attribute, there is an inherent danger in converting a large information base, especially about controversial and intricate subjects, into a series of images. The real-world situation may be portrayed in too simplistic a manner. Or the graphics and animations could be manipulated to promote a particular point of view—possibly a distorted view of the facts.

Because we tend to believe what we see, a production that effectively uses mixed media could be misleading. We also generally do not have an opportunity to refute the information on a point-by-point basis, as is the case in a courtroom situation. In court, expert witnesses could be called, and the information could possibly be challenged and rebutted.

Finally, the potential to present an altered view of the real world becomes an even more pressing concern in light of a virtual reality system's capabilities, as covered in the next section. A computer-generated world, designed and controlled by a human operator, could serve as an analog for real-world situations. Closer to home, news pictures and other images can be manipulated and altered with a computer.

## VIRTUAL REALITY

### On the Threshold of a Dream

Virtual reality (VR) is a footnote to many of the book's topics. The implications also cut across technological, social, and ethical lines.

In brief, VR can be defined as a display and control technology that can surround a person in an interactive computer-generated or computer-mediated virtual environment. Using head-tracked head-mounted displays, gesture trackers, and 3-D sound, it creates an artificial world of visual . . . and auditory experience. With a digital model of an environment, it creates an artificial place to be explored with virtual objects to be manipulated.[12]

As just described by Michael W. Mc-Greevy, a virtual reality pioneer, you can use a system to enter into and interact with a computer-generated environment. The passage to this realm can be through different VR platforms.

In the configuration we're probably the most familiar with, you wear special "goggles" to view the computer-generated images and a glove to navigate through and to interact with this environment.[13] Realistic sound may also be a key element, and head motions are tracked so the scene shifts as you move your head.

Outfitted with this equipment, the real world can be blocked out as you are fed images of a room. You can move through this space with a gesture of your hand. You can also grab and move objects and even turn wall switches on and off. The objects can act as they would in real life. But in this reality, the action is taking place in a virtual world, the world generated by the computer.

## Applications

As of this writing, there has not been a general consumer/business VR killer-app—that is, an application that would make virtual reality systems an everyday tool, much like spreadsheet and word processing programs helped establish the PC as a mainstream business tool. Nevertheless, numerous applications have been developed and supported. Representative examples include the following:

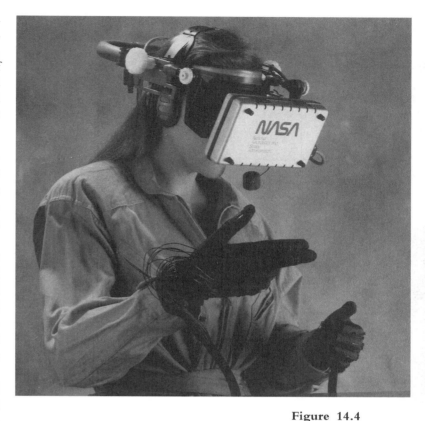

**Figure 14.4**
*A virtual reality outfit.*
*(Courtesy of NASA,*
*Ames Research Center.)*

1. **Telepresence.** Telepresence is the "ability to interact in a distant environment through robotic technology."[14] In one example, NASA has experimented with remote robotic systems, controlled by distant operators, for space-based construction and repairs. Instead of simply pressing buttons to carry out commands, the robotic device becomes a natural extension of its human operator.

2. **Teleconferencing.** With conventional teleconferencing, which enables you to hold an electronic meeting, you may only see another individual on a monitor.[15] In a virtual reality setup, your virtual image could shake hands with your counterpart's image in Yankee Stadium, a tropical beach, or another computer-generated environment. The Internet2, as described elsewhere, may help extend this concept to an online setting, particularly for collaborative work.

PRODUCTION TECHNOLOGIES

3. **Architecture and Prototyping.** You could walk through a building before it is constructed, so you could judge the design's effectiveness, and if necessary, make modifications. This may be one of the closest sensations to living in a building, to try it out, before it is built. A VR system could also be used to help design buildings that better accommodate the needs of individuals with disabilities (e.g., someone who uses a wheelchair).[16] Similarly, VR could be used in car design and other virtual prototyping. In one example, an engineer would

sit on a real seat with a steering wheel in front of him. Wearing a head-mounted display (HMD), she can see the interior's layout and evaluate visibility and accessibility of various components. The cup holding positioning may be tested by having a real coffee cup in hand . . . By tracking the tester's head and hand position, it's obvious if she can easily see and reach the cup holder.[17]

If there's a problem, it could be changed and retested. This capability speeds-up the design process.

4. **Medicine.** The medical applications are numerous. A surgeon could practice a procedure with a virtual patient while a medical student could take a tour through the human body. In a more specialized application, VR has been used to study a disease's molecular structure and "to design drug molecules which could shut down its defense systems."[18] The setup allowed the researchers to see, and in a sense, to experience the data in new ways. The data were not simply viewed on a flat, conventional monitor.

5. **Education.** Applications include the medical tour, meeting historical figures, and investigating an ancient city or even the Solar System.

6. **Personal Freedom.** This is a potent VR application. You could assume a new identity or explore another world and other

environments that you may never be able to experience in real life.

7. **Other Applications.** Other application areas include military war games, enhanced scientific visualization, art, and virtual sex. If you like video games, you could be in for the ride of your life since you could be immersed in the action.[19] The same scenario may also hold true for sports—as you pedal a bike, your representation would follow suit in a virtual environment.[20]

While not strictly VR, virtual sets have been adopted by the broadcast industry. Instead of building elaborate and expensive sets, computer-generated environments are used. You can also tap into effects and capabilities you may not be able to match with a "real" set. CBS news, in fact, touted this feature in its 1996 presidential election coverage.

Like the book's other topics, virtual reality raises ethical and social questions. In one case, a VR configuration could be used for torture. In other examples, some individuals have claimed that VR may have a powerful addictive and "intelligence-dulling" power as well as an insulating quality.[21] If we use realistic simulations, for instance, could we become desensitized to a war in the real world where real people die? In essence, could war turn into a VR game?[22] In another situation, will people be able to strike a balance between the *virtual* and the *real* world? Other questions and concerns exist as well.

*Summary.* Like other technologically driven fields, VR's tools will continue to mature. The equipment should become lighter, more mobile, and less expensive. The graphics quality should also improve, in response to a current complaint, while feedback developments should keep pace. You may experience a physical response similar to the one you get when you pick up a real ball or other object.

Building on contemporary systems, you may also enter a room, much like the holodeck in the *Star Trek* television series. In this setting, you would be unencumbered by equipment. As portrayed in different episodes, a fine line may also emerge between what is reality and what is a virtual reality.[23]

If VR is used appropriately we may all, as is stated in the Moody Blue's album title that opened this section, be on the "threshold of a dream."

## ELECTRONIC MUSIC

In the entertainment field, the proliferation of electronic musical instruments has led to its own controversy. Musicians, for instance, used a *New York Times'* ad to protest a budgetary and/or profit-related move in which a synthesizer replaced eight string players. The musicians had played for a Broadway production, and the ad read, in part:

If the producers . . . really believed that a synthesizer sounds as good as real string instruments, they could have used the synthesizer to play the string parts from the beginning! But they knew that real strings sound better, so they hired eight top-notch string players to enhance the production—that's what they wanted theater critics to hear![24]

This incident is not isolated and, in fact, somewhat resurfaced in 2003. This time, producers wanted to set a lower minimum figure for the number of musicians that would be required for a given show.

The key points for our discussion are the sound's quality and the human presence. In the former, the musicians argued that real strings sound better than a synthesizer. A similar argument has been waged between CD (digital) and high-end LP (analog) proponents.

**Figure 14.5**
*The STAR★TRAK RF wireless motion capture system. It can capture the motion of two performers in three-dimensional space. Interfaces of this type enhance our human–machine capabilities. In one application, you could capture a golf swing or dance steps for various applications. (Courtesy of Polhemus, Inc.)*

In the latter, we all have expectations for different events. When you see a play where an orchestra is central to the production, the expectation almost calls for a conventional orchestra. It is part of the ambience, at least, as we currently perceive it.

Yet this perception could change. Although we may object to electronic instruments in certain circumstances, this situation may be the norm for the next generation.

Until that time, technological changes will continue to conflict with traditional standards. But a balance may eventually be

**Figure 14.6**
*An example of a scene you can now generate, and more importantly, can explore with your PC: Olympus Mons on Mars. (Software courtesy of Virtual Reality Laboratories; Vista Pro.)*

reached. In the music world, both electronically and traditionally generated music have value. They can coexist, when used appropriately.

## PAPERLESS SOCIETY

A paperless society is a product of different technologies and applications. They run the gamut from optical disks to the Internet to multimedia productions. In brief, in a paperless society information is increasingly created, exchanged, and stored in an electronic form. It is also a society that is no longer dependent on paper for writing and storing memos, checks, and other documents.

### Benefits

A switch to a paperless society could have some benefits:

1. We would witness the creation of an enhanced communications system. Information could be read, stored, and retrieved with the accuracy and control afforded by a computer.

New software tools are helping to make this goal a reality. For example, documents can be transported across multiple computer platforms and used without an arcane conversion process. This makes the technology somewhat transparent to the user.

2. Physical space would be conserved because information would no longer be saved on bulky paper. Environmentally, fewer trees would be cut down for paper, and less waste material would be generated to further clog our landfills.

3. Electronic publications could be quickly revised. This is an important factor for industries with rapidly shifting technology bases that necessitate the continual updating of manuals.

4. The advent of cost-effective optical recording systems opens up electronic publishing to more people. If you have a mass storage demand, but not a mass distribution requirement, a DVD or CD recorder may prove useful. We can also establish our own web sites. These developments have, in turn, made it possible to exchange graphically-rich information.[25]

5. A paperless society could save money. Checks could potentially be eliminated, saving banks handling and associative charges.[26] The Internet has accelerated this process through e-commerce and other functions.

6. Information in an electronic form is elastic. It can be stored diversely and communicated through different retrieval mechanisms. Digital clips and other information can also work together, as in a multimedia presentation.

7. A PC's processing power can be utilized to conduct data searches. Instead of flipping through hundreds of pages to find a Sherlock Holmes' quote, use a PC. Load in the appropriate CD-ROM, type in a keyword, and the PC can find the line.

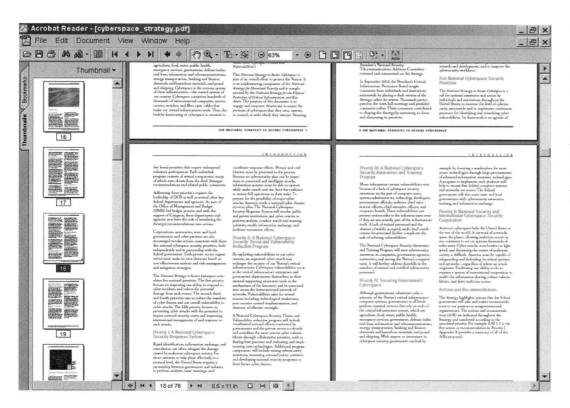

**Figure 14.7**
*New software has made it possible to exchange documents across different platforms. Acrobat helped pioneer this application, particularly at the PC level. The full document (right) and individual pages, represented by icons (left), provide easy access to the information. The document is a White House report to help secure cyberspace, as documented in the Internet chapter. (Software courtesy of Adobe Systems, Inc., Acrobat.)*

Depending on the interface, you may also discover complementary information.

Or you can use the PC to find a term. Click on the word and bring up the definition in a window. You could also search through an on-screen table of contents, bring up a passage in one window, and retrieve and then view a graphic or even a digital clip in another.[27]

8. A paperless society affords you access to an almost unlimited information resource. By utilizing the Internet, you can explore this ever expanding data pool and learn about King Arthur, view the latest space telescope images, and communicate with a friend who lives half a world away.

9. A paperless society can speed up your work. You may be conducting research about privacy. In this electronic environment, you could use a search mechanism to locate relevant information. If available online, you could have it in a matter of seconds.

**Pitfalls**

Despite the benefits, a paperless society does have potential pitfalls. Paper is a better choice for some uses. Newspapers, books, and other paper documents are cost effective, easily mass produced, and can be read almost anywhere. People are also used to working with paper.

Paper can also speed up certain jobs, such as quickly scanning through a stack of books and magazines. While you can somewhat duplicate this process with a computer, you may face hardware and software constraints, especially if you have a small monitor. You may only be able to read small chunks of information rather than the broad sweeps you can take-in by quickly scanning a page(s) in a book(s).

Electronic information systems, for their part, call for machine-dependent operations to store, retrieve, and exchange information. If standards are not adopted, technical

PRODUCTION TECHNOLOGIES

roadblocks may impede this information flow.[28]

Paperless or electronic resources can also be relatively expensive. Like a thirsty swimmer in an ocean, you may be surrounded by a sea of information, but may not be able to use it. You may lack the prerequisite knowledge, skills, or money.

This scenario brings up two questions. If information is reduced to an electronic form, will it be readily accessible? What is a realistic timetable for the technology to trickle down to all levels of society? A paperless society also has privacy implications. E-mail concerns, as later described, may only be the tip of the proverbial iceberg.

## THE DEMOCRATIZATION AND FREE FLOW OF INFORMATION—MAYBE

The production technologies have contributed to the democratization of information. With the correct outfit and skills, you can publish a newspaper, magazine, video, or multimedia project. This capability implies that information cannot be controlled by only a select group of people or by the government.

Established media organizations could be censored or even shut down. But it may be impossible to completely shut off the flow of information. There may be too many desktop publishing and video systems, copy machines, VCRs, the Internet, and other production and distribution mechanisms.

This concept has been put to the test on different occasions, including the aborted August 1991 coup in the Soviet Union when Mikhail Gorbachev was deposed. Boris Yeltsin and his support group used desktop publishing and other communications tools to keep the Russian citizens and the world community apprised of events.

The free flow of information, during this time, contributed to the coup's eventual collapse.[29]

The democratization of information also has a personal connotation. Using sophisticated communications tools, you could launch your own company for, say, desktop publishing or serving Internet customers.

You could also present your personal ideas in the open marketplace via a letter, political broadside, or video production. You may only attract a small audience. No one may even agree with you. But in the context of our discussion, is this important?

## Potential Problems: Media Concentration and Ownership

While there are more tools and communications channels to "voice" your opinion, some organizations have great concern over increased *media concentration*—fewer companies and individuals now own or control more newspapers, television stations, and other major media outlets. If there are fewer owners, the diversity of programming choices and viewpoints could be affected. This is particularly crucial for news coverage since the mass media are still a staple news source. This situation would be exacerbated through cross-ownership where different types of media companies may have a single owner.[30]

Michael Powell, an FCC chairman, was a major proponent for relaxing rules that governed such options. At an April 2003 meeting, he stated

Change is now inevitable. A digital migration has begun taking us from the old world—marked by analog technologies, narrowband infrastructure, and the monopoly regulation model—to the new world; marked by digital technologies, broadband infrastructure and a broader minded view of regulation, informed by listening to technology more than lobbyists. The changes

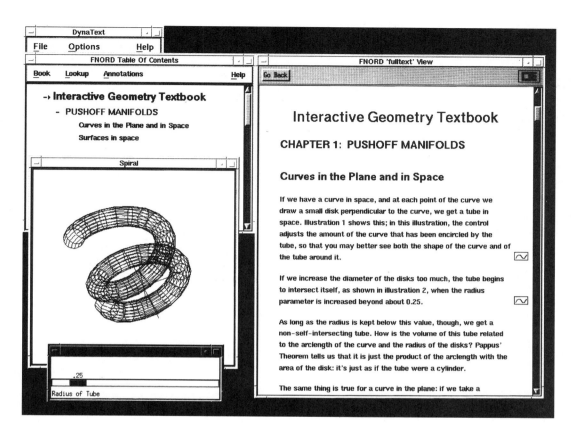

**Figure 14.8**
*A sophisticated electronic book. Multiple views and hypertext links may be supported, including in this case, a link to a program that illustrates the theorem mentioned in the text. It may also be possible to create your own electronic notes. (Courtesy of Electronic Book Technologies, Inc.)*

brought forth by digital life will finally force a change in the decades old outlook of communications policy.... As I have said, in the old world governments are central and monopolists are central. The most important thing technology is doing is placing consumers and citizens more squarely in the driver's seat of choosing and controlling their personal communications space.[31]

In this new information order, "old style" regulation may no longer be beneficial, thus, the push to relax the rules.

But as stated, critics contend that despite our new information and communications options, the diversity of news offered by the mass media may suffer. In the words of one publisher, "People ought to be scared to death when you have a handful of big businesses that are getting bigger, that are going to control all the conduits of information ..."[32]

## CONCLUSION

To sum up, all the topics have common threads. They:

- are the offspring of video, image, optical, printing, and audio processing/manipulation technologies;
- have influenced society; and
- have raised important questions. The most pressing issues are related to news image manipulation and the democratization of information.

For the democratization of information, the communication revolution has presented us with powerful tools. We have become information producers and are now more ensured of the free flow of information. You can censor a newspaper, but can you shut down thousands of DTP/video systems?

There is the old joke about the censorship police knocking on your door. With the new technologies, some believe there may be too many doors. But by tapping into the new communications and information technologies, can more of these doors now be watched?

For the news media, does increased media concentration and cross-ownership reduce the diversity of information? For a democratic society, if the diversity of voices is potentially curtailed, what are the implications?

Finally, technological developments have raised eyebrows about another application and its implications. Can people still trust your reporting? How do you know what is real and what is fabricated? Once you lose the public's trust, can it ever be fully recovered?

This has also raised an incident with another application. As outlined in the editorial from *Broadcasting & Cable*, it has serious implications. "Box of Virtual Chocolates," by John Eggerton (Deputy Editor) and Harry Jessell (Editor), *Broadcasting & Cable*. Reprinted with permission by *Broadcasting & Cable*. Originally appeared in the January 7, 2000 issue, p. 150.[33]

On the subject of believing. It sure isn't seeing anymore.

When *Forrest Gump* first came out, some of us didn't know actor Gary Sinise. The filmmakers didn't show him from the waist down until after his character's legs had been amputated, so we though the actor was likely an amputee. He isn't, of course. It was computers that removed his legs, just as they had put Tom Hanks into George Wallace's schoolhouse door speech and Kennedy's White House. In an age when computers can put anything into a picture or take it out, the distinction between reality and special effects is as blurry as the virtual images are sharp. That is why we reacted so viscerally to the quote from the director of the *CBS Evening News* to the effect that virtual insertion technology has "applications that I think are very valid and lend themselves perfectly to news, such as *obscuring things you don't want in the frame* [my emphasis]."

Now CBS News executives are saying that they are going to be careful in how they use the video insertion technology and so far they have been. They have used it only to post their logo in strategic places. . . . They haven't distorted events in any substantive way, although they have the distinction of being the only news organization to show Times Square as it wasn't at the dawning of the millennium.

Nonetheless, we believe that the best policy regarding video insertion is not to use it during newscasts. Viewers must be able to trust that video, especially live video, is the real thing. . . . No less a CBS executive than former president Frank Stanton had this to say on the subject, on the occasion of presenting a First Amendment award to Walter Cronkite in 1995:

"Digital technology opens a Pandora's box. The options are startling and tempting. No longer will it necessarily be that what you see is what you get. Ultimately it will be easy and inexpensive to fake the picture. And virtually undetectable. Consider the temptations and the burden these developments will put on the producer, the television reporter, and his organization. The Forrest Gumps of the evening news could have a field day. And the public could be the loser. The audience will not know who or what to believe."

We join Mr. Stanton in urging news departments everywhere to consider the temptations and their consequences. It may be impossible to put the digital genie back in the bottle, but when the news business becomes about obscuring things, we're all in trouble.

## REFERENCES/NOTES

1. The shades look too similar.

2. The processed image is a false color image.

3. An analogy can be drawn with the Sistine Chapel restoration. Some critics claimed this process altered Michelangelo's original work; others stated it was a cleaning and restoration process.

4. Steve Ciarcia, "Using the ImageWise Video Digitizer: Part 2: Colorization." *Byte* 12 (August 1987), 117. *Note:* As part of the process, a new print has been made, which may actually help preserve a copy of the film.

5. You may notice a label that appears on a colorized product: "This is a colorized version of a film. . . . It has been altered without the participation of the principal director, screenwriters, and other creators of the original film." From *Village of the Damned,* aired on TBS, 1995.

6. Jane Hundertmark, "When Enhancement Is Deception," *Publish* 6 (October 1991), 51.

7. Ibid. Film cameras add an additional step to computer processing. The film must be developed and the negative, slide, or print (produced, for instance, from the negative) must be scanned.

8. Don Sutherland, "Journalism's Image Manipulation Debate: Whose Ethics Will Matter?" *Advanced Imaging* 6 (November 1991), 59.

9. Charles Rubin, "Multimedia on Trial," *NewMedia* 1 (April 1991), 27. *Note:* This case was settled out of court. Opposing lawyers would also have objected to the use of the presentation.

10. Jon L. Roberts, "The Digital Image Processing in Courtroom Forensics: Increasingly Admissible at Last," *Advanced Imaging* 15 (May 2000), 25. The article also covers admissibility guidelines and the shifting role of the judge as a "gatekeeper" for determining a computer-based image or sequence as evidence.

11. Carrie McLean, "Houston Lawyer Makes a Case for Computer Animation in the Courtroom," *Presentation Products* 5 (May 1991), 18.

12. Michael W. McGreevy, "Virtual Reality and Planetary Exploration," paper presented at 29th AAS Goddard Memorial Symposium, March 1991, Washington, DC, 1.

13. "The Future Is Now," *Autodesk* (1992), 32.

14. Linda Jacobson, "Virtual Reality: A Status Report," *AI Expert* 6 (August 1991), 32.

15. Teleconferencing is discussed in another chapter.

16. Ben Delaney, "Where Virtual Rubber Meets the Road," *AI Expert,* Virtual Reality 93: Special Report (1993), 18. *Note:* This special edition of *AI Expert* focuses on the virtual reality field/applications.

17. Ben Delaney, "Virtual Prototyping: VR Goes to the Factory." www.gate.com (July 2000), 26.

18. Elliot King, "The Frontiers of Virtual Reality and Visualization in Biochemical Research," *Scientific Computing & Automation* (July 1996), 20.

19. Early video game users had a taste of this potential through Mattel's Power Glove for the Nintendo system. Please see Howard Eglowstein, "Reach Out and Touch Your Data," *Byte* 15 (July 1990), 283–290 for more information; other articles described how to interface the Power Glove to a PC.

20. In one setup, as you pedal a stationary bike, you would travel through the virtual environment. Please see Ben Delaney, "VR Applications Take the Tedium Out of Indoor Exercise," *Computer Graphics World* 24 (July 2001), 48–52.

21. Howard Rheingold, *Virtual Reality* (New York: Summit Books, 1991), 355.

22. For example, many of the bombing runs in Operation Desert Storm, as portrayed on television, took on the appearance of a video game. See Rheingold, *Virtual Reality*, 357–362, for some of the implications.

23. Tom Reveaux, "Virtual Reality Gets Real," *NewMedia* 3 (January 1993), 34.

24. Advertisement, "Grand Hotel: The Rip-Off," *New York Times* (November 3, 1991), Section 4, 7.

25. In the past, such information may have generally been text-based.

26. This includes check maintenance and delivering checks, if relevant, to customers.

PRODUCTION TECHNOLOGIES

PRODUCTION TECHNOLOGIES

27. Louis R. Reynolds and Steven J. Derose, "Electronic Books," *Byte* 17 (June 1992), 268.

28. There is also an archival issue. If you use a proprietary system, and the company goes out of business, what do you do if your system breaks?

29. See Richard Raucci, "Overthrowing the Russians and Orwell," *Publish* 6 (November 1991), 21; Howard Rheingold, "The Death of Disinfotainment," *Publish* 6 (December 1991), 40–42, for additional information about this event and similar incidents, such as the 1989 Tiananmen Square uprising. Rheingold's article also touches on other implications.

30. Consumer Federation of America, "Democratic Discourse in the Digital Information Age: Legal Principles and Economic Challenges at the Millennium."

31. Remarks of Michael K. Powell, Chairman Federal Communications Commission, "Hear Ye, Hear Ye Read all About It!" At the Associated Press Annual Meeting and General Session of the Newspaper Association of America Annual Convention, April 28, 2003.

32. Broadcast Engineering, "FCC's Powell Says Newspapers Will 'Fare Well' Under Ownership Changes," downloaded from http://editorial1.industryclick.com/microsites/index.asp?srid=11266&pageid=7018&siteid=15&magazineid=158&srtype=1#cap.

33. John Eggerton and Harry Jessell, "Box of Virtual Chocolates," Reprinted with permission by *Broadcasting & Cable*. Originally appeared in the January 7, 2000 issue, p. 150.

## SUGGESTED READINGS

Baxes, Gregory A. *Digital Image Processing*. New York: John Wiley & Sons, 1994; Richard Berry and James Burnell. *The Handbook of Astronomical Image Processing*. Richmond, VA: Willmann-Bell, Inc., 2002. Two excellent books about image processing. The Berry book covers astronomical image processing but many of the concepts are applicable for other application areas.

Benedikt, Michael, ed. *Cyberspace: First Steps*. Cambridge, MA: The MIT Press, 1991; Howard Rheingold. *Virtual Reality*. New York: Summit Books, 1991. Though older, excellent introductions/discussions about the topic.

Berry, Richard. "Image Processing in Astronomy." *Sky and Telescope* (April 1994), 30–36; Chris Scher and John O'Farrell. "Improve Your Astrophotos by Combining Images." *Sky and Telescope* (October 1999), 135–139. Overviews of image processing, PC-based astronomical applications, and relevant software.

Bove, Tony. "On Location." *NewMedia* 7 (September 1997), 40–45; Mark Hodges. "Virtual Reality in Training." *Computer Graphics World* 21 (August 1998), 45–52; Diana Philips Mahoney. "Better than Real." *Computer Graphics World* 22 (February 1999), 32–40; NASA. "Alternative Engineering: How Virtual Reality and Simulation Are Changing Design and Analysis." *NASA Tech Briefs* 25 (June 2001), 16–24; Elizabeth M. Wenzel. "Three-Dimensional Virtual Acoustic Displays." NASA Technical Memorandum 103835, July 1991. Various VR applications and issues.

Chouthier, Ron. "Glossary of Image Processing Terminology." *Lasers and Optronics* 6 (August 1987), 60–61; Susan Wels. *Titanic*. New York: Time Life Books, 1997. Image processing issues: a glossary of image processing terms and the history/fate of the *Titanic*; includes information about imaging techniques used to capture pictures of the ship.

Emerson, Toni et al. "Medicine and Virtual Reality: A Guide to the Literature." HITL Technical Report No. B-94-1, downloaded from http://kb.hitl.washington.edu/kb/medvr/medvr.html; Elliot King. "The Frontiers of Virtual Reality and Visualization in Biochemical Research," *Scientific Computing*

& Automation (July 1996), 20–22; M. Russell Taylor II and Vernon L. Chi. "Take a Walk on the Image with Virtual-Reality Microscope Display." *Laser Focus World* 30 (May 1994), 145–149. VR medical applications; the online document is a comprehensive and excellent bibliography on the topic.

Ganzel, Rebecca. "Digital Justice." *Presentations* 13 (November 1999), 36–48; Jerry W. Jackson. "Case-Making Courtroom Car Crash Visualization." *Advanced Imaging* 7 (July 1992), 48–51; Andrew Lichtman. "Forensic Animation." *Amazing Computing* 6 (January 1991), 42–44, 46–47; Samuel R. Rod "3D Measurement & Reconstruction for Crime & Accident Investigation." *Advanced Imaging* 15 (July 2000), 15–17; Erica Schroeder. "3D Studio Gives Crime-Solving a New Twist." *PC Week* 9 (March 9, 1992), 51, 58. Court-room imagery, types of other imaging applications, and animation use (in a courtroom).

Kuchwara, Michael. "Strike Shuts Down Broadway Musicals." AP, March 9, 2003, downloaded from www.yahoo.com. Information about the musician's strike of 2003.

Reveaux, Tony. "Virtual Set Technology Expands." *TV Technology* 17 (July 28, 1999), 14; David A. Tubbs. "Virtual Sets: Studios in a Box." *Broadcast Engineering* 40 (November 1998), 90–92. Overviews of virtual set systems.

Wylie, Philip. *The End of the Dream.* New York: DAW Books, Inc., 1973. See pp. 162–165 for a precursor look at a virtual sex-type setup. The book, a science-fiction novel, primarily focuses on environmental issues and the Earth's potential future.

## GLOSSARY

*Colorization:* The process by which color or "colorized" versions of black-and-white movies are produced.

*Democratization of Information:* New communications tools contribute to the free flow of information in society. The implications? More people can now be information producers, and it may be harder to institute censorship on a broad scale. But there are potential limiting factors, such as media concentration and cross-ownership.

*Image Processing:* The field and technique in which an image (for example, created by a video camera) is digitized and manipulated. Typical operations include image enhancement and correction.

*Paperless Society:* A society in which information is increasingly created, stored, and exchanged in an electronic form.

*Photograph and Image Manipulation:* The manipulation of photographs for, in one example, electronic retouching. A potential problem is the alteration of news images.

*Virtual Reality (VR):* A computer-generated environment in which a user can interact with and/or manipulate various elements.

PRODUCTION TECHNOLOGIES

# V

## INFORMATION, ENTERTAINMENT, & COMMUNICATION SYSTEMS

# 15  The Cable and Telephone Industries and Your Home

As described, fiber-optic (FO) lines may serve as the backbone of digital-based entertainment and information systems. They could feature interactive and conventional television programming, stock market quotes, and additional services. These offerings can be supported by cable and telephone companies (telcos). The satellite industry is also a player in this field.

## VIDEO-ON-DEMAND

A goal of the communications industry is to provide viewers with more control over programming choices, whether it is a television show or a new optional service. A term associated with this development is video-on-demand (VOD).

In brief, VOD, which comes in different flavors that sport different features, and can be thought of as enhanced pay-per-view (PPV). With PPV, you're charged a set fee to view a movie, concert, or special event.

VOD expands this capability by supporting a more diverse and tailored programming pool. In one configuration, VOD could function much like a video rental store. You select the programming you want to see when you want to see it. There could also be a library of movies, old television series, and other programming. Interactive controls, which would function much like those on a conventional VCR, could also be

featured.[1] This also means you can save and view the program at your own leisure not just when the program is initially delivered (e.g., via a cable system).

VOD also reflects a fallout of the new communications technologies. As described in other chapters, rather than serving the mass audience, individual needs could now be satisfied. In one sense, the communication revolution could be viewed as a *personal* revolution.

Another key issue centers on the players in this arena. Cable and satellite companies are not the only options, and the telephone industry has emerged as a major contender. But it was a somewhat tortuous legal path that led to this development since the telcos could not initially offer VOD and other advanced services.

For example, as part of the ongoing evaluation of the consent decree that led to the divestiture of AT&T, Judge Harold Greene, the architect of this decision, barred the seven regional holding companies from offering various information services. The companies were created after the breakup of AT&T, and the restrictions were adopted to prevent the telephone industry from dominating the new, emerging information infrastructure.

But political pressures and other factors led to a relaxation of the original decision in the late 1980s and early 1990s. This opened the door for the telcos to support information services, such as electronic

INFORMATION, ENTERTAINMENT & COMMUNICATION SYSTEMS

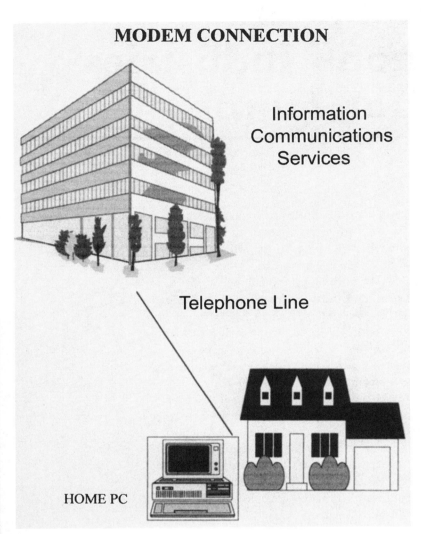

## MODEM CONNECTION

Information
Communications
Services

Telephone Line

HOME PC

**Figure 15.1**

*When first implemented, the typical connection between a home user and an information/communications service was the telephone line. As described in earlier chapters and in this chapter, competing communications channels, with a range of new services, have entered the picture.*

Yellow Pages and, potentially, to enter the video market.[2]

Another major step was video dialtone (VDT).[3] In 1992, the FCC modified its rules to allow telephone companies to compete in the video arena. One provision called for the support of a "basic platform that will deliver video programming and potentially other services to end-users" on a common carrier, that is, a nondiscriminatory basis.[4] In essence, the telcos could now carry video programming, with certain restrictions.[5]

Since the telcos had deep financial pockets and subscribers bases (homes with telephone connections), competing indus-

tries did not relish this decision. Other developments followed that further expanded the telcos' options, including the 1996 Telecom Act.[6] For this discussion, four points merit our attention:

1. Support was provided for the proliferation of set-top boxes described in the next section.
2. Cable companies could enter the telephony business.
3. Existing VDT policies and rules were repealed.
4. Rules for an Open Video System (OVS) were outlined. They could be considered an extension to the original VDT concept.[7]

The last provision was designed to help relax the guidelines that restricted the broad telcos entry into the video field.[8]

## THE ENTERTAINMENT-INFORMATION MERGER

### Qube

The idea of providing customers with more control over their viewing choices did not begin in the 1990s. The Qube cable television service, launched in Columbus, Ohio, in the 1970s, let subscribers tailor their cable service to match their viewing requirements.

Its most unique characteristic was an interactive capability. After a speech delivered by then-President Jimmy Carter, for instance, Qube subscribers participated in a televised electronic survey. The subscribers registered their answers to questions via keypads and the results were tabulated.[9]

Similar viewer response programs were produced in other areas, and the system had the capability to support a PPV option, information services, and electronic transactions. The latter could encompass shopping and banking at home. Even though Qube later abandoned its ambitious plans, it actually helped define the concept of an *integrated entertainment and information utility*.

## Implications

For our purpose, an integrated entertainment and information utility implies that a single company may provide you with entertainment and information services. You may use your television for conventional viewing, VOD, retrieving an electronic television viewing guide, and other activities. The same delivery mechanism may also support an Internet hookup. These broad service categories could be grouped under the interactive television umbrella where the viewer interacts with the service through the television set. This would be analogous to using a computer keyboard as you surf the Internet when trying to find and retrieve certain information.

This type of system could become a reality through cable, telephone, or other links. For example, experimental VOD operations were launched, and during the early-to mid-1990s, the cable and telephone industries rushed in to support different levels of this application. At the same time, they experimented with new information delivery methods.

It is also important to stress digital technology's role in this endeavor. It will help supply the television programming to potential customers through, in one case, video servers.

Another element is the convergence factor. Numerous companies, as well as technologies and applications, will have an impact on this infrastructure's growth. For example, new technological developments and a competitive marketplace have played a role in inter- and intra-industry mergers. Software companies have also joined with broadcast entities, and cable and computer

**Figure 15.2**
*An advantage of a fiber-optic line is its great information carrying capacity. New developments in this field, led by companies such as Corning, have revolutionized the way we use and relay information. (Courtesy of Corning, Inc.)*

companies have becomes partners with television equipment manufacturers.

This sector of the communications industry is a complex mix of services and companies. The rest of our discussion focuses on potential players, the key issues that must be addressed, and the questions that are, as of this writing, still unanswered.

## Issues and Questions

1. True or real-time VOD and other interactive services require either:

- Enhanced television converter boxes. The new generation of set-top boxes (STBs) resemble computers in terms of their sophistication and user interfaces, and various computer companies have contributed to this development. The planned operations call for STBs that could handle data, digital video, and other information. In this scenario, the STB functions as an interface to these enhanced services.

A related factor is the ability of manufacturers to create a flexible technology base—an STB should be readily extensible by accommodating new services without major retooling. In one case, could a new service be initiated through a software update delivered by, for instance, the cable company to the box?[10] If STBs have to be regularly replaced, this would raise the cost of subsequent services.

- Your television set could potentially handle enhanced services without an STB. The Cable and Consumer Electronics Industries filed a Memorandum of Understanding that outlined an agreement for the manufacturing of a new generation of plug and play television receivers. Unlike current systems, consumers would now "be able to plug their cable directly into their digital TV without the need of a set-top box."[11] As envisioned, this development would simplify the adoption of new services, including digital television operations.

For our discussion, the focus is on the potential viability of enhanced services. It does not matter if it is delivered through an STB or a new digital receiver. In fact, it should also be noted the adoption of a new generation of television receivers might not eliminate the need for external hardware in specific cases and for certain services.[12]

2. A cable company must have the channel capacity to support new entertainment and information services. In one setup,

- A fiber/coaxial hybrid system could be constructed. Fiber could form the trunk line with coaxial cable being used in other parts of the system.
- Fiber could be directed to a curbside terminal where multiple subscribers would then be connected by standard cable.[13]

- Sony and other companies are developing new delivery systems, which tap the current cable infrastructure and STBs, to deliver enhanced services.[14]

3. Even though it appears the trend may be for the eventual implementation of all-fiber systems, the timetable is still unknown. The cost for a full switchover would be expensive and, as described, fiber/coaxial hybrids and/or data compression may prove suitable, at least in the interim.

Digital compression can also increase the number of channels a cable system could support. Consequently, instead of building a new FO plant, a coaxial plant may be able to handle more advanced services.

But this switchover, be it based on telephone or even cable technologies and systems, may mandate the adoption of either new STBs, television receivers, or other conversion devices. The cost can vary, and to achieve a wide market penetration, it must be somewhat competitive with current pricing structures.

4. The development of an integrated entertainment and information utility hinge on the growth of information services. This could mean videoconferencing, support for multimedia presentations, and Internet access.

As described in Chapter 17, the Internet is an ever-widening information pool that subscribers can tap into via various communications channels. It is also one of the wild cards or unknown factors in our discussion.[15]

Some important issues include

- Will a company, which is your Internet service provider, also support your television entertainment needs?
- Will an Internet "hook" convince subscribers to switch to a full-service company that may offer a high-speed Internet link, television programming, and

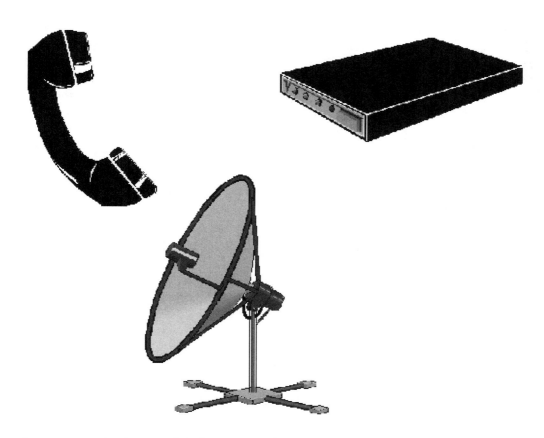

**Figure 15.3**
*Three of the players that want to deliver information and possibly entertainment programming to your home: satellite, cable, and telephone companies.*

other communications and information options?

- Will we generally continue to use multiple companies for television, telephone, and information service deliveries?
- What role will the Internet ultimately play in this development? As will be covered in other chapters, the Internet can be used for delivering information ranging from music clips to video programming. As the technology base is enhanced in tandem with the information delivery systems (e.g., DSL-based operations), will the Internet emerge as the VOD medium of choice? What of other information and entertainment services?

5. New and emerging technologies will continue to shape the way we gain access to and use this rich information mix. In one example, digital television systems can de-liver enhanced pictures and sounds to our homes. When combined with VOD, your living room could be converted into a movie theater. Digital television has also promoted the merger between television and computer-based technologies. In essence, could your computer serve as your television-viewing device?

In another example, consumers have purchased personal video recorders or PVRs. Much like a computer stores data, a PVR records and stores television programs for replay. But unlike the ubiquitous VCR, a disk-based PVR has some advantages, including the ability to simultaneously record and playback programming. On the flip side, PVRs are more expensive than VCRs and have some other shortcomings, including, as of this writing, a limited market penetration.[16] Nevertheless, could a future and enhanced PVR enable us to store hours

INFORMATION, ENTERTAINMENT & COMMUNICATION SYSTEMS

of high-quality video and audio programming that we could view at our own leisure, archive for future use, or exchange with our friends? What of the copyright implications for the use of such information?

It should also be noted that PVR functions could be incorporated in an STB. In one example, Motorola introduced such a system in the early 2000s.[17]

6. What are the television and satellite industries' roles in this endeavor? Television stations may be positioned to offer consumers multiple entertainment channels, data, and information services. Satellite operators, on the other hand, have already surpassed the cable and telephone industries in certain respects. DirecTV has relayed digitally compressed television programming to its subscribers, and it can support an Internet link.

The satellite industry also has two major advantages. First, it may be easier to upgrade a service. Once a subscriber owns a dish, technical changes may be software based and/or through STB enhancements. Cable and telephone companies may have to additionally retool their physical plants.

Second, like DirecTV, a satellite-based company does not have to be tied to an existing, and possibly antiquated, communications infrastructure. Thus, it could take advantage of new technological developments. Cable and telephone companies that have serviced a given region for years, and may have a sizable investment in current facilities, may not have the same luxury.

7. A related question is that of software—in this context, the television/movie programming. If your company plans to offer a VOD service, do you have access to a range of program options? As of this writing, the telephone companies are not as well placed as their competitors. There is also a question of public perception. Cable and satellite companies have established consumer bases for television services. Will this factor influ-

ence consumers as they choose their VOD platform of choice? Or if a subscriber is comfortable with using the Internet through a DSL-based operation, will this individual turn to a telephone company for service if the programming options were sufficient?

8. Another factor may be the initiation of more cooperative ventures between different industries.[18] New technological developments and accompanying applications will also be introduced.

9. Technically, how many years will it take for the necessary technologies to reach smaller markets? Are the changes inevitable, as is the case with most industries where a physical plant is modernized over time, or will some companies balk at the additional expense?

10. Although specific demographic groups may quickly adopt VOD or other entertainment and information services, the question about broad consumer acceptance is still unanswered. How much money are people willing to spend for such services? Can they be sustained by targeting select subscriber groups via narrowcasting?[19]

11. The concept of a VOD system has been extended to other environments. In one case, VCRs used for playing movies in certain hotels have been replaced by a hybrid digital/analog system. A similar or all digital system could also be used for corporate training, linking information kiosks in departments stores, and delivering college lectures.[20] What future applications could be accommodated? Will they help create a niche market for VOD if the general consumer market fails to materialize?

## CONCLUSION

The communications industry is undergoing a series of rapid changes. As indicated, companies and individuals must contend with a complex mix of current and potential services. We could gain access to VOD and the Internet through our cable, telephone, or satellite connection. But there are still questions, especially for the consumer market.

Will people be willing to spend additional money for these services? If broad consumer acceptance is not achieved, should projects be abandoned? Will it take time for people to accept and use these services, that is, will it be a slow, but sure, growth pattern? Or should companies focus more on niche markets?

There are also new developments and applications that could have an impact on this field. In one example, the FCC initiated an inquiry about delivering broadband services via power lines—*Broadband Over Power Line* (BPL). According to the FCC, BPL can provide consumers with the freedom to access broadband services from any room in the house without adding or paying for additional connections by simply plugging a BPL device into an existing electrical outlet. BPL may be able to provide an additional means for "last-mile" delivery of broadband services and may offer a competitive alternative to digital subscriber line (DSL) and cable modem services. This will also enable access to communications services in rural and remote areas of the country.

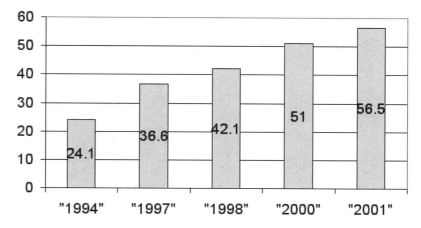

In addition, BPL systems can be used by electric utility companies to more effectively manage their electric power networks.[21]

Basically, you could tap into broadband services by plugging a device into an electrical outlet. Demonstrations have already been held, and if widely adopted, BPL could have a major impact in the telecommunications industry. But as has typically been the case with many new services, particularly one in a highly competitive market, it also may unleash a flood of policy, technical, and marketing questions.[22]

**Figure 15.4**
*The percent of U.S. households with computers by designated years. (Source: NTIA and ESA, U.S. Department of Commerce, using U.S. Bureau of the Census Current Population Survey supplements.)*

## REFERENCES/NOTES

1. Richard L. Worsnop, "Pay-Per-View TV," *CQ Researcher* 1 (October 4, 1991), 743.

2. Steve Higgins, "NYNEX, Pacific Bell Tout Electronic Information Services," *PC Week* 9 (March 9, 1992), 67.

3. VDT, used in the context of computer systems, is a video display terminal. For this specific discussion, though, it represents video dialtone.

4. FCC, CC Docket 87–266, July 16, 1992, *Local Telephone Companies to Be Allowed to Offer Video Dialtone Services*; Repeal of Statutory Telco-Cable Prohibition Recommended to Congress, downloaded from CompuServe, October 1992. *Note:* There had been a call for a VDT-type option for a number of years.

5. Harry C. Martin, "Telcos Offer Video Services," *Broadcast Engineering* 34 (September 1992), 8.

6. The Telecom Act is discussed in different chapters of this book.

7. Chris McConnell, "FCC Begins Work on Telecom Act," *Broadcasting & Cable* (February 19, 1996), 15.

8. If a telco opted for this framework, it would be free of certain regulations, but would

INFORMATION, ENTERTAINMENT & COMMUNICATION SYSTEMS

INFORMATION, ENTERTAINMENT & COMMUNICATION SYSTEMS

have other restrictions. See "OVS: Other Door to Cable Entry," *Broadcasting & Cable* (February 12, 1996), 54, by Michael Katz, for additional information.

9. Edward Meadows, "Why TV Sets Do More in Columbus, Ohio," *Fortune* (October 6, 1980), 67.

10. Please see "Future-Proofing Set-Top Boxes," by David Barringer, downloaded from www.nab.org/conventions/nab2002/proceedings/dtvreceivers3.pdf.

11. FCC Press Release. "FCC Seeks Comment on Cable-Consumer Electronics Agreement on 'Plug and Play'." January 10, 2003, downloaded from www.fcc.org.

12. Sanjay Talwani, "Cable, CEA Agree on Plug-and-Play." *TV Technology* 21 (January 8, 2003), 15.

13. Larry Aiello, Jr., "Bring Fiber Home," *Guidelines* (1992), 7.

14. "Rites of Passage: Sony Lays Down Cable's Future," *eMedia* 16 (May 2003), 10.

15. Please see Chapter 2 for a discussion about high-speed communications systems/Internet access issues.

16. Mark Schubin, "Taper in Gout," *Videography* (December 2002), 18.

17. "Motorola Launches HD, PVR-Integrated Digital Cable Set-Tops," *TV Technol-ogy*, downloaded from www.tvtechnology.com/dailynews/one.php?id=1102.

18. For example, New York Telephone and Liberty Cable Television, a wireless operator in Manhattan, developed a plan to deliver programming while Blockbuster Video teamed-up with two companies to develop a system to deliver VOD programming via the Internet. Please see Rich Brown, "New York Connects to Video Dialtone," *Broadcasting* (October 12, 1992), 38 and Edward B. Driscoll, Jr., "Video On Demand," *Smart TV* 5 (spring 2001), 53–56, for the respective details.

19. Narrowcasting targets defined subscriber groups in contrast to producing, for example, a television program geared for a broad-based—a general—audience.

20. Claire Tristram, "Stream On: Video Servers in the Real World," *NewMedia* 5 (April 1995), 50.

21. FCC, "FCC Begins Inquiry Regarding Broadband Over Power Line (BPL)," Press Release, April 23, 2003, downloaded from http://hraunfoss.fcc.gov/edocs_public/attachmatch/DOC-233537A1.doc.

22. This may include potential interference issues.

## SUGGESTED READINGS

Anderson, Lesley. "Interactive TV's New Approach." *The Industry Standard* 2 (October 11, 1999), 60, 62; Richard V. Ducey. "A New Digital Marketplace in Information and Entertainment Services: Organizing Around Connectivity and Interoperability," downloaded from www.nab.org; Michael Grotticelli and Ken Kerschbaumer. "Slow and Steady." *Broadcasting & Cable* 131 (July 9, 2001), 32–42. Excellent overviews of interactive television services and various issues. The last article includes sidebar features, such as the necessity of the broadcast industry to conduct surveys to monitor audience behaviors vis a vis interactive programming wants and desires.

"Breakup of the Bell System." *Communications News* (September 1984), 98–99; Kevin Tanzillo. "Flood Gates Open." *Communications News* (January 1989), 48–51, 65. The first article describes the divestiture of AT&T; the second looks at the industry five years later.

*Broadcasting & Cable. Broadcasting & Cable* magazine has chronicled the telcos' entry in the video market. Sample articles, prior to the 1996 Telecom Act include: "Bell Atlantic Sees Television in Its Future." (May 8, 1989), 30;

"Telco's Army Poised for Assault on TV Entry." (October 3, 1988), 38–47.

Brown, Peter. "Digital Set-Top Boxes—Slow to Shine." *Digital Television* 2 (May 1999), 1, 28, 49; John Healey. "All Set for Advanced Set-Tops?" *Broadcasting & Cable* (April 27, 1998), 48–54; Joe McGarvey. "Competition Heats Up Early Digital Set-Top Market." *Inter@ctive Week* 2 (January 16, 1995), 26–27; Kyle Pope and Scott Collins. "The New Killer App." [Inside] (December 12. 2000), 63–68. An overview of set-top boxes, PVRs, and related issues.

FCC. CS Docket No. 96–46. March 11, 1996. *Implementation of Section 302 of the Telecommunications Act of 1966.* Open Video Systems. Everything you want to know about OVS. The introduction also provides a good overview of video dialtone and its repeal by the 1996 Telecom Act.

Gall, Don, and Mitch Shapiro. "Fiber-to-the-Home—Why Not Now?" *Lightwave* (May 2000), 18; Robert Israel. "Video Un-Demand." *Telecom* 7 (April 2, 2001), 47–54; Reed Majors. "Television from the Telephone." *Broadcast Engineering* 44 (August 2002), 24, 26. The articles explore related topics—issues concerning the use of fiber-optic lines to connect with home users; DSL issues including its role as a VOD carrier; the Internet as an information/entertainment programming vehicle and the telcos' role in these ventures.

Telecommunications Act of 1996, 104th Congress, 2nd Session, January 3, 1996. Sections of the Act have an impact on some of this chapter's applications.

## GLOSSARY

*Entertainment and Information Utility:* A company that will deliver entertainment programming and information services. A precursor of such a system was Qube, a cable system in Columbus, Ohio.

*Telecommunications Act of 1996 (1996 Telecom Act):* Comprehensive legislation that has influenced the communications industry.

*Two-Way Cable:* An interactive, two-way cable system.

*Video Dialtone (VDT):* VDT permitted telcos to participate in the video marketplace. It was superseded by the 1996 Telecom Act.

*Video-On-Demand (VOD):* VOD can be viewed as enhanced pay-per-view. At one level, you could gain access to and view a movie or program from a central library at your convenience.

*Video Server:* A high speed and high capacity information processing/storage device geared for a video relay. Applications include VOD and commercial playbacks (by a television station).

# 16    Teleconferencing and Computer Conferencing

## TELECONFERENCING INTRODUCTION

In Chapter 1 you wrote a note to a friend about the New York Yankees. Now, you want to celebrate another World Series win.[1] Do you write another note, use the telephone, or hold a personal videoconference so you can also see each other?

The latter falls under the umbrella term *teleconference*. A teleconference is an electronic meeting between two or more sites. It can range from an audioconference, where you hold an interactive conversation, to a videoconference, where video information can also be exchanged.

Another type of meeting, a computer conference, is an extension of the systems presented in the Internet chapter, Chapter 17. A computer conference can range from an exchange of pictures between two people to dedicated networks. It is placed in this chapter for organizational purposes, and for our discussion, the focus is on its electronic meeting characteristics.

## VIDEOCONFERENCES

There are different types of videoconferencing. In a two-way videoconference, you can hear and see each other through cameras, monitors, microphones, and speakers. In a one-way videoconference, the information exchange is a one-way audio and video delivery system. But there's an option for interaction through a telephone or a fax connection. In a typical meeting, a party receiving the information could ask questions by telephone.[2]

Videoconferences have also been either motion or nonmotion. Motion implies movement, but the capabilities vary. You can appear on a television screen in a lifelike manner. Or your motions may be jerky, and the picture quality may be poor.

In a nonmotion videoconference, a series of still images is relayed. Even though the visual element is not lifelike, an audio hookup could support a conversation.

Regardless of its form, a meeting can be held over satellites, telephone lines, and other communications channels. A company may also own a private teleconferencing network, or a public room could be rented for the occasional meeting.

### Two-Way Videoconference

In a two-way videoconference, the various parties can see and hear each other, and its primary advantage lies in the replication of a face-to-face meeting. The participants or conferees can react to each other's body language; valuable visual clues in interpersonal transactions.

A room can also be designed with this goal in mind. For example, the table where

INFORMATION, ENTERTAINMENT & COMMUNICATION SYSTEMS

**Figure 16.1**
*A teleconferencing setup. Note participants in this room (foreground) and the shot of a conferee at another location (background screen). (Courtesy of U.S. Sprint.)*

the conferees sit can be shaped to maximize their view of the room's monitors. The object is to promote good eye contact between the different parties.[3] Strategically placed cameras can produce various shots of the conferees, and an additional camera(s) can shoot graphics and other supporting materials. Proper lighting and acoustics are also important.

Before the development of dedicated facilities, a meeting may have been hampered by inadequate equipment and design considerations. A large office may have been converted into a videoconferencing room by simply adding the necessary communications equipment. Thus, lighting and other production considerations, the human element that would have made people more comfortable, may have been overlooked.

Another past and current problem has to do with the notion of appearing on camera. Some people get nervous or may think a videoconference is an unnatural way to communicate.

Finally, two-way videoconferences had generally been used for point-to-point meetings, uniting two sites. But advances made it possible to take advantage of a multisite capability, and this has become a

popular option. You can see and talk with people scattered across, in one case, the country.

## One-Way Videoconference

The one-way videoconference, which has also been called business television among other names, is typically relayed via satellite from one location to multiple sites (point-to-multipoint). One-way videoconferences have generally been analog, but digital compression techniques, when combined with satellite delivery, have made for effective relays. In these situations, though, the picture quality may be more of a concern than with a two-way setup, where a lower quality image, but a higher compression ratio, may be more acceptable.[4]

In a typical application, the audio and video information can be a one-way stream from a corporate headquarters to its branch offices. These sites may communicate back with a telephone or other hookup for question and answer periods and discussions.

In light of its point-to-multipoint capabilities, dedicated videoconferencing networks have supported the one-way videoconference and organizations that lease satellite time and the prerequisite facilities.[5] One-time or occasional videoconferences, also called special events or ad hoc videoconferences, have also been held.

This electronic meeting's power has been vividly demonstrated in medical videoconferences. Through this real-time process, doctors can view a new surgical technique performed at another site.

## Nonmotion Videoconference

In a nonmotion videoconference, still images are delivered over voice-grade or faster channels. The relay time can vary, depending on the communications line and the image's resolution, and either dedicated or PC-based configurations have been used.

A nonmotion videoconference has some advantages. A system can be relatively inexpensive, and the transmission cost could be reduced to that of a telephone call. It is also suitable for situations where visual information must be exchanged, but not necessarily moving pictures. These have included X rays, other medical imaging output, and illustrations for the medical and educational fields.

The former, an element of the *telemedicine* market, is particularly well suited for remote geographical locations. If a doctor is not available, pertinent information can be relayed to a distant hospital for review. In another scenario, an aging population and too few doctors and specialists have made telemedicine a potential ally. Applications may include setting-up a monitoring system in a patient's home as well as high-speed relays where complex data could be rapidly exchanged.[6]

It should be noted that telemedicine also embraces motion videoconferencing, among this chapter's other categories. The military is also experimenting with systems, originally pioneered by NASA, to remotely monitor the medical status of their personnel.[7] This type of development may be extended to civilian applications.

## AUDIOCONFERENCES

An audioconference can be thought of as an extended telephone conversation. But instead of talking with only one person, you may be talking with several or more people. In this operation, multiple sites can be connected through a teleconferencing bridge.

An audioconference is a satisfactory communications tool in many situations. It is relatively inexpensive and is supported by its own array of equipment, including a micro-

phone that may serve several people through a 360-degree pickup pattern.

Audioconferencing can also handle numerous applications. It can create an inexpensive communications link between a journalism class and a sports reporter, who may live and work 200 miles away. It can also support business meetings and national political campaigns.

The primary criterion for adopting audioconferencing, or one of the visual systems, is an organization's needs. In many cases an audioconferencing link may suffice.

## OTHER CONSIDERATIONS

The teleconferencing field has grown over the years. In fact, it surged in the early 1990s when product and service sales for the videoconferencing sector alone rose from $127 million to $510 million in 1987 and 1991, respectively.[8] An accelerated growth pattern has been projected through the twenty-first century, particularly in the audio and Internet-based arenas. Some of the reasons for this growth are in the following subsections.

### Standards

A standard, H.261, also known as Px64, has resulted in a level of compatibility between manufacturers to help facilitate the flow of information.[9] Other standards promoted videoconferencing's growth over communications channels ranging from the ISDN (H.320) to the "plain old telephone system" (POTS).[10] Even though we're hurling toward a digital world, the analog telephone infrastructure is still important. By videoconferencing over standard telephone lines, you could tap the large, existing network.[11]

The most dramatic growth, however, is fueled by H.323, which "uses the Internet to send and receive data . . . H.323, a set of standards for multimedia conferencing over

INFORMATION, ENTERTAINMENT & COMMUNICATION SYSTEMS

packet-based networks, describes the . . . equipment and services required for multimedia conferencing."[12] This development has also fueled the use of a company's internal network for such operations, with some possible system changes and enhancements.

However, a problem can arise once you leave your internal (e.g., a company) network and connect with the public Internet. The company-run network typically has a greater quality-of-service control (QoS). QoS refers to the "ability to define a level of performance in a data communications system."[13] In essence, while a videoconference may look great within your company, once you leave this environment, the quality could deteriorate. Other factors also play a role, and by using special third party networks and options, the overall quality could be enhanced.

Another important factor has been the introduction of more flexible, dual-mode systems, that are H.320 and H.323 compliant. In one case, a company with an investment in ISDN-based systems could also extend its capabilities with this type of equipment.[14]

### Transmissions

The industry has benefited from lower transmission and equipment costs. It is less expensive to hold a videoconference, and the quality has improved.

It is also easier to hold a meeting. New equipment, standards, and an improved communications infrastructure have made this application more user accessible and transparent. These include faster communication channels and systems that are easier to set-up and use than their predecessors. Enhanced compression techniques also play a vital role in this development.

### Flexibility

Teleconferences can handle more types of information. In one case, computer and conventional videoconferencing capabilities have been married. In a multimedia videoconference, graphics and computer data can be exchanged, documents can be electronically annotated, hard copies can be printed, and simplified interfaces can be set up.[15] The goal is to provide users with the same tools they might use in a face-to-face meeting.

Teleconferencing is also becoming more flexible as the industry continues to mature. Organizations can rent public rooms to experiment with videoconferencing or for occasional use. Satellite and fiber-optic relays are now widely available and networks accommodate domestic and international relays.[16]

A company can also use a portable and less expensive roll-about unit to hold a videoconference. This unit can be moved from room to room and eliminates the need for a dedicated facility. However, its aesthetic elements could be enhanced if the unit is used in a board or conference room with the proper lighting and acoustical treatment.

### The Internet

The Internet is expanding at a dizzying pace and, as stated, this environment can support teleconferencing. One pioneering product, CU-SeeMe, offered users a low-cost videoconferencing option via a PC.[17] The quality was poor, compared to television standards, but it did provide the Internet community with an inexpensive teleconferencing tool.

More important, this type of product pointed the way to more sophisticated applications. In short, the Internet has emerged as the world's largest teleconferencing environment. This is particularly true as new techniques, that support real-time audio-video relays, are developed and widely adopted.

The Internet will also emerge as a more potent teleconferencing tool with the widespread implementation of Internet2.[18] This environment supports enhanced audio-video relays and conferences and, at some future date, could help make videoconferencing as ubiquitous as making a telephone call. But as described in the next section, this development in the personal videoconferencing arena, which for our purposes focuses on personal/home rather than business use, will also depend on a human factor.

## PERSONAL VIDEOCONFERENCES

Videoconferencing has come to the desktop. In the early 1990s, a number of desktop configurations were introduced. These devices can trace their origins to AT&T, which experimented with the Picturephone transmission standard and the Picturephone, its hardware component.

The Picturephone was a compact desktop unit that incorporated a small television monitor, camera, and audio components. Even though it was a pioneering service, it did not catch on in either the business or private sectors. The transmission and hardware costs, a Picturephone user's possible reluctance to appear on camera, and the inadequate reproduction of printed documents hampered sales.

Since 1964, when AT&T publicly demonstrated a video telephone at the World's Fair, newer products have been introduced.[19] These include consumer-oriented nonmotion units, which relayed still black-and-white images over telephone lines, and newer motion systems.

The latter have ranged from PC-based to stand-alone desktop models. Analog and digital transmission schemes have been supported, and consumers and businesses have been targeted.

While personal videoconferencing may be inevitable, the time frame for its broad acceptance is unknown. The quality, as of this writing, may still be unacceptable to some users, and the human factor must be considered.[20]

New transmission lines, as outlined in previous chapters, may support faster relays, and equipment costs may continue to fall. But systems geared for the consumer may run into the same problem faced by the original Picturephone—many people may not want to appear on camera. Think of the way you use the telephone. When talking, are you always attentive? Or are you also working on a computer, eating, watching TV, or as indicated in research that evaluated the Picturephone, reading, doodling, or even yawning?[21] While you could use a system in a voice-only mode, would the person you're talking to be insulted?

This type of system calls for a different mind-set. It may only be fully accepted when a generation of children actually grows up with the technology, and it is a part of their everyday lives. At that time, appearing on camera will just be old hat.

## ADVANTAGES OF TELECONFERENCING

As a communications tool, teleconferencing has advantages, regardless of its form.

1. It can promote productivity. By linking an organization's offices, meetings can be held either as needed or on a regular basis. With a private setup, a company could also meet on short notice to respond to crisis situations.[22]
2. Employee morale can be boosted. Management can use a teleconference to keep employees informed about relevant events.
3. Guest lecturers have electronically met with classes while businesses have engaged consultants, especially those who may be unable to travel to a given site.
4. Reducing travel can save time and money.

INFORMATION, ENTERTAINMENT & COMMUNICATION SYSTEMS

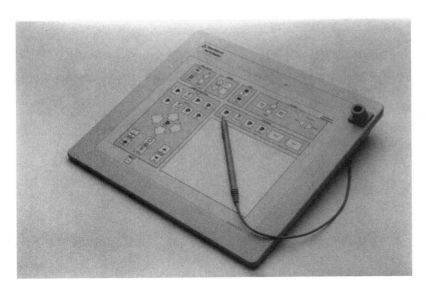

**Figure 16.2**

*More transparent and easier-to-use control interfaces, such as the one depicted, has simplified teleconferencing operations. (Courtesy of VideoTelecom Corp.)*

Your job may call for extensive traveling, which normally entails a drive to the airport, waiting and flight time, another drive to the meeting point, and the return flight. This itinerary may then be repeated. In today's world, this wasted time and energy are two resources that can never be recouped. The financial cost can also be prohibitive.

Teleconferences can help alleviate these problems. If an electronic meeting could be arranged between East and West Coast offices, the same information could potentially be exchanged through a teleconference, rather than a series of trips.

But teleconferencing will not eliminate all travel. Face-to-face communication is still important. For example, if you are learning how to use a complicated piece of equipment, an on-site seminar may be appropriate. Some people also prefer face-to-face meetings. Or at the very least, the first meeting is conducted in person, which could be followed up by teleconferences.

Another advantage is that, as with business travel, you could save time and money by conducting a job interview via a teleconference. You may also be able to interview more candidates.[23]

In sum, the teleconference is emerging as another tool that organizations and individuals can tap into to improve their communication capabilities. These advantages also extend to the international market. Since many companies are now global in scope, teleconferencing can help unite these geographically distant sites. This capability cannot be overemphasized because international considerations continue to play an integral role in domestic operations.

The development of PC-based and inexpensive teleconferencing outfits, which may become standard office and home fixtures, may also change this meeting from a special to an everyday event. Faster and more cost-effective communications channels, with new hardware/software releases, may accelerate this trend.

## COMPUTER CONFERENCING

The term computer conferencing can describe a computer-based meeting. For our discussion, a conference can range from an exchange of pictures to dedicated networks that may link multiple domestic and international users (e.g., via the Internet).

In one application, a PC-based configuration could unite national real estate offices. An office in Denver could send pictures of houses to the New York branch for review by a client. A telephone conversation could supplement this exchange.

Computer-based systems can also support conferencing. An advantage of this operation is its availability. Scattered groups of people can gain access to the system, and the company provides the prerequisite organizational and communications infrastructures. In a different setting, an organization may opt to create its own internal network. Meetings can be rapidly convened and a variety of information can be exchanged.

In one growing application area, individuals can collaborate on a document through data conferencing. People share the same file for editing. When working on complex projects, this capability could speed up and help support the writing process.[24] In this light, it could be viewed as an interactive extension of e-mail.

Hardware and software have also supported computer conferencing in real- and nonreal-time situations. Real-time, in this context, implies that information can be sent and received as you interact with the system and other people. The nonreal-time elements may encompass a series of longer messages, a central database of information, and a record of comments all the conferees can see.

## Telecommuting

Computer conferencing can also be held between an individual and an organization, as is the case with telecommuting. As defined by Osman Eldib and Daniel Minoli, telecommuting is "the ability of workers to either work out of their homes or to only drive a few minutes and reach a complex in their immediate neighborhood where, through advanced communication and computing support . . . they can access their corporate computing resources and undertake work."[25] As further described by Eldib and Minoli, telecommuting can also be associated with the virtual office, covered in another chapter.

For our present discussion, if you are a telecommuter, you can work at home and maintain contact with the office by computer. You can be an employee, a consultant, or a chief executive officer.

Telecommuting is becoming a popular work option. It can, for example, be cost and time efficient. Think about your own commute, especially if you have to travel on an overextended highway or mass transit system.[26] Telecommuting would reduce the frequency of, or possibly eliminate, this daily subway, bus, or car trip.

Telecommuting also gives you more options. You could choose to live further away from work or take a geographically inaccessible job.

Telecommuters may also be more productive and experience a higher level of job satisfaction. There are also corresponding social benefits, such as fuel savings, air-pollution reduction, and potential child care improvements.[27]

Several factors have combined to make telecommuting a viable alternative. These include faster communications lines, the proliferation of PCs, and the information age. A job in the information sector could potentially be completed at home.

Yet everyone is not happy with this development.

You may be uncomfortable with this work environment, and there is a potential for abuse—you could be treated as a contract worker instead of a salaried employee with its attendant benefits. Lower wages may also be earned. You could be cut off from the "informal communication channels" and be "less well integrated into an organization's structure and culture."[28]

Another issue flared-up concerning work at home safety conditions. The U.S. Labor Secretary wrote a letter to an employer stating that "companies have the same responsibilities for employees working at home as in the office."[29] The letter was withdrawn after criticism from the business community. But it did draw into focus two contrary viewpoints:

1. An employer should be liable for safe, working conditions in an employee's home. The goal is to essentially provide these employees with the same rights as those who work in a traditional setting.

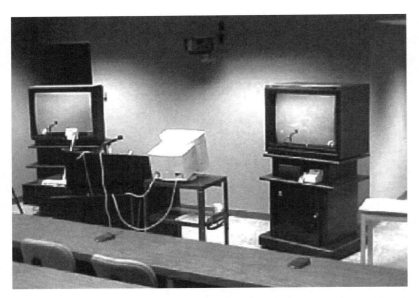

**Figure 16.3**
*An example of a videoconferencing room used for applications ranging from holding courses to interviews to meetings. The monitors are used to display the participants at other sites; microphones are placed on the tables for each participant. (Courtesy of Marywood University.)*

2. Since any employer does not have direct control over an employee's home, such a regulation would be unfair and impossible to enforce. It could also curtail the field's growth.[30]

A final question focuses on another issue—does telecommuting help reduce human communication to a machine-dominated format? This could have an adverse effect on the socialization process since the interpersonal relationships we develop at work may play an important role in our lives.

Isaac Asimov, the late science fiction writer and science authority, painted an interesting portrait of such a society. Communications technologies have contributed to the birth of a rigid social structure where personal human contact is avoided. In Asimov's world, even casual meetings are conducted through lifelike 3-D visual relays.[31]

Could our own society be similarly affected? What if telecommuting is weighed with other technologies that enable us to shop at home, to create sophisticated home entertainment centers, and to view a museum's collection via the web?

## Education

The concept of telecommuting has also been extended to education. You can register for a distance education course, which may be delivered by computer or a videoconference, among other options. Like "real" school, you are assigned a teacher and textbooks. But instead of taking notes in class, lectures can be downloaded, and your teacher may be available for electronic office hours.

Students can also expand their course options since geography and time do not limit them. If you do not live near a school, and work during the day, it may still be possible to pursue a degree. Instructors for their part, can reach a broader student body and can tap into the Internet and other electronic resources when designing a course.

Similarly, the Internet has emerged as an important conduit for this application. Various tools, including the WebCT software package, support real-time interaction in meeting rooms as well as a range of data types ranging from documents to digital media clips.[32]

Courses can also be delivered by compressed video. This set-up more accurately duplicates a classroom situation since you can see and interact with an instructor. But it does carry a price. Much like a videoconferencing room, you have to set-up a suitable environment with monitors, microphones for student responses, and possibly, a document camera and computer. An instructor may also have to learn how to teach to a camera/monitor in addition to other prep work.

This concept has also been applied to the Internet. You can utilize this environment to deliver audio and video information to students located at distant sites through inexpensive webcams.

• Connect the webcam to your computer.
• Go online to the appropriate site.

- You can now see the instructor on your monitor as he or she views you.

One of the downsides, as of this writing, may be the image's small size and limited quality. But like similar applications, faster communications lines, enhanced compression methods, and hardware/software improvements will improve a relay's quality.

The use of these tools is not solely limited to educational institutions. Companies have adopted the equipment and techniques to support, in one application, employee training. Besides reaching what may be a geographically scattered sales force, this environment can help ensure that individuals in different locations can "train simultaneously, eliminating knowledge lag between various far-flung sites."[33] In essence, a company's personnel will receive the same training at the same time.

## CONCLUSION

Teleconferences and computer conferences have emerged as potent communications tools. They serve organizations, and through personal videoconferencing, may have an impact on our daily lives.

We may also have direct contact with such systems through telecommuting and by taking electronic courses. Although both applications are not without their faults, they may enable us to work at a geographically removed site and to attend a class that may, just a few years ago, have been impossible to take.

## REFERENCES/NOTES

1. FYI—As of this writing, the Yankees have won 26 World Championships; a major league baseball record.

2. Bill Dunne, "Screening the Confusion Out of the Different Types of Videoconferencing," *Communications News* (November 1985), 34.

3. David B. Mensit and Bernard A. Wright, "Picturephone Meeting Service: The System," in *Teleconferencing and Electronic Communications: Applications, Technologies, and Human Factors* (Madison, WI: Regents of the University of Wisconsin, 1982), 174.

4. Clarke Bishop et al., "Compressed Digital Video for Business TV Applications," *Communications News* (December 1991), 40.

5. A network can support educational and business applications, among other options. See the International Teleconferencing Association, "ITCA Teleconferencing Definitions," Guide to Membership Services, brochure.

6. Laura Robinson, "Japan's New Telemedicine Momentum: Imaging Opportunities?," *Advanced Imaging* (May 1999), 15.

7. John Rhea, "Telemedicine Ties into Tactical Communications Infrastructure," *Military & Aerospace Electronics* 8 (April 1997), 11–13.

8. The International Teleconferencing Association, "Business Television Private Network Market Reaches $606 Million," press release, February 1992, from a survey by TRI and IFC Resources, Ltd., of 100 companies responding to the impact of videoconferencing on their operations.

9. "Video Communications Comes of Age with H.261 Standard," *Communications News* (February 1991), 10.

10. Andrew W. Davis, "Videoconferencing via POTS Now: Proprietary Codecs and Emerging Standards," *Advanced Imaging* 10 (June 1995), 32.

11. *Note:* The quality may be lower.

12. Catherine Murphy, "The Megaconference," *Syllabus* (August 2000), 27.

13. TechEncyclopedia. QoS, downloaded from www.techweb.com/encyclopedia/defineterm?term=QoS.

INFORMATION, ENTERTAINMENT & COMMUNICATION SYSTEMS

14. Two manufacturers of such equipment, as of this writing are: Vtel and Polycom at www.vtel.com and www.polycom.com, respectively. You can visit their web sites for more information.

15. Conversation with VideoTelecom Corporation, summer 1992, developer of such a teleconferencing system.

16. Sprint Communications, "Sprint Meeting Channel-Rates and Services," brochure.

17. Davis, "Videoconferencing Via POTS Now," 36.

18. Please see the Internet chapter, Chapter 17, for more information about Internet2.

19. Patrick Portway, "The Promise of Video Telephony Made in 1964 Finally Is Being Fulfilled at Reasonable Prices," *Communications News* (February 1988), 35.

20. Elliot M. Gold, "Trends in Desktop Video and Videophones," *Networking Management* (May 1992), 46.

21. Howard Falk, "Picturephone and Beyond," *IEEE Spectrum* (November 1973), 48.

22. The International Teleconferencing Association, "Business Television Private Network Market." Brochure.

23. Since you saved time and money.

24. Herman Mehling, "The Ties that Unbind," *Information Week* (November 6, 1995), 58.

25. Osman Eldib and Daniel Minoli, *Telecommuting* (Norwood, MA: Artech House, Inc., 1995), 1.

26. Ibid., 4.

27. Since a parent works at home.

28. Robert E. Kraut, "Telecommuting: The Trade-Offs of Home Work," *Journal of Communication* (summer 1989), 43.

29. Fran Calpin, "Work at Home Injuries," *The Sunday Times*, Section C, (February 20, 2000), 1.

30. Ibid.

31. Isaac Asimov, *The Naked Sun* (New York: Ballantine Books, 1957), 46.

32. WebCT is used to create interactive, online courses.

33. Tracy Mayor, "E-Learning: Does It Make the Grade," CIO 14 (January 15, 2001), 114.

## SUGGESTED READINGS

Carr, Sarah. "Is Anyone Making Money on Distance Education?" *The Chronicle of Higher Education* (February 16, 2001), A41–A43; Peter A. Concelmo. "Business Television Blossoms with Distance Learning." *Satellite Communications* 21 (October 1997), 20–30; Kim Kiser. "Ready for Lift Off." *Online-Learning* 5 (April 2001), 38–42. Distance learning applications; the first article also examines how educational institutions have measured their distance education programs' financial success (or not).

DeMaria, Michael J. "Making Your Bandwidth Count." *Network Computing* 13 (October 21, 2002), 85–88; Bob Doyle. "How Codecs Work." *New Media* 4 (March 1994), 52–55. Technical information about holding a videoconference and a description of codecs/compression, respectively.

Fritz, Mark. "Light Waves." *EMedia Magazine* (November 2001), 44–52. An excellent and comprehensive review of videoconferencing trends and technical implementations.

Halhed, Basil R., and Lynn D. Scott. "Videoconferencing Terminology." *Video Systems* 18 (May 1992), supplement, 8–10. A list and accompanying definitions of the more important terms in the videoconferencing field.

Kames, A. J., and W. G. Heffron. "Human Factors Design of Picturephone Meeting Service." A paper presented at Globecom '82, sponsored by the IEEE, in Miami, FL, November 29 to December 2, 1982. A review of an AT&T videoconference service. The paper also discusses the tests that were conducted to judge the system's effectiveness as a communications tool.

Lundsten, Apryl, and Robert Doiel. "Digital Video Growing." *Syllabus* (August 2000), 12–16; "Collaborative Video Conferencing Services over Internet2: Concept and Preliminary Service Description." White Paper of the Internet2 Digital Video Conferencing Subcommittee; November 9, 2000. Internet 2 and teleconferencing applications and issues.

Nisenson, Kyle. "Tune in to IP Videoconferencing." *Network World* 15 (September 21, 1998), 43–45; Christine Olgren. "Trends in the Use of Teleconferencing Illustrate the Wide Variety of Applications That It Offers." *Communication News* (February 1988), 22–26. Two articles about teleconferencing; the second article presents an interesting view of the state of the industry in the late 1980s.

Schooley, Ann K. "Allowing FDA Regulation of Communications Software Used in Telemedicine: A Potentially Fatal Misdiagnosis?" *Federal Communications Law Journal* (May 1998) 50 Fed. Com. L.J. 731, downloaded from LEXIS; Delbert D. Smith. "Distant Doctors." *Satellite Communications* 22 (May 1998), 32–40; Jon Surmacz. "Long Distance Medical Call," downloaded from www.darwinmag.com. Telemedicine and legal implications.

Videoconferencing Cookbook, V. 3. This online document is an excellent source for videoconferencing information and related topics. It can be accessed at www.videnet.gatech.edu/cookbook.

## GLOSSARY

*Audioconference:* A form of a teleconference. In an audioconference, individuals at two or more sites can speak to and hear each other.

*Computer Conference:* A meeting conducted through computers that can support the exchange of information ranging from text to graphics.

*Motion Videoconference:* A motion videoconference can duplicate a face-to-face meeting. In a two-way, motion configuration, for example, the participants can see and hear each other.

*Nonmotion Videoconference:* A nonmotion videoconference supports the relay and exchange of still pictures.

*Personal Videoconferencing:* Personal videoconferencing is a general term that describes desktop videoconferencing systems.

*Telecommuter:* An individual who works at home and maintains contact with the office by computer.

*Teleconference:* An umbrella term for the various categories of electronic meetings and events, ranging from audioconferences to videoconferences.

*Videoconference:* A teleconference wherein, as implied by the name, video or visual information is exchanged.

INFORMATION, ENTERTAINMENT & COMMUNICATION SYSTEMS

# 17 Information Services: The Internet and the World Wide Web

## INTRODUCTION

Jules Verne's *Around the World in Eighty Days* describes Phineas Fogg's momentous journey. Today, we can complete the same trip in seconds via the Internet. In brief, the Internet can be described as a global data highway. You can travel on this electronic road to exchange information with sites scattered across the globe.

The Internet can trace its roots to the late 1960s. It started as a U.S. government project, the Advanced Research Projects Agency Network (ARPANET). Designed, in part, to experiment with and to demonstrate decentralized computer networking, ARPANET eventually evolved into the Internet structure.[1]

The Internet also remains a decentralized entity. It can be viewed as a collection of independent computer systems that no one individual or organization owns. It is almost like the Wild West; an information frontier without boundaries that is primarily governed by technical standards.[2]

The Internet is also an evolving network. New information pools become available as additional computers are linked to it. Internet contributors include government agencies, profit and nonprofit organizations, educational institutions, and individuals. You can read government documents, send and receive e-mail, browse through the Library of Congress, join discussion groups, and visit personalized information sites.

Internet-based operations have also virtually eliminated geographical and time-based constraints. You can retrieve information from around the world, 24 hours a day. In an earlier demonstration of this capability, we witnessed the comet Shoemaker-Levy 9 as it collided with Jupiter during the summer of 1994. This once in a lifetime event created a scientific and public sensation that was particularly felt on the Internet.

As comet fragments slammed into the planet's atmosphere, images were made available and were retrieved from Internet sites. The demand was so great that primary and even mirror sites, which stored duplicate image files, were overwhelmed by requests. One estimate placed the number of downloaded images at the two million mark in a relatively short time.[3]

Prior to the Internet, this level of accessibility was not possible. But now you can assemble your own image library and participate in the event in almost a real-time sequencing, that is, as it unfolds.

The collision also demonstrated how the Internet could work with more traditional communications venues. The PBS television station WHYY ran a live program that tied an Internet link with "satellite dishes and a videophone to bring together multiple images and experts for an interactive program shared with other PBS stations around the country."[4] The different media provided viewers with a broad over-

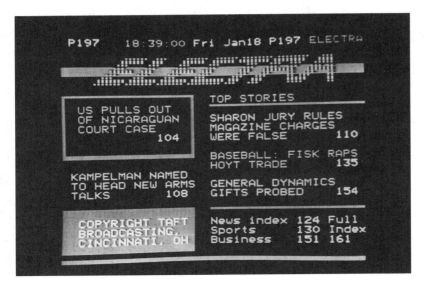

**Figure 17.1**

*An index page of a teletext magazine. (Reprinted by permission of ELECTRA, a part of Great American Broadcasting; Electra.)*

INFORMATION, ENTERTAINMENT & COMMUNICATION SYSTEMS

view of this event in addition to expert commentary.

## OTHER SYSTEMS

Before we explore the Internet, it is important to note that it is not the first or only electronic information resource. Two precursors were teletext magazines and videotext systems.

A *teletext magazine* is an electronic publication delivered to a television set via a television signal.[5] Composed of text and graphics, it typically featured news, sports, and other stories. Subscribers generally received the same information, and the only option was the page you viewed at any given time.

Teletext services only achieved limited success in the United States. Competing technical standards, cost factors, and other issues contributed to a flat market.[6] They were, however, more successful in England and other countries.

In contrast with a teletext service's one way information stream, an interactive system performs as the name implies. You can interact with the system to request and receive specific information and services.

This family can also be divided into two general classifications: videotext and PC-based operations. For our discussion, PC-based systems describe both dedicated and more consumer-oriented, AOL-type operations. The former also targets narrow subscriber groups with specialized information pools.

## Videotext and PC-Based Operations

A videotext service is a graphics-oriented, interactive operation. You could tap a database of thousands of frames or pages of information through its two-way capabilities and a telephone line, the prevalent communications channel for such two-way systems.

An easy-to-use interface and hookup have distinguished a videotext service. Consumers were the primary target, and different terminals were developed for this connection. A typical configuration consisted of a keypad, decoder, and television set.[7]

These services also helped pioneer simplified user interfaces, gateway options to gain access to computers outside of the host system, and the inclusion of *Information Providers*, an organization that contributed information to a database. An airline, for instance, typically provided flight, seating, and pricing information.[8]

Like the teletext industry, the U.S. videotext industry failed to mature. But the systems found wider acceptance in other countries. These included Prestel (England) and Telidon (Canada).

In contrast, AOL, CompuServe, and other contemporary PC-based counterparts were more successful. A partial explanation may lie in the targeted user groups. For example, earlier PC-based systems were geared toward more focused audiences, people who were willing to pay for a service for specific reasons. This principle was particularly true for dedicated systems that were designed for lawyers and other select groups with desirable information pools.[9]

***Evolution.*** As of this writing, PC-based systems coexist with the Internet for various reasons, including the following:

- PC-based systems typically support proprietary information pools and may make it easier to navigate the Internet to locate specific information.
- Certain information may be available via the Internet and not through these systems. Consequently, you may use both to meet your information needs.
- The Internet's popularity made it the center of a fierce competition in the mid-1990s. AOL and other companies subsequently rushed to offer their subscribers Internet access, especially access to the World Wide Web (WWW), or web, for short. In these cases, you could "have your cake and eat it too"—you could gain access to the proprietary services/information and the Internet.

During the same general time frame, an electronic cottage industry was also born. Entrepreneurs recognized that some people actually wanted Internet access without the other offerings provided by AOL-type organizations. Thus, they launched companies that primarily served as communications conduits. If you owned a PC and modem, they provided the hookup and necessary software.

These companies also helped open up cyberspace to more people. *Cyberspace*, a term popularized by William Gibson's science fiction work *Neuromancer*, can be described as a computer-generated environment or world where you can interact with other people and work and play. The creation of cyberspace also spawned a collection of complementary words and activities. A March 1994 *CompuServe Magazine* article presented such a list, most of which are self-explanatory. A sample includes Cyberart, Cybergames, Cybersex, and Cyberschool.[10]

Finally, one of the most successful interactive information systems to date has been France's Teletel. To promote this national service and to lower printing costs, the telephone directory was installed on the network. The government subsequently distributed free Minitel terminals to French citizens. These factors contributed to its impressive growth.

For more information about teletext, videotext, and PC-based services, please see the references in the Suggested Readings section of this chapter.

## WORLD WIDE WEB

### Introduction

The web is a product of the Swiss-based CERN research center. Pioneered by Tim Berners-Lee to help facilitate the exchange of information, its popularity had soared by the mid-1990s.

As outlined by Eric Richard, the web could be described as "a collection of protocols and standards used to access the information available on the Internet . . . [which is] the physical medium used to transport the data." The web has been primarily defined by three standards:

URLs (Uniform Resource Locators), HTTP (HyperText Transfer Protocol), and HTML (HyperText Markup Language). These standards are used by WWW servers and clients to provide a simple mechanism for locating, accessing, and displaying information available through other common network protocols. . . . However, HTTP serves as the primary protocol used to retrieve information via the Web.[11]

In essence, the web can be viewed as an overlying net that, as described by Richard, is used to mine or gain access to the Internet's information pools.[12] You actually travel across the web via a *browser*, software used to visit different sites and to conduct other activities. These can include engaging in business, searching online catalogs, conducting financial transactions, retrieving scientific data, and reading movie reviews.

Prior to the web, you typed a series of commands to gain access to information and to conduct other Internet-based operations. In the case of the File Transfer Protocol (FTP), you initiated a sessions by typing *ftp*, followed by an electronic address. Once connected, you could use keywords to retrieve files, such as the aforementioned comet images. Additional tools include telnet and gopher.[13]

For some, these tools made the Internet a readily traversable environment. For others, the Internet was still a hostile territory that you could only cross with a handful of instruction books, a series of arcane commands, and a prayer.

But the second perception greatly improved with the web's growth. Unlike most traditional Internet tools, you could explore the information universe with software that sported an intuitive interface. As a typical user, you could throw the instruction books away.

***HTML and URL.***    The HTML specification is used for document formatting and design. It employs tags to format text and to delineate a document's appearance. In two examples, the ⟨P⟩ and ⟨H⟩ tags, respectively, mark a new paragraph and a text heading level (on-screen text size).

HTML is also used to embed links to other documents and Internet services.[14] These links, and the web's underlying structure, are based on a concept you are already familiar with, hypertext. You navigate across the web, retrieving information and visiting new sites, via this tool.

The URL, for its part, is an addressing system. It can identify a document's loca-

tion. A typical URL is http://www.ac. anyschool.edu/newcomm.html. When you activate its link, you will retrieve the newcomm.html document at the ac.any school.edu site.[15] Three of the letters specify, in this case, an educational (edu) rather than a commercial (com) or other type of organization/institution. As you create a web page, you can forge links to other information by using URLs embedded in your document.

***Browsers and Search Engines.***    As indicated, you explore the web with a browser, a graphical rather than a text-based tool. Analogous to a graphical user interface (GUI), you use a mouse to point to and select different functions. But with the web, you can retrieve information or travel around the world. It is a visual interface that also supports graphics and other visual information.

Different browsers have been released. Mosaic, a product of the National Center for Supercomputing Applications (NCSA), helped define this software field. It also helped make the web the Internet's hot spot. As of this writing, Netscape and Microsoft's Internet Explorer dominate the industry.[16]

But regardless of the system you use, they generally share some common elements:

- Drop-down menus with commands.
- Hypertext link ID system. Hotspots or links are color coded for easy identification.
- A point and shoot interface like GUIs. To complete an operation, move the cursor and click a button.
- Navigational buttons you can activate to go, for instance, back to the previous link (e.g., a different document and/or web site).
- An option to view a page's underlying HTML code. This is a good way to learn how to create your own documents.

- Extensibility: As the web evolves and new media types are introduced (e.g., audio format), your browser should be able to accommodate them through "plug-ins"—software that extends your browser's capabilities.

Using this tool, you can electronically wander around the web for days looking for specific information. Web search mechanisms can help by identifying and locating resources. You can find government documents or locate a site that covers Phil Ochs or another folk singer.

One of the most popular search mechanisms has been Yahoo. A comprehensive service, Yahoo features a keyword search option as well as broad subject categories that can facilitate this operation. Other systems have supported queries in the form of a question. You type a question, and based on its criteria, potential web sites that "answer" the question are listed for your perusal.[17]

Another popular search engine is Google. It sports an enhanced searching capability and user interface that can speed up a search.

## INTERNET AND WEB GROWTH

### Some Factors

Several reasons for the Internet's growth are listed in the following subsections. They also provide a framework and an historical perspective for tracking this infrastructure's technological maturity.

*Interface.* The web has provided users with a familiar and intuitive interface. PC owners are already familiar with GUIs and, as indicated, graphical browsers extend this metaphor to the Internet.

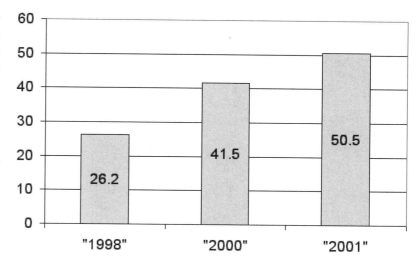

Browsers also help make Internet access transparent. Like a telephone, you do not have to think about or fully comprehend the web's underlying technologies or structure to use it.

*Real-Time Interaction.* The web's real-time interaction can save time. Prior to the web, you generally relied on file descriptions to identify data. The problem? If this information was inaccurate or imprecise, you may have spent valuable time to retrieve a file that was actually unsuitable for your purpose.

With the web, you may be able to listen to and view specific data types, such as a graphic, before downloading. Most sites use small, low-resolution "thumbnails," pictures that are linked to larger, higher resolution images, as visual guides. You look at a thumbnail, and if it suits your needs, you retrieve its counterpart with the click of a mouse button.

Similarly, audio and video retrieval became a real-time event. Real-time, for this discussion, signifies you do not have to download an entire file before you can hear or see it (e.g., video).[18]

Various products support this *streaming* of video and audio information over the Inter-

**Figure 17.2**
*Percent of U.S. households with Internet access for 1998, 2000, and 2001. (Source: NTIA and ESA, U.S. Department of Commerce, using U.S. Bureau of Census Current Population Survey supplements.)*

INFORMATION, ENTERTAINMENT & COMMUNICATION SYSTEMS

net. They range from streaming appliances, or self-contained units that may simplify an operation, to PC-based systems. As of this writing, Apple Computer, Microsoft, and RealNetworks are dominant players in the PC market. Their respective players provide users with the software to view this information, among other media types. The other major components in this process are a server, which "delivers streams to audience members," and an encoder that "converts raw audio and video files into a format that can be streamed."[19]

The actual content, the programming, could be stored on a computer and subsequently retrieved and viewed by multiple users. In this case, as described in a prior chapter, audio and video depositories could be created and subsequently tapped by users on an on-demand basis.

In another setting, a meeting could be viewed live—as it actually occurs. The latter is also emerging as a new production form with new rules of the game. As an independent producer, you will also have to be computer savvy and work through some of the medium's potential shortcomings. This may include a limited channel capacity that, in turn, may have an impact on the videos perceived quality.

Regardless of the system, the capability to deliver real-time audio and video is an important achievement. Additional applications include the following:

- The launching of Internet-based radio stations. In one scenario, a radio station can simultaneously distribute its programming via the Internet. Although the audience is limited to online users, you can reach the international community with this mechanism. Web surfers, for their part, can pick and choose the materials they want to hear and view.[20]
- Advertising and running movie and television promotions.

- Online demonstrations.
- Delivering electronic news and entertainment.
- Online training/education.
- Teleconferencing.

In a related application, the Internet has been utilized for "telephone" conversations. You can talk with a friend with your PC, microphone, and special software, all without long-distance telephone charges. Although this capability was a boon for many users, telephone companies, as you might guess, were not as enthusiastic about the application.

It is also important to note the parallel between audio and video distribution via the Internet and early PC-based digital movies. Some video producers may have initially wondered what the fuss was all about with QuickTime and other products. The on-screen size was small and the picture quality was poor.

However, as the technology-base matured, they became important video and multimedia production tools. The same scenario may play out for real-time web services. Even though the video quality can be somewhat limited, particularly when a modem/telephone line connection is used, the potential remains. High-speed communications channels *and* enhanced compression schemes will accelerate this trend.

More sophisticated tools have also been developed to take advantage of this Internet capability. Besides dedicated software/hardware that can convert a video program to a streaming format, other components, including newer generation NLE systems, can handle this process. There typically is an option to export an edited production for web-based distribution. You may also be able to convert your program for recording on a DVD. As stated elsewhere, the same software may support multiple distribution venues— you can reach a wider audience.

INFORMATION, ENTERTAINMENT & COMMUNICATION SYSTEMS

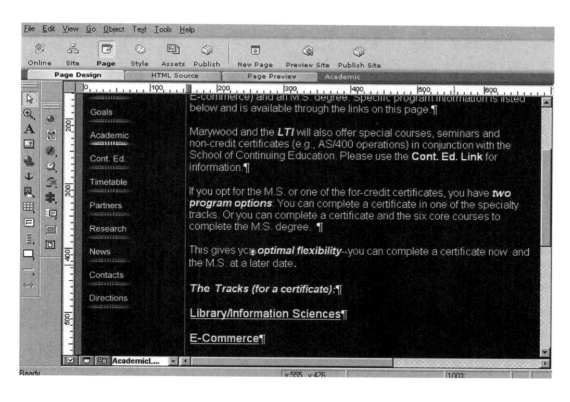

**Figure 17.3**
*An example of a web design program. Note the toolbar, to gain access to different functions, on the left side of the screen. (Software courtesy of Websitepros; NetObjects Fusion.)*

Finally, an intellectual property question remains with regard to using audio and video clips (e.g., Internet radio and television stations). How do you monitor the legal use of this information?

***HTML Authoring and Java.*** Most people are able to create HTML-based documents. If you're comfortable with macros for spreadsheets and have access to a word processing program with an ASCII option, you can create web publications. All you need is the prerequisite Internet link.[21]

New tools have also simplified web page construction. Dedicated authoring programs have been introduced and modules have been released for desktop publishing and presentation programs. Essentially, with the right software, your project can be published and distributed across the print and electronic domains.

Additional Internet enhancements, including those spearheaded by Sun Microsystems, Apple, Microsoft, Netscape Communications, and others, have also made the web a more flexible environment. One development has been Sun Microsystem's Java programming language.

Using Java, you can write applets, programs that "can be included in an HTML page, much like an image can be included. When you use a Java compatible browser to view a page that contains a Java applet, the applet's code is transferred to your system and executed by the browser."[22] Applications range from animations to interactive ads.[23]

Although Java may have required programming skills when introduced, newer products have simplified the development process. Java and related products also signaled an important stage in the Internet's maturation. Starting as a text-based information and communications channel, it gradually emerged as a robust and flexible system. Much like the first PCs, which

INFORMATION, ENTERTAINMENT & COMMUNICATION SYSTEMS

**Figure 17.4**

*An example of a web design program. As highlighted you can also examine a page's underlying structure and components. (Software courtesy of Websitepros; NetObjects Fusion.)*

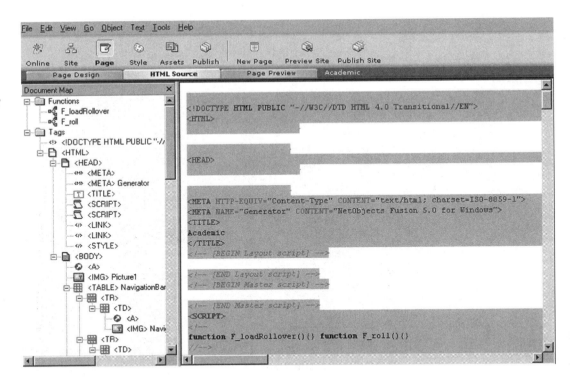

evolved into today's sophisticated machines, the Internet has evolved into a sophisticated multimedia engine.

Similarly, other tools have extended a web page's visual design. *Flash*, for one, is used to create and distribute web-based animations and movies. The latest releases also support video playback.

Other software products, which may have originally targeted CD-ROMs or other media for distribution, may now also support the Internet. As indicated in the multimedia chapter, Chapter 11, this includes Director, a popular multimedia development tool.

But this capability does raise a production question. Based on their file size, the same graphics you may use for a CD-ROM may not be suitable for Internet distribution.

Many individuals still use a telephone/modem link to the Internet. Thus, a graphic that may load quickly on a computer via a CD-ROM may take too long to load via the Internet. The upshot? You can use the same graphics but may have to reduce the file sizes through compression. More pointedly, when you work with graphics and other information, it's important to keep final output(s) in mind—are you creating a project solely for the Internet? Will it also be distributed on a DVD?

***Virtual Reality Modeling Language.*** The Virtual Reality Modeling Language (VRML) introduced virtual reality concepts to the Internet. As discussed elsewhere, conventional virtual reality systems could be used to enter into computer-generated environments. You could explore computer-generated cities, another planet, the human body, or a museum's collection.

Geared for limited-capacity communications channels, VRML extended this framework to the Internet. While there are limitations versus a more conventional setup, it has provided the Internet with a visual third dimension.

This capability can accommodate different applications. Real estate agents can take prospective clients on a tour of a building. You can also set up a museum with links. As you pass through a door to another room, you could be hyperlinked or connected to a new document, and for this type of application, a new virtual environment.[24]

Exhibits could also be rotated to highlight different pieces, and new "rooms" could be added. Unlike a conventional museum, which has limited exhibition space, a virtual museum could readily extend its display capacity.

A VRML-type interface, when combined with the web's ubiquitousness, could also support other operations. In one potential application, photographs from a future space probe could be used to produce a flight over Mars or other planetary body. Using a VRML-type system, these data could be quickly released to thousands of interested armchair astronomers and explorers via the web and could possibly be complemented by a more immersive environment.[25] VRML and similar systems can also help make the Internet a richer multimedia environment.

*Cottage Industry and Advertising.* The web has promoted the growth of an electronic cottage industry. Participants include service providers who may offer web access, web page designers, and advertisers.

As a visually driven environment, the web can be an ideal advertising platform. Small ads may greet you when you navigate to a site, and an ad could tap into the web's interactive capabilities. In the latter, a computer company's home page may serve as a launching point to retrieve additional product data. You could also utilize Java, VRML, and other capabilities to produce exciting ads—you are not restricted to a static, print layout.

Other options include support for a deeper information pool, potentially reaching a larger audience, and setting up feedback mechanisms through forms. So, if you are the New York Yankees, besides listing home games, you can present team information and audio-video sequences.

Yet despite these benefits, the web does pose a unique challenge. Unless an ad is incorporated on a regularly viewed page, like an ad placed on a search mechanism's home page, people may not get to see it. When you read a magazine, you may scan the ads as you flip through the pages. The same scenario generally does not play out on the web. You typically have to seek a site. Thus, one suggestion for advertisers has been to make their sites interesting stop points and to offer interactive games and other incentives.[26]

On a more positive note, the web does enable large and small companies to engage in this form of advertising. This can help level the playing field for reaching potential clients. But this field can shift with the ebb and flow of technology.

Larger companies also have a funding advantage. They can load their web sites with the latest technological advancements. More in line with traditional advertising, they can also place ads on other strategic web pages and hire the best graphic artists.

But a key fact remains unchanged: The Internet provides individuals with a platform to launch and advertise their own companies. Like other fields, technological developments can also trickle down to make them more accessible to the average user.

For example, for a modest fee, or even for free, as has been the case with various "hosting" companies, you can establish a personal web site. Whether or not someone actually visits is another matter. It depends on your site's contents and perceived value. Even if it only has a few "hits" or visitors,

**Figure 17.5**

*A web design program may feature templates—much like a desktop publishing program—to facilitate a web page and site's design. In the case of NetObjects Fusion, you can select one of many such templates. (Software courtesy of Websitepros; NetObjects Fusion.)*

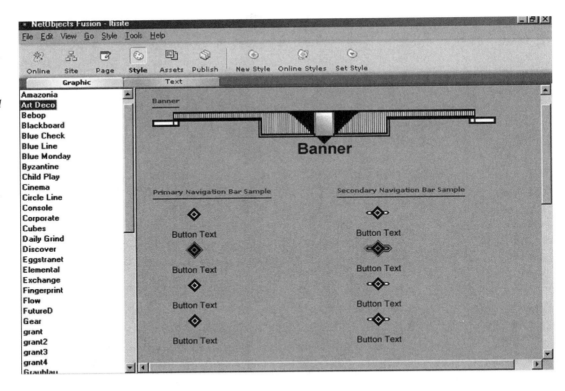

INFORMATION, ENTERTAINMENT & COMMUNICATION SYSTEMS

you can still utilize this communications and information tool to present your ideas.

## OTHER CONSIDERATIONS

The Internet and web have raised additional considerations. These range from heavy traffic concerns to security to censorship.

### Channel Capacity and Implications

The availability of high-speed communications lines is a critical issue for Internet users. Many are tied to standard modem/telephone line connections. This has implications.

Relay times are increased. While surfing the web, you may find a QuickTime file you want to retrieve. You click a mouse to start the procedure, and with a conventional modem, you can probably finish lunch by the time the file is downloaded. A faster modem might help, but depending on the file's size, you may still have time for at least coffee.

There's also an impact on the type of data you can use. As outlined, the web supports real-time audio and video. But the communications channel still imposes restrictions, including a reduced frame rate and a lower quality image. You could, however, alleviate these problems with a DSL, cable hook-up or other high-speed line.[27]

By the mid- to late-1990s, you may also have experienced a general Internet slowdown. The cause? Overburdened equipment and communications lines, more frequent graphical and audio-based relays, and an expanding user base.

As outlined in an early report, it was estimated that Internet traffic exceeded 30 terabytes of data a month, equal to "30 million 700 page novels."[28] The end result? Longer retrieval times and possibly not getting access to specific sites. Various solutions were offered. These ranged from the creation of a two-tiered Internet structure, one of which would have supported an access fee for enhanced services, to providing infrastructure support through advertisements.

Another solution has been to create an enhanced communications and information network. Internet2, among other initiatives, is a consortium using high performance networks to help develop "advanced network applications and technology . . ." that may eventually find their way to the broader Internet community.[29] While not designed to replace the Internet, Internet2 may advance the way we use this type of tool, particularly in the realm of digital video, developing sophisticated online collaborative environments, and deploying advanced display techniques. The latter may include virtual reality-type systems and applications.[30]

## Security

Security is another important Internet issue. This has encompassed providing security for companies to prevent break-ins and for financial transactions.

### Firewalls and Disruptions.

Firewalls and other techniques can help safeguard your data. A firewall is "a barrier placed between your network and the outside world to prevent unwanted and potentially damaging intrusion of your network."[31] Firewalls can block out unwanted visitors and help maintain a system's integrity. But like any other protection scheme, it should be viewed as only a component of a broader security system. You may also determine who gets access privileges to specific data pools.[32]

Firewalls also protect a company from the potential intrusion of its own employees. BellSouth, for one, set up an intranet environment, an internal Internet-like system that could be used for data sharing. To help protect this data, firewalls have separated its different operating units.[33]

Organizations face another potential security breach. While not new, the Internet community received a warning shot across it electronic bow in early 2000. Some of its "hottest" sites were temporarily disabled. Individuals using special software essentially overloaded different computer-based systems with information, thus causing the downtime. Legitimate users could not enter the sites, causing revenue losses, including advertising revenue. This is analogous to losses experienced by a radio or television station if it goes off the air. In response, companies tightened their security and the FBI was called in to help.[34]

This type of security breach has repercussions—the Internet has become a commercial boomtown. It was a time when electronic business ventures exploded, and the Internet served as a vital communications and information conduit. What would happen if this conduit was disrupted? Would certain industries be disrupted, even temporarily? If so, what would be the cost?

### Other Security Issues.

In the wake of the World Trade Center attack, more individuals called for tighter security for the Internet and our overall communications and information infrastructure. While there were new initiatives, it must be noted the government and industry had been working on this problem for years.

For example, as stated in a U.S. General Accounting Office (GAO) Report,

To address concern about protecting the nation's critical computer-dependent infrastructures from computer-based attacks and disruption, in 1998, the President issued Presidential Decision Directive (PDD) 63. A key element of the strategy outlined in that directive was establishment of the National Infrastructure Protection Center (NIPC) as "a national focal point" for gathering information on threats and facilitating the fede-

INFORMATION, ENTERTAINMENT & COMMUNICATION SYSTEMS

ral government's response to computer-based incidents.[35]

It is a daunting task. It requires a well-trained technical staff and a coordinated effort, potentially across multiple government agencies, private industry, and the international community.

This concept of protecting our electronic infrastructure was extended by the *National Strategy to Secure Cyberspace*.[36] It is a part of a broader plan that called for the protection of critical national assets and fell under the *National Strategy for Homeland Security*.[37] Its strategic objectives were to

- Prevent cyber attacks against America's critical infrastructures;
- Reduce national vulnerability to cyber attacks; and
- Minimize damage and recovery time from cyber attacks that do occur.[38]

The plan called for a combined government and private effort to secure cyberspace and proposed

I.   A National Cyberspace Security Response System;

II.  A National Cyberspace Security Threat and Vulnerability Reduction Program;

III. A National Cyberspace Security Awareness and Training Program;

IV.  Securing Governments' Cyberspace; and

V.   National Security and International Cyberspace Security Cooperation.[39]

The last point, for instance, called for heightened international cooperation to secure the United States and the global network.

While applauded in some quarters, questions were raised about the plan's potential effectiveness and impact. In one case, security experts agreed the document height-ened the general awareness about security. But the recommendations for certain provisions were ambiguous, among other issues.[40]

A broader question, which concerns this topic as a whole, focuses on the actual damage *cyberterrorism* could cause.[41] While a terrorist attack could physically destroy a power plant or lead to a loss of life, a cyber-attack would, as previously described, most likely result in a disruption of service. It is a matter of the degree of impact. In fact, "many dislike the term 'cyberterrorism.' Ambiguity over its definition—and, therefore, which threats are real . . . —has confused the public and given rise to countless myths."[42]

In essence, the term cyberterrorism and the specter of a threat may be overblown in contrast to actual physical attacks. The hyperbole may also hide a more potent threat, the use of the Internet as a communications tool to support terrorist activities.[43] It may also be a new variation of the old saying, "the pen is mightier than the sword." *But in this case, it could be an electronic pen that could be used to guide a more potent sword.*

What it comes down to in the end is this—the communication revolution has influenced what we now consider to be vital assets. As discussed in Chapter 1, the production, manipulation, and transmission of information drive our world. The Internet and the other roads and tools we use to sustain this system have become key elements of our national and international fabric and infrastructure. Yet as raised by some individuals, this concern should also be weighed against other, more physical threats, which may actually cause a greater loss of life.

## Sales and E-Commerce/E-Business

*Commercial Transactions.*   For our discussion, the Internet serves as an infrastructure

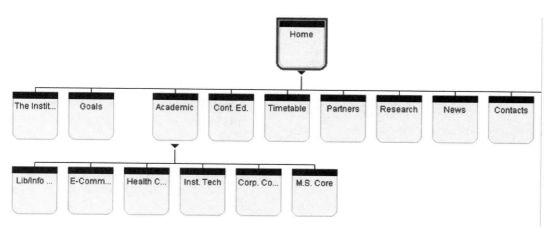

**Figure 17.6**
*NetObject Fusion supports a web site management option. Each icon represents a page, providing the designer with a visual map of the overall site. The icons, and related pages, can also be rapidly edited, moved, or deleted. (Software courtesy of Websitepros; NetObjects Fusion.)*

that can support transactions ranging from stock trading to sales. It is also the world's largest shopping mall and communications and information network.

Large and small companies offer Internet users a vast array of products and services. The Internet's accessibility, ubiquitous nature, cost-effectiveness, and freedom from geographical and time constraints, play key roles in this development.

Banks also became players in this field. A traditional bank could reduce its overhead and reap substantial profits by creating a virtual bank, in essence, an electronic bank or a bank without walls. But individuals who conduct business over the Internet have to be assured their credit information will remain secure and private.[44] Other issues include digital cash and signature verification schemes.

Internet-based auction companies, such as Ebay, have also flourished. As a subscriber, you could electronically bid on products ranging from cars to ancient coins to toys. You could also sell your own products for a fee. The cottage industries described in the chapter are also part of this emerging world.

Two terms are associated with the use of the Internet for business, e-commerce and e-business. E-commerce "generally refers to buying and selling over the Internet.... E-business has a wider meaning, encom-

passing e-commerce but generally using the Internet and the technology behind it to connect business processes over the Internet."[45] Some relevant activities include the following:

- providing service,
- promoting brand awareness,
- extending market reach,
- distributing information,
- delivering distance learning,
- managing business partners, and
- launching a web-based business.[46]

Besides a business selling services/products to individuals and consumer(s) (*B2C*), which is the way we normally think about Internet business, operations include business to business (*B2B*) functions—companies selling their services/products to other companies. Together, they help compose a rapidly growing sector of the international financial market, and one in which revenues are expected to be measured in the trillions of dollars during the early years of the twenty-first century.[47]

Finally, it may be appropriate to wrap up this section with a quote that defines how these transactions may affect us all—some noticed, others unnoticed, on a daily basis.

INFORMATION, ENTERTAINMENT & COMMUNICATION SYSTEMS

. . . real economic and social transformation is not just about far-reaching multimillion or billion-dollar investments or grand schemes for generating new products and new wealth. It is also made up of . . . *everyday events that often go unnoticed* [my emphasis]. It is about connection, discourse, and commerce that occurs in what sociologist Manuel Castells describes as "grass-rooted networks" and the "back alley of society."[48]

Take a moment to think about this the next time you buy a camera, magazine, or Yankees baseball card from Ebay and the Internet.

### Monty Python and Privacy

Another Internet item may be familiar to Monty Python fans: spam, or more precisely, *spamming*. An important online attraction has been Usenet newsgroups, basically special interest or discussion groups. By 1995, there were approximately 14,000 such groups, and the Internet has provided a door to these discussions.[49]

In April 1994, a law firm posted an ad for its services to thousands of newsgroups, the practice of spamming. According to many users, this action violated Internet etiquette or "Netiquette." This triggered an avalanche of flames or hostile electronic responses.

Spamming has a number of implications. First, the messages can contribute to the ever-growing Internet traffic jam. Second, it has led to the development of cancelbots, programs that automatically erase messages that are posted to multiple newsgroups. Cancelbots, in turn, raised First Amendment questions. Did anyone have the right to curtail spamming, or as some individuals have claimed, to hamper free speech?[50] Third, the 1994 incident highlighted other Internet-related issues. These included the use of mail bombs, large messages designed to clog-up your electronic mailbox, and other censorship topics.

Privacy issues also became a primary concern:

- How do you ensure that other people are not monitoring the sites you visit? In one example, your computer may be infected with *spyware*—software that you are not aware of but may be sending back information about the sites you visit. While browsing the Internet, spyware may "infect" your computer without your knowledge or consent. Thus, as you continue site hopping, this information is collected and may be sold to third parties (e.g., as an advertising tool). It is analogous to a survey, but in this case, your browsing preferences are recorded. You can actually remove spyware from your system by using other software that identifies and cleans your system—the spyware is eliminated.[51]

- Another privacy tool is to use a service that provides you with *anonymous browsing*. One such operation has been Anonymizer.[52] You could select various privacy options, including one that would shield your Internet identity. The actual URLs you visit may also be scrambled so they are unintelligible to other individuals.

- You can permanently erase Internet-based information that details the sites you visit and other data. When you erase information from your hard drive, it could potentially be recovered through software and/or hardware tools. A browser may also store information about your Internet travels that you may not even be aware of—in various directories or folders.

Consequently, software has been written to locate and permanently erase this information as well as help ensure that other data cannot be recovered once erased.[53] The latter is, for instance, particularly crucial for government agencies and organizations that work with sensitive information.

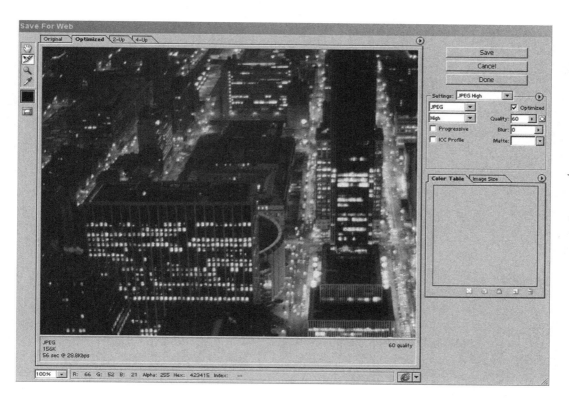

**Figure 17.7**
*A graphics program may help you produce optimized images for the web. Photoshop has an option that depicts, in one case, an image's appearance in an appropriate web-based format using different compression levels. (Software courtesy of Adobe Systems, Photoshop.)*

Much like a commercial enterprise, a firewall is also an important tool for individual users, particularly if a computer is connected to the Internet through DSL or cable-based systems. Besides providing a high-speed connection, these systems may be on all the time, making the computer an attractive target to individuals who may hijack the system for their own use. Basically, your Internet connection can provide someone with an open door to your computer. It may be used without your knowledge for various activities, and a firewall can help prevent this from happening.[54]

In sum, when you roam the Internet, you run a gauntlet of challenges. These range from spyware to viruses. But there is help in the form of hardware/software that can help protect your privacy. It should also be noted that legislative initiatives have been proposed with this goal in mind.

## CONCLUSION

The Internet remains an evolving system. It is also a challenging topic to cover since it literally changes every day. Taking these factors into consideration, several generalizations can still be drawn.

The Internet will continue to grow in importance as a premiere information and communications system. It has provided us with a tool to conduct research, to launch new business ventures, and to exchange information.

It's continued growth should also be fueled by faster communications channels. This enhanced capability will enable us to communicate faster with a wider variety of data.

It is also important to view the Internet as a focal point for the convergence of different communications companies/channels and technologies. The former includes the

INFORMATION, ENTERTAINMENT & COMMUNICATION SYSTEMS

telephone, satellite, and cable industries. The latter range from streaming audio and video data to VRML to applications that have not yet left the drawing board.

It is this last feature that makes the Internet so exciting. As its driving technologies are refined, it becomes possible to support new applications.

It may also be possible to receive and view your television programming via the same communications channel and device that you use to surf the Internet. As an element of this diverse information mix, the Internet may contribute to the creation of an integrated information and entertainment system.

In one example, the Internet may become the world's largest distribution center for music and video. As described, MP3 and other compression schemes have made it possible to relay music over the Internet. As these technologies are refined, we may be ordering our music via the web in addition to going to music stores. The same scenario may play out for movies—they may be retrieved over the web and/or other distribution tools.

We may also be gaining access to Internet-based resources through wireless technologies. In one case, a cell phone may now offer you wireless Internet access, and a key development was the Wireless Application Protocol (WAP). Using a cell phone or other WAP-ready device, you could tap into this resource. While the options were rudimentary when first introduced, the promise offered by 3G-based services would greatly extend these capabilities. Please see the wireless chapter, Chapter 8, for more information about 3G services.[55]

Another factor, covered in a later chapter, is the Internet and censorship. While the Internet is the world's largest information conduit, censorship in the United States and abroad could limit the access to certain types of information. In the United States, for example, provisions in the 1996 Telecom Act would have had an impact on the information we could receive. Cuba, China, and other countries, for their part, have tight Internet controls in place.[56]

## REFERENCES/NOTES

1. See Howard Rheingold, *The Virtual Community* (New York: Addison-Wesley Publishing Co., 1993), for an in-depth look. *Note:* ARPANET was also a reflection of the Cold War. The United States was concerned that a nuclear conflict would sever traditional communications/information ties, thus the impetus for an ARPANET-type environment.

2. It could almost be categorized as an organism or artificial life. It is growing and is host to other organisms (e.g., cancelbots described in a later section).

3. J. Kelly Beatty and Stuart J. Goldman, "The Great Crash of 1994: A First Report," *Sky and Telescope* 88 (October 1994), 21.

4. Adam Gaffin, "Comet's Impact Felt on Internet," *Network World* 11 (July 25, 1994), 65.

5. They have been distributed as part of a television signal's Vertical Blanking Interval (VBI). *Note:* During the television scanning cycle, the electron beam initially scans the odd-numbered lines and when it returns to scan the even-numbered lines, a period of time elapses in which the beam is blanked or turned off. This is known as the VBI. All the lines in a television frame do not contain picture information. Teletext magazines have been inserted in a portion of the approximately 21 "empty" lines, the VBI, of a television signal.

6. Joseph Roizen, "Teletext in the USA," *SMPTE Journal* 90 (July 1981), 603. U. S. standards were World System Teletext (WST), based on British operations, and the North American Broadcast Teletext Standard (NABTS) developed through the combined efforts of CBS, the Canadians, and the French. Domestic supporters included Electra and Keyfax (WST) and EXTRAVISION (NABTS). The British services were Ceefax and Oracle.

7. Stand-alone terminals were also developed; systems in England and Canada, for example, were also geared toward business.

8. Prestel, press release, August 29, 1983.

9. LEXIS is an example of such an information pool. It is a legal database with court cases, law journal articles, and other information.

10. Cathryn Conroy et al., "All Things Cyber," *CompuServe Magazine* (March 1994), 10–21. *Note:* Cybersex encompasses a range of activities, including interactive conversations with other people.

11. Eric Richard, "Anatomy of the World Wide Web," *Internet World* (April 1995), 28.

12. Peter Kent, "Browser Shootout," *Internet World* (April 1995), 46.

13. The telnet tool enables you to log on to a remote computer. You're not limited to file transfers and can fully interact with the system as though you're on-site. Gopher is a menu-driven system that combines, in part, FTP and telnet capabilities. Telnet and FTP are still used as of this writing.

14. Robert Jon Mudry, *Serving the Web* (Scottsdale, AZ: The Coriolis Group, 1995), 109. *Note:* These other services have included FTP, gopher, and telnet. The protocols that drive the web also have applications on other types of networks.

15. The http designation indicates the document is on a web server. Other addresses might be, for example, for FTP sites.

16. *Note:* Other browsers also exist.

17. One such popular site has been ask.com.

18. Real-time also implies that a radio station's programming, for example, is simultaneously distributed and can be listened to over the air and on the web. The term live may also be associated with the function.

19. Steve Mack, *Streaming Media Bible* (New York: Hungry Minds, 2002), p. 33.

20. A web surfer is analogous to a cable television surfer.

21. For more sophisticated operations, though, you may need programming expertise. For example, you can create a document with a form, but you may have to write a cgi script to actually tap into this information. Or you may be able to use a premade script that could be downloaded from the Internet.

22. 1995 Sun Microsystems web site, downloaded from http://java.sun.com.

23. See Dan Amdur, "The Scene Is Set for Multimedia on the Web," *NewMedia* 5 (November 1995), 44, for such an example (fantasy baseball).

24. A paper by David Raggett, Hewlett-Packard Laboratories, "Extending WWW to Support Platform Independent Virtual Reality," downloaded.

25. An immersive environment would more closely model a traditional VR set-up.

26. Ron Maclean, "New Game Rules," *American Advertising* 2 (spring 1995), 9.

27. This includes satellite. Please see Frank Beacham, "DirecTV Offers On-Ramp to the Net," *TV Technology* 13 (September 1995), 1, for an earlier look at this application.

28. Mark John, "Cybergridlock Feared," *Communications Industries* Report 12 (August 1995), 9.

29. "Frequently Asked Questions About Internet2," downloaded from http://www.internet2.edu/about/faq.html. This includes the use of a new protocol, IPv6. Please see the Suggested Readings section for more information.

30. Internet2 Health Sciences, downloaded from http://www.internet2.edu/resources/infosheethealth.pdf. *Note:* A virtual reality system could also be used for online collaboration.

31. John Bryan, "Build a Firewall," *Byte* 20 (April 1995), 91.

32. Peter Kolsti, "Firewalls Keep Users from Kicking Your Apps," *Network World* 12 (November 13, 1995), 61.

33. Steven Vonder Haar, "Firewalls Address the Threat Within," *Inter@ctive Week* 3 (January

INFORMATION, ENTERTAINMENT & COMMUNICATION SYSTEMS

INFORMATION, ENTERTAINMENT & COMMUNICATION SYSTEMS

29, 1996), 23. *Note:* Intranets became a growing market in their own right.

34. Rutrell Yasin, "Cyberterrorists Crash Web Party," *InternetWeek* (February 14, 2000), 66.

35. GAO, "Critical Infrastructure Protection Significant Challenges in Developing National Capabilities." Report to the Subcommittee on Technology, Terrorism, and Government Information, Committee on the Judiciary U.S. Senate. GAO-01–323.

36. Drafted and released in late-2002/early-2003.

37. Please see the National Strategy for the Physical Protection of Critical Infrastructures and Key Assets; www.whitehouse.gov/pcipb/physical.html.

38. The National Strategy to Secure Cyberspace, downloaded from www.whitehouse.gov/pcipb/executive_summary.pdf.

39. Ibid.

40. Dennis Fisher, "Cyber Plan's Future Bleak," *eWeek* (February 24, 2003), 18.

41. Cyberterrorism in this context refers to an attack on the Internet or other network for malicious purposes.

42. Robert Lemos. "What Are the Real Risks of Cyberterrorism." Special to ZDNet. August 26, 2002, downloaded from http://zdnet.com.com/2100–1105–955293.html.

43. Ibid.

44. Peter Noglows, "Virtual Tellers: Internet Bank to Open," *Inter@active Week* 2 (October 9, 1995), 53.

45. Scott Bury, "The e-Business Explosion," *Electronic Publishing* 24 (February 2000), 24.

46. Martin C. Clague, "Understanding e-Business," In *Renewing Administration* (Boston, MA: Anker Publishing Co., Inc., 1999), p. 47.

47. Ibid., p. 59.

48. Blake Harris, "Training in the New Society," *Government Technology* (May 2001), 8.

49. Joel Furr, "The Ups and Downs of Usenet," *Internet World* (November 1995), 58.

50. Laurie Flynn, "Spamming on the Internet," *New York Times* (October 16, 1994), Section F-9.

51. As of this writing, Ad-aware by Lavasoft is one such free spyware tracking/disinfecting tool. The URL is www.lavasoft.nu.

52. For more information, the web site is www.anonymizer.com.

53. An example of such a program has been CyberScrub. CyberScrub has offered options to erase Internet-based tracking information as well as to permanently delete data so it could not be recovered; www.cyberscrub.com.

54. As of this writing, there are commercial/free software firewalls. An example of a popular free program has been ZoneAlarm; a commercial version is also available; www.zonelabs.com.

55. For more information about WAP devices, see "How Wireless Internet Works," by Jeff Tyson, downloaded from http://www22.verizon.com/about/community/learningcenter/articles/displayarticle1/0,4065,1087z1,00.html and www.cnet.com. In Cnet, search for cell phones or WAP for this information.

56. Michael Hatcher, Jay McDannell, and Stacy Ostfeld, "Computer Crimes," *American Criminal Law Review*, summer, 1999 (36 Am. Crim. L. Rev. 397), downloaded from LEXIS.

## SUGGESTED READINGS

### Teletext, Videotex, and PC-Based Systems

Connelly, Terry. "Teletext Enhances WKRC's Local News Image." *Television/Broadcast Communications* (October 1983), 52–58. An overview of the Electra teletext magazine.

France Telecom Intelmatique. Teletel traffic data, downloaded from France Telecom Intelmatique.

Harrison, David. "Lexis-Nexis Moves to the World Wide Web." *Inter@active Week* 4 (October 13, 1997), 42–43; Kevin Jones. "The French Evolution Comes to the Web—Slowly." *Inter@active Week* 5 (November 30, 1998), 25. Two traditional services and Internet implications.

"Knight-Ridder Pulls Plug on Viewtron." *Broadcasting* (March 24, 1986), 45.

Martin, James. *View Data and the Information Society*. New York: Prentice-Hall, 1982; C. D. O'Brien et al., "Telidon Videotext Presentation Level Protocol: Augmented Picture Description Instructions: CRC Technical Note 709-E." Ottawa, Canada: Department of Communications, 1982. Excellent sources about earlier interactive/videotext systems.

Morgenstern, Barbara, and Michael Mirabito. "Educational Applications of the Keyfax Teletext Service." *Educational Technology* 8 (August 1984), 46–47. Examines some of the educational applications of a teletext magazine.

Rayers, D. J. "The UK Teletext Standard for Telesoftware Transmissions." In *International Conference on Telesoftware*. London: The Institution of Electronic and Radio Engineers, 1984. Explores telesoftware technical issues—using a teletext magazine to distribute computer software.

Tenne-Sens, Andrej. "Telidon Graphics and Library Applications." *Information Technology and Libraries* 1 (June 1982).

**The Internet**

*American Advertising*. Vol. 2 (spring 1995). This issue covers advertising and the Internet. Includes Ron Maclean. "New Games New Rules," 9–12.

Baldasserini, Mario. "Playing with Fire." *Computer Shopper* (June 1995), 606–608; Steven Vonder Haar. "Firewalls Address the Threat Within." *Inter@ctive Week* 3 (January 29, 1996), 23. Coverage of firewalls. The last article examines firewalls within an organization.

Bielski, Lauren. "Image-Enabling E-Commerce: TrueSpectras's Mission and Technology." *Advanced Imaging*, 28–30; Eric Chabrow. "E-Services." *Information Week* (November 13, 2000), 46–62; Eugene Grygo. "B-to-C Companies Are Extending Their Reach." *InfoWorld* (July 31, 2000), 37; Marcelo Halpern. "Not All E-Signatures Are Equal." *CIO* 14 (January 15, 2001), 54, 56; NASA. "Power Struggle." *NASA Tech Briefs* 25 (June 2001), 14; David Thompson. "The Social Engineering of Security." *eWeek* 18 (June 11, 2001), 25, 28; Debby Young. "Meter Readers." *CIO* 14 (January 15, 2001), 100–104. Various e-business/commerce issues; the NASA article concerns new power supplies, a critical issue in respect to the Internet/computer users (who are major electrical consumers).

Bowen, Cathy. "Security Takes Center Stage." *Card Technology* (November 2001), 39–47; James Cope. "Distributing IT Resources Across Multiple Locations Could Make It Easier to Recover from a Disaster." *Computerworld* 36 (July 15, 2002), 38–39; Robert Lemos. "Security Pros: Our Defenses Are Down." Special to ZDNET News. September 10, 2002, downloaded from http://zdnet.com.com/2100–1105–957219.html; U.S. Government. "National Strategy to Secure Cyberspace." For the executive summary, please see www.whitehouse.gov/pcipb/executive_summary.pdf (you can also search the site for the complete document). Various security issues.

Bridis, Ted. "Virus Overwhelms Global Internet Systems." Yahoo Press, AP, downloaded from www.yahoo.com; *The Industry Standard* covered the "attacks" on Internet sites. Two articles from the February 21, 2000 issue are: Elinor Abreu. "The Hack Attack," pp. 66–67 and Polly Sprenger. "Go Hack Yourself," pp. 68–69.

Coco, Donna. "A Web Developer's Dream" *Computer Graphics World* (August 1998), 54–58; S. D. Katz. "This Is Deep Virtual Reality Comes to the Web." *Millimeter* 28 (November 2000), 96–102. Presenting *A Midsummer Night's Dream* over the Internet via VRML and the Web and 3-D graphics, respectively.

Cox, John. "It May Be Immature, But Fans Claim Java's No Jive." *Network World* 13 (March 18, 1996), 1, 14–15. A look at Java and an early evaluation.

Dodge, John. "Can the Internet Let Freedom Ring in China?" *PC Week* 16 (June 14, 1999), 3; Leslie Pappas. "The Long March to Cyberspace." *The Industry Standard* 3 (February 21, 2000), 134–139. Internet issues and China.

Duffy, Daintry. "Why Do Intranets Fail." *Darwin* 1 (November 2001), 57–62; Jeffrey Schwartz and Richard Karpinski. "Intranets Grow Up." *InternetWeek* (December 21, 1998), 17. Intranets; Daintry's article also cover Intranet costs.

INFORMATION, ENTERTAINMENT & COMMUNICATION SYSTEMS

Flynn, Laurie. "Spamming on the Internet." *New York Times* (October 16, 1994), Section F-9. Spamming and a Q&A session with Laurence Canter and Martha Siegel; Joshua A. Marcus. "Commercial Speech on the Internet: Spam and the First Amendment." *Cardozo Arts and Entertainment Law Journal* 1998. (16 Cardozo Arts & Ent Lj 245), downloaded from LEXIS. Censorship and the Internet.

John, Mark. "Cybergridlock Feared." *Communications Industries Report* 12 (August 1995), 9; Brock Meeks. "Commerce Dept. Study Highlights Unequal Access." *Inter@active Week* 2 (July 31, 1995), 12. Earlier Internet views about too much traffic, a possible two-tier system, and access issues.

Kening, Dan. "A Connected Electorate." *CompuServe Magazine* 11 (September 1992), 35; John McKeon. "Examining the Net's Effect on the '96 Elections." *Communications Industries Reports* 12 (August 1995), 1, 6. A PC-based system/the Internet and political applications/implications.

Mack, Steve. *Streaming Media Bible*. New York: Hungry Minds, 2002; Gus Venditto. "Instant Video." *Internet World* (November 1996), 85–101; Jeff Sauer. "Streaming Appliances Do the Dirty Work." *EMedia Magazine* 14 (September 2001), 36–45; Ben Waggoner. "Streaming Media." *Interactivity* 4 (September 1998), 23–29. Media streaming, earlier systems, and new developments. Mack's *Streaming Media Bible* is highly recommended and provides a comprehensive examination of this application.

McCandish, Stanton. "EFF's Privacy Top 12," downloaded from www.eff.org/Privacy/eff_privacy_top_12.html. Overview of steps to protect your identity/privacy in regard to the Internet.

Mulqueen, John T. "Latin America Online: Overcrowded Field?" *Inter@active Week* 7 (August 14, 2000), 56; NTIA. *A Nation Online: How Americans Are Expanding Their Use of the Internet.* Washington, DC, February 2002, downloaded from www.ntia.doc.gov/ntiahome/dn/index.html; Robert N. Wold. "Europe's Pursuit of Internet and Broadband Revenues." *Via Satellite* VI (September 2001), 36–44. World and U.S. Internet expansion and use.

Plsern, Florence. "Internet2's Network Adopts New Protocol for Addressing and Packaging Data," downloaded from http://chronicle.com/free/2002/08/2002081601t.htm; www.internet2.edu. Internet2 information and the Internet2 web site.

## GLOSSARY

*Browser:* Software used to visit different web sites.

*Cyberspace:* A term popularized by William Gibson's science fiction work *Neuromancer*, cyberspace can be described as a computer-generated environment (or world) where you can interact with other people and work and play.

*Firewall:* A barrier typically placed between your network and the outside world to prevent unauthorized entry.

*HyperText Markup Language (HTML):* The basic language for creating web documents. If you are comfortable with macros for spreadsheets and have access to a word processing program with an ASCII option, you can create web publications.

*Java:* A language that extends the web's capabilities.

*Real-Time Audio and Video Retrieval:* The ability to hear/see a file without having to completely download it.

*Spamming:* If you are a Monty Python fan, a funny skit (spam, spam, spam . . .). If not, the act of posting an ad, for example, to thousands of newsgroups or sending e-mails advertising a product. This can trigger an avalanche of flames—hostile electronic responses.

*Streaming:* Delivering real-time audio and video information via the Internet.

*Virtual Reality Modeling Language (VRML):* Introduced virtual reality applications to the web.

*Web Search Mechanisms:* Software tools used to identify and locate web-based resources (e.g., keyword search).

*World Wide Web (WWW):* A product of the Swiss-based CERN research center, it was pioneered by Tim Berners-Lee to help facilitate the exchange of information. You use the web to travel across the Internet and to retrieve and exchange information ranging from text to digital video.

INFORMATION, ENTERTAINMENT & COMMUNICATION SYSTEMS

# VI

## THE LAW

# 18     E-Mail and Privacy

The United States has a rich tradition of protecting an individual's privacy rights. This protection can be traced to the Fourth Amendment, which guarantees

the right of the people to be secure in their persons, houses, papers, and effects, against unreasonable searches and seizures, shall not be violated, and no warrant shall be issued, but upon probable cause, supported by Oath of affirmation, and particularly describing the places to be searched, and the persons and things to be seized.[1]

California, Washington, and some other states also provide privacy guarantees within their constitutions. In California, for example, "All people by nature are free and independent and have inalienable rights. Among these are enjoying and defending life and liberty, acquiring and possessing, and protecting property, and purchasing and obtaining safety, happiness, and privacy."[2]

Based on the intent found within such documents, legislation has been enacted to protect our privacy rights. But over the past several years, new communications technologies have led to a potential deterioration of these privileges. In response, federal and state courts have been asked to examine and interpret relevant law. Of particular concern is the area of privacy and human-to-computer and computer-to-computer communication, which includes e-mail privacy.

## FEDERAL CASE LAW

The Fourth Amendment provides protection to individuals in narrowly defined situations, but it does not guarantee employee privacy in the workplace or in their e-mail. Fourth Amendment protection is only afforded if the employee has a reasonable expectation of privacy. This expectation must be balanced against the employer's right to control, supervise, and maintain workplace efficiency.

According to the reasonableness standard, the search and seizure of an employee's private workplace property must also be reasonable at its inception and in scope. An e-mail search is justified at inception if there is a reasonable belief the search will provide evidence of employee workplace misconduct or if the search is required to retrieve noninvestigatory work-related material— for example, your supervisor accesses some routine work reports from your e-mail on a day you are sick.[3]

An e-mail search is reasonable in scope if "the measures taken are reasonably related to the objectives of the search."[4]

### Two-Prong Test

A two-prong test must also be conducted. First, a subjective expectation of privacy must exist. This is your belief that you are entitled to a reasonable expectation of privacy with regard to your briefcase, personal mail, and other workplace personal items, including e-mail. Second, there must be an objective expectation of privacy. Society must agree that your perception is reasonable.[5]

The reasonableness of a search must also be judged by the search's intrusiveness. In one example, the federal courts have ruled that no reasonable expectation of privacy exists in workplace situations where there are extensive security precautions (e.g., at a military site).[6] In contrast, a warrantless body cavity search for suspected workplace misconduct is not reasonable because it is very intrusive.[7] Therefore, a search of employee items, including their e-mail, is a complex issue.

### Precedent and *Stare Decisis*

The paucity of federal e-mail cases has led privacy advocates to cite other, relevant Fourth Amendment cases.[8] The cases discuss whether employees have a reasonable expectation of privacy in their desks, file cabinets, wastebaskets, and similar workplace items. If they are afforded a reasonable degree of privacy protection, this could establish precedent for e-mail.

Precedent is defined as an established rule of law set by previous courts, deciding a case involving similar facts and legal issues.[9] Another legal concept that comes into play is *stare decisis*. *Stare decisis* ensures that similar cases will be decided in a similar way.[10] So precedent sets the standard—privacy law protects your private wastebasket. *Stare decisis* then guarantees that similar wastebasket privacy cases will be decided in the same way.

Based on these two legal concepts, e-mail privacy advocates claim the following:

- A line of Fourth Amendment cases serve as precedent for extending workplace privacy protection to an employee's e-mail.

- Workplace e-mail should be afforded the same privacy protections other workplace items currently enjoy.

### Cases

*O'Connor v. Ortega (480 U.S. 707.1987).* One such case, which could serve as a precedent, was *O'Connor v. Ortega*. The U.S. Supreme Court ruled that Dr. Ortega, a physician who was a state hospital employee, had a reasonable expectation of privacy in his desk, file cabinets, and office. They were not shared with other employees and only contained personal items.

The Court decided this case on Fourth Amendment grounds. It ruled that state hospital administrators had violated Ortega's privacy rights. They failed to meet the requirements of the reasonableness standard while investigating possible work-related misconduct.[11]

In brief, hospital administrators were concerned about Ortega's management of the hospital's psychiatric program. They subsequently placed him on leave and launched an investigation. At this time, the administrators searched Ortega's office, removed some items, and subsequently used them in an administrative hearing that resulted in Ortega's firing. He responded with a lawsuit against the hospital.

In deciding the case, the Court recognized the need to fairly balance the employee's legitimate right to privacy against the employer's need to supervise, control, and maintain the workplace's efficient operation. However, the court also ruled that a search must be reasonable at its inception and in scope. In the Ortega case, the Court ultimately ruled that Ortega had a reasonable expectation of privacy in his workplace items.[12]

THE LAW

Following this logic, an employer would violate an employee's rights if private, workplace e-mail was accessed without any prior suspicion of work-related misconduct or if the search was more intrusive than necessary. Any evidence of work-related misconduct/criminal behavior, which had been obtained in this manner, would be inadmissible in an administrative or judicial proceeding (e.g., to fire the employee).

### Richard Schowengerdt v. General Dynamics et al. (823 F. 2d. 1328. 944 F. 2d. 483.1987).
Richard Schowengerdt was a civil engineer employed by the U.S. Navy to work on high security weapons and projects. In light of their sensitive nature, random workplace searches and other extensive security precautions were taken. Nevertheless, following a routine search of his office and desk, Schowengerdt filed suit claiming he had an objective reasonable expectation of privacy in his workplace.

The court disagreed stating that employees knew about the "tight security procedures and the ongoing surveillance."[13] Therefore, the measures taken to guarantee the facility's security would negate any objective expectation of privacy.

Based on this case, it appears that if the employer has a policy informing employees that their e-mail may be read, accessing it would only be legal for monitoring purposes. It would be illegal in narrow circumstances, such as access by an unauthorized person.

### Michael A. Smyth v. The Pillsbury Company (914 F. Supp 97. 1996).
In 1996, the federal courts finally heard cases specifically involving workplace e-mail privacy. In Smyth v. Pillsbury, the U.S. District Court for the Eastern District of Pennsylvania ruled that an employee who used the company's private, internal e-mail system to communicate "inappropriate and unprofessional comments" to his supervisor, did not have a reasonable expectation of privacy in his e-mail messages. He had been terminated for this action.[14]

The court also ruled that the company's right to prevent inappropriate, unprofessional, or illegal comments and activity, outweighed the employee's workplace e-mail privacy interest.[15] This decision serves as a particularly strong endorsement of employer rights—Pillsbury's company policy states that employee e-mail communication is confidential and privileged. Nevertheless, the court still held that Pillsbury had the right to monitor internal employee e-mail communication.[16]

### John Bohach et al. v. The City of Reno (932 F. Supp. 1232. 1996).
Another federal case specifically involving workplace e-mail privacy is Bohach v. the City of Reno. Unlike Smyth v. Pillsbury, the Bohach case involved a public employer, the Reno, Nevada Police Department.

Two police officers used the department's Alphapage message system to exchange information that was later used in an internal investigation against the officers.[17] They filed suit against the police department in federal court on the grounds that the storage, retrieval, and disclosure of the messages violated their right to workplace e-mail privacy.[18]

The court ruled that while the two officers may have believed their messages were private, no reasonable expectation of privacy in their workplace e-mail messages existed.[19] This decision was based on

1. The system automatically recorded and stored all messages.
2. Officers were informed that all messages would be logged and that some types of messages were prohibited on the system.
3. Message recording was considered a part of the "ordinary course of business" for a police department.

THE LAW

4. As the service provider, the police department was free to access its system messages and use them as they saw fit.[20]

Thus, the court held the officers did not have a reasonable expectation of privacy in their public, workplace e-mail messages.

### *Richard Fraser v. Nationwide Mutual Insurance Co. (135 F. Supp. 2d 623. 2001).*

In *Fraser v. Nationwide*, the court examined the legality of accessing workplace e-mail communication that has been retrieved by an employer after being received and discarded. Nationwide Insurance agent Richard Fraser filed suit against his employer claiming the retrieval of his workplace e-mail from post-transmission storage violated the Pennsylvania Wiretap Act as well as the Federal Wiretap Act and the Stored Communications Act.[21]

Nationwide executives searched the company's server for Fraser's e-mail. They believed he had written and sent false and libelous communications to insurance regulatory agencies and to Nationwide's competitors. The search produced such information, and Fraser was subsequently fired. Upon termination, Fraser sued Nationwide on numerous grounds, including the accusation that retrieving his e-mail violated the Federal Wiretap Act which protects unauthorized "interception" of electronic communication and the Stored Communications Act which protects "access" to electronic communications while in electronic storage. Nationwide responded by claiming the Acts did not protect e-mail that had been sent, received and discarded.[22]

Judge Anita Brody of the U.S. District Court for the Eastern District of Pennsylvania ruled in favor of Nationwide on the grounds that Fraser's e-mail was in post-transmission storage. Judge Brody ruled the Wiretap Act and Stored Communications Act only protected electronic communications that was in the process of being transmitted. Therefore, no objective reasonable expectation of privacy in Fraser's e-mail existed.[23]

### *Case Law Summary as It Relates to E-Mail.*

The courts provide the employer with certain advantages in the employer-employee relationship. If an employer can demonstrate the place of business, including office files and computer systems, are easily accessible to coworkers, customers, and others, the employee does not enjoy a reasonable expectation of workplace privacy. This includes e-mail.

However, if an employer accesses an employee's e-mail, this action would constitute a search and, thus, the employer must meet the reasonableness standard. Since an employer has the right to supervise a business and see that it is efficient, a search warrant does not have to be obtained to gain access to routine, noninvestigatory business information.

In essence, although the Fourth Amendment provides protection in certain narrowly defined situations, it does not guarantee your privacy in the workplace, or as of this writing, your e-mail.

## FEDERAL LEGISLATION

Beyond federal case law, privacy advocates have been concerned about legislative initiatives. One problem has been the legal system's inability to keep pace with technological developments.

For example, Congress enacted Title III of the Omnibus Crime Control and Safe Streets Act in 1968 to protect certain privacy rights. But it became apparent, especially in the 1980s, that the act was not comprehensive enough. It only provided protection for wire and oral communication and the aural interception of voice communication.

Human-to-computer communication and computer-to-computer communication were not protected.[24]

Recognizing the gap, Congress passed the Electronic Communications Privacy Act of 1986 (ECPA) to eliminate some of the loopholes.[25] But while the ECPA does apply to public e-mail, private e-mail is not protected.

A public e-mail system functions much like a public telephone company and enables you to send and receive messages.[26] If such a company illegally discloses a message's content to unauthorized individuals, it can be sued. In contrast, a private, internal e-mail system, such as one set up by a company, only has to guarantee that "unauthorized parties cannot gain access to your e-mail."[27]

## STATE LEGISLATION AND CASE LAW

A body of state legislation and case law exists that specifically addresses e-mail workplace privacy.[28] In California, for instance, section 630 of the California Penal Code states:

The Legislature hereby declares that advances . . . have led to . . . new devices and techniques for the purpose of eavesdropping on private communications and that the invasion of privacy resulting from the continual and increasing use of such devices . . . has created a serious threat to the free exercise of personal liberties and cannot be tolerated in a free and civilized society.[29]

E-mail that was illegally accessed could fall into this category. Code violations would render any illegally obtained information inadmissible. However, the California code supports the legitimate use of the latest technologies, which could include accessing an employee's e-mail, if a legal search warrant had first been obtained.[30]

With regard to case law, one relevant example is *Alana Shoars v. Epson American*

(No. 073234. Super. Ct. No. SWC 112749.1994). A court examined state law and the state constitution to determine if an employer could legally access, print out, and read an employee's workplace e-mail.[31]

Alana Shoars was hired by Epson to teach employees how to use the company's e-mail system to communicate with external parties via an MCI setup. Shoars issued personal passwords to each employee and told them their e-mail would be considered private communication. But Epson management did not share Shoar's viewpoint about confidentiality. Employee e-mail was frequently accessed, printed out, and read.

When discovered, Shoars confronted her supervisor and demanded the practice be stopped. He threatened to have her fired if she continued to pursue the matter. Shoars refused and was fired. She then filed suit claiming that Epson's actions violated the California Penal Code and the California State Constitution.

Epson management claimed they did not violate the law. Employees were provided with computer tools to tie their internal e-mail system with MCI's operation. This meant that messages were automatically logged and downloaded to Epson's computers (for example, for troubleshooting).

More pointedly, Epson claimed its supervisors had to scan and read through this information as a part of the support process. Thus, Epson argued the law had not been violated.

After hearing the arguments, the court ruled in Epson's favor. The company had not engaged in illegal tapping or unauthorized connections with a telephone line. The court further held the act of downloading messages, as practiced by the company, did not constitute a violation. It similarly ruled in favor of Epson on other grounds.

A more recent state workplace e-mail privacy case, *Restuccia v. Burk Technology, Inc.* (5 Mass. L. Rptr. 712. 1996), was heard by

THE LAW

the Massachusetts State Court in 1996.[32] This case involved two employees who were fired by a private employer, Burk Technology. The company contended the firings were based on the content of their internal workplace e-mail messages and for the excessive use of the company's e-mail system[33] The employees subsequently filed suit in state court on the grounds that accessing and reading their workplace e-mail messages violated the Massachusetts wiretapping statute.[34]

The Burk Technology e-mail system had been created for internal communication between employees and required the use of a password. Employees had not been told their messages were automatically saved or that e-mail messages could be accessed by company personnel. Furthermore, Burk Technology did not have a company policy prohibiting communication of nonbusiness messages via the e-mail system. But it did have a written policy prohibiting excessive system use.[35]

The Massachusetts State Court eventually ruled that an employer's accessing and reading of e-mail messages, which had automatically been saved, did not violate the state wiretapping statute. The court further held this practice fell under the "ordinary course of business" exemption clause and as such constituted protected company action.[36]

Finally, in a 1999 case, *McLaren v. Microsoft* (No. 05-97-00824-CV, Tex. App. LEXIS 4103), a Microsoft employee was suspended and subsequently fired for sexual harassment and inventory irregularities. Bill McLaren subsequently sued Microsoft for invasion of privacy after company officials decrypted his private password, and retrieved and read information in his "personal" file folders. McLaren claimed an objective reasonable expectation of privacy in this case. Even though the files in question were on his work computer owned by Microsoft, the information was obtained from material placed in his e-mail "personal files folders."

The court ruled that Microsoft's interest in preventing inappropriate, unprofessional or even illegal activity over its e-mail system outweighed McLaren's privacy claim. Therefore, the court ruled that McLaren did not have an objective reasonable expectation of privacy in his workplace e-mail.[37]

## CONCLUSION

Online privacy is afforded protection under various federal and state constitutions and legislation, but it is not fully extended to the workplace. Once you become an employee, much of this protection is lost or at least significantly diminished.

For example, most employer privacy guidelines prohibit "unauthorized parties" from accessing and disclosing employee e-mail and other personal workplace effects. The employee is also protected from an unreasonable and warrantless search. An employer, however, has sweeping powers to monitor and disclose an employee's e-mail without prior knowledge or consent.

While an employee's expectation of e-mail privacy is shaky at its best, two avenues of protection may be passwords and addresses. In the former, employees are assigned private passwords; thus the expectation of privacy exists. In the latter, if e-mail is addressed to a specific individual, the sender believes the message will be sent to that person alone.

But employers have disputed these contentions. The hardware and software required to use e-mail are company owned—these tools are subject to company review. An employer also has a legitimate right to supervise employee work, including e-mail, for quality control.

## Federal and State Response

E-mail privacy supporters have filed suit in federal and state courts on the grounds that employer monitoring of e-mail violates the law. Specifically, federal courts have been asked to decide if this practice violates the Fourth Amendment.

To-date, the federal courts have declined to extend such protection to workplace e-mail. Most constitutional scholars predict that future Fourth Amendment protection would only be extended to workplace e-mail in cases where obvious abuses exist.

State courts have generally followed suit. Like federal court, most scholars maintain that privacy protection would only be afforded in extreme circumstances. Workplace privacy advocates also claim that federal and state legislation should offer some protection. However, as of this writing, they have provided limited protection at best.

Federal and state law have also made a distinction between a public employer (e.g., the government) and a private employer. Generally speaking, if you work for a public employer, you have slightly more privacy protection. A private employer has more leeway and can provide significantly less protection.

## Employer Code

If an employer has expectations for employee use of company property, including e-mail, formal guidelines should be drafted and distributed (e.g., each party's legal/ethical responsibilities). When designing such policy, the employer should also define the workplace privacy that will be afforded to the employee, regardless of the communication form.

For example, does existing company policy provide privacy protection to an employee's paper files, desk drawers, and voice mail? If so, how does it compare with the privacy afforded to e-mail? If e-mail has been singled out for special treatment, the employee should be informed.[38]

Workplace privacy should also meet the needs of both the employer and the employee. But it can be a very fine line. An employer has a right to monitor a business. But the employee has a right to expect a positive and secure workplace. If, for instance, e-mail is going to be monitored, the employee should be warned. The least intrusive method of obtaining the desired information should also be used. This access should be limited in scope, and the employer should preserve as much of the employee's privacy as possible.

In conclusion, individuals have privacy protection on the federal and state levels. But in reality, much of the protection is illusionary once you become an employee. Until a body of law is created that is more favorable to employee privacy rights, your workplace e-mail will be subject to employer monitoring.

## REFERENCES/NOTES

1. Constitution, Amendment IV.
2. California Constitution, Article I, Section I.
3. *New Jersey v. TLO.* 469 U.S. 325. 1985.
4. *United States v. Charles Katz.* 389 U.S. 347. 1967.
5. Ibid., 359.
6. *Richard Schowengerdt v. General Dynamics et al.* 823 F. 2d. 1328. 944 F. 2d. 483. 1987.
7. *Gerald W. Shields v. David Burge et al.* 874 F. 2d. 1201. 1989.
8. Cases have, however, been heard (unlike the time of this book's third edition).

THE LAW

9. *Black's Law Dictionary* (St. Paul, MN: West Law Publishing, 1979), 1059.

10. Ibid., 1261.

11. *O'Connor v. Ortega.* 480 U.S. 717.

12. Ibid., 717.

13. *Richard Schowengerdt v. General Dynamics et al.* 823 F. 2d. 1328. 944 F. 2d. 483. 1987.

14. *Michael A. Smyth v. The Pillsbury Company.* 914 F. Supp. 97. 1996, downloaded from www.Loundy.com.

15. Ibid.

16. Ibid.

17. *John Bohach et al. v. the City of Reno.* 932 F. Supp. 1232. 1996, downloaded from LEXIS.

18. Ibid.

19. Ibid.

20. Ibid.

21. *Richard Fraser v. Nationwide Mutual Insurance Co.* 135 F. Supp. 2d 623 at 1/2. 2001, downloaded from LEXIS.

22. Ibid., 5.

23. Ibid., 1.

24. Steven Winters. "Do Not Fold, Spindle or Mutilate: An Examination of Workplace Privacy in Electronic Mail," *Southern California Interdisciplinary Law Journal* 85 (spring 1992), 45–46.

25. Electronic Communications Privacy Act of 1986. 100 Stat 1848. Section 2510, 1.

26. Electronic Communications Privacy Act. viii.

27. Ibid.

28. California Constitution, Article I, Section I.

29. California Penal Code, Section 630.

30. California Penal Code, Section 631.

31. *Alana Shoars v. Epson America, Inc.* No B 073234, Calf. Super. Crt. No. SWC112749.

32. Jenna Wischmeyer. "E-Mail and the Workplace," 1998, downloaded from raven.cc.uKans.edu.

33. Ibid.

34. Ibid.

35. Ibid.

36. Ibid.

37. *Bill McLaren v. Microsoft.* No. 05-97-00824-CV, 1999. Tex. App. LEXIS 4103. 1999.

38. Electronic Mail Association. *Access to and Use and Disclosure of Electronic Mail on Company Computer Systems: A Tool Kit for Formulating Your Company's Policy* (September 1991), 1.

## SUGGESTED READINGS

### Case Law

*Bonita Bourke v. Nissan Motor Corporation in U.S.A.* (B 068705. Super. C. 1993).

*Bonita Bourke v. Nissan Motor Corporation in U.S.A.* (No. YC003979).

*Cameron v. Mentor Graphics* (No. 716361. Cal. Spr. Ct.).

*Francis Gillard v. Schmidt* (570 F. 2d. 825. 1978).

*Fraser v. Nationwide Mutual Insurance Co.* (135 F. Supp. 2d 623. 2001).

*Lacrone v. Ohio Bell Co.* (182 N.E. 2d. 15. 1961).

*McLaren v. Microsoft* (No. 05-97-00824-CV, 1999 Tex. App. LEXIS 4103).

*National Treasury Employee Union v. William Von Raab* (489 U.S. 656. 1989).

*O'Connor v. Ortega* (480 U.S. 709. 1987).

*Ray Oliver v. Richard Thornton* (466 U.S. 170. 1984).

*Rhodes v. Graham* (37 S.W. 2d. 46. 1931).

*Roach v. Harper* (105 S.E. 2d. 564. 1958).

*Alana Shoars v. Epson American, Inc.* (No. 073234. Super. Ct. No. SWC 112749. 1994).

*Samuel Skinner v. Railway Labor Executives Association* (489 U.S. 602. 1989).

*Watkins v. Berry* (704 F. 2d. 577. 1983).

### Federal and State Legislation

Computer Security Act of 1987. 101 STAT. 1724. P.L. 100-235. 1988.

Electronic Communications Privacy Act of 1986. 100 STAT. 1848. P.L. 99-508. 1986.

Privacy for Consumers and Workers Act. Hearing. S. 984. 1993.

Title III Wiretapping and Electronic Surveillance. 82 STAT. 211. P.L. 90-351. 1968.

**Journals, Magazines, and Online Documents**

American Management Association International. "Electronic Monitoring and Surveillance: 1997 AMA Survey." 1997, downloaded from www.cybersquirrel.com.

Barlow, John Perry. "Electronic Frontier: Bill O' Rights." *Communications of the ACM* (March 1993), 21–23.

Baumhart, Julia. "The Employer's Right to Read Employee E-Mail: Protecting Property or Personal Prying," *LABLAW* 8 (fall 1992), 1–65.

Boehmer, Robert. "Artificial Monitoring and Surveillance of Employees: The Fine Line Dividing the Prudently Managed Enterprise from the Modern Sweatshop." *Dickinson Law Review* (spring 1992), 593–666.

"Current Legal Standards for Access to Papers, Records, and Communications." 1999, downloaded from www.cdt.org.

CyberSpace Law Center. "Workplace Privacy Bibliography." 1998, downloaded from www.cybersquirrel.com.

Electronic Mail Association. *Access to and Use and Disclosure on Company Computer Systems: A Tool Kit for Formulating Your Company's Policy.* September 1991.

Electronic Mail Association. *Protecting Electronic Messaging: A Guide to the Electronic Communications Privacy Act of 1986.*

Tuerkheimer, Frank. "The Underpinnings of Privacy Protection." *Communications of the ACM* (August 1993), 69–73.

Witte, Lois. "Terminally Nosy: Are Employers Free to Access Our Electronic Mail?" *Dickinson Law Review* (1993), 545–573.

## GLOSSARY

*Electronic Communications:* Any transfer of signs, signals, writings, images, sounds, or data transmitted via wire, radio, electromagnetic, photoelectric, or photo-optical systems.

*Electronic Communications Privacy Act (ECPA):* This act was passed in 1986 to close some regulatory loopholes that existed in the Omnibus Crime Control and Safe Streets Act. This protects the acquisition of the contents of any wire, electronic, or oral communication via an electronic or mechanical device.

*Objective Expectation of Privacy:* Society agrees with an employee that he or she has a reasonable expectation of privacy in workplace items.

*Precedent:* An established rule of law set by a previous court deciding a case involving similar facts and legal issues.

*Stare Decisis:* Similar cases should be decided in a similar way.

*Subjective Expectation of Privacy:* The right to expect that personal workplace items such as a briefcase, personal mail, desk, files, and wastebaskets will not be subject to an unreasonable employer search.

THE LAW

# 19 The First Amendment and Online Obscenity

The new communication technologies and their applications have raised constitutional questions, including the First Amendment implications of regulating online obscenity. Relevant legislation and case law concerning various aspects of this issue are discussed in this chapter.

For online obscenity, the main focus is the Communications Decency Act (CDA) and the Child Online Protection Act (COPA). Background information is presented first, followed by an examination of relevant case law.

## THE COMMUNICATIONS DECENCY ACT

In early 1995, U.S. Senators James Exon (D-Nebraska) and Slade Gorton (R-Washington) sponsored the CDA to regulate online sexual content. According to Senator Exon, this legislation, which was later incorporated in the 1996 Telecom Act, was necessary. Some individuals were using online services to communicate with children about inappropriate sexually oriented topics. Some adults were also offended by the sexually explicit material that was available over the Internet.[1]

In the past, sexually explicit speech was regulated on the basis of consent. For example, the 1934 Communications Act only regulated nonconsensual sexual speech that was unwanted by one of the parties.[2] But the CDA made no distinction between consensual and nonconsensual speech. It criminalized all obscene, lewd, lascivious, filthy, and indecent communication distributed via a telecommunications device. This law also placed the regulatory burden on providers and services.[3]

### The CDA's Constitutionality

The CDA was challenged in court on constitutional grounds. In 1996, two lower courts held portions of the CDA unconstitutional. CDA opponents indicate the act would have a chilling effect on computer communication. Chilling implies that a threat of regulation may make you think twice about providing what may be inappropriate information. In this case, you may not post some questionable material online because you thought it might be illegal—communication is stopped before it can actually take place. In this light, it could be viewed as a form of censorship.

Critics also contend the CDA would create a new "measuring stick" to determine if a message was obscene. Under the CDA, only material judged appropriate for children would be permitted online.[4]

Another concern was the targeting of obscene as well as indecent speech. Indecent speech is defined as nonobscene language that is sexually explicit or includes profanity.[5] Although most First Amendment scholars agree that obscene speech is not protected, there is no precedent for denying constitutional protection for indecent speech. In fact, the Supreme Court has ruled that indecent speech is protected.

In the precedent-setting case, *Miller v. California*, nonobscene speech was ruled constitutional. Only language that appeals to prurient interests is patently offensive, and lacks serious literary, artistic, political, and scientific value is prohibited.[6] Indecent speech does not fall within this definition and has been protected by the First Amendment.

Furthermore, in *Sable Communications of California Inc. v. Federal Communications Commission*, the court ruled that sexual expression, which is indecent but not obscene, was protected. For example, the ability of a child to potentially retrieve a sexually explicit telephone message does not render such speech unconstitutional.[7]

***Internet Regulation and the Broadcast Model.*** When a new communication form is born, Congress and other bodies must determine if and how it will be regulated. In this process, established communication systems may be examined as regulatory models.

For the Internet, the broadcast industry served as one example. Broadcast regulation has been based, in part, on the spectrum scarcity theory—since the spectrum is finite, only a limited number of individuals could receive a license. Thus, FCC involvement in this arena could, depending on your viewpoint, be considered a legitimate use of power.

But when applied to computer communication, critics contend the scarcity argument is not valid. As described in previous chapters, the Internet and online services constitute a vast and rapidly growing infrastructure. Essentially, "whenever you add a computer to the Internet, you increase the Internet size [sic] and capabilities."[8]

A similar principle applies to the pervasive medium theory, which argues that the broadcast industry could be regulated since messages are "thrust" on a sometimes unsuspecting audience. But like spectrum scarcity, this regulatory stance may not be valid in computer-based systems.

Computer communication on the Internet, for instance, requires you to take a number of affirmative steps to retrieve information. "Exposure to the content is primarily driven by choice with little risk of unwitting exposure to offensive or indecent material."[9] Consequently, the broadcast model may not be a valid Internet regulatory model even though it was used to explore such options.

### Internet Concerns

One key factor, which provided impetus for Internet regulation, was a concern that some Internet material may be inappropriate for children. But CDA critics indicated the online community was developing blocking and screening software and options. Parents, teachers, and other responsible adults would be able to control the material children could retrieve.[10]

In one case, Cyber Patrol functioned by producing a "CyberNOT" list of objectionable sites, and it organized questionable content into categories. In the violence and profanity entry, obscene words and communication that encouraged extreme cruelty against any animal or person were included. Sites with these characteristics would be placed on the list and minors would be blocked from entering.[11]

### The Case: *American Civil Liberties Union v. Attorney General Janet Reno*

The CDA gave the U.S. Attorney General the authority to investigate and prosecute any party in violation of the act's indecency and patently offensive provisions. Specifically, the indecency provision (Section 223) states:

. . . any person in interstate or foreign communication who, "by means of a telecommunications device, knowingly . . . makes, creates, or solicits" and "initiates the transmission" of "any comment, request, suggestion, proposal, image or other communication which is obscene or indecent, knowing that the recipient of the communication is under 18 years of age," shall be criminally fined or imprisoned.[12]

The patently offensive provision (Section 223)

. . . makes it a crime to use an "interactive computer service" to "send" or "display in a manner available" to a person under age 18, "any comment, request, suggestion, proposal, image, or other communication that, in context, depicts or describes, in terms patently offensive as measured by contemporary community standards, sexual or excretory activities or organs, regardless of whether the user of such service placed the call or initiated the communication."[13]

In response, the American Civil Liberties Union (ACLU) and other plaintiffs filed suit in federal court to halt the provisions' enforcement. They requested a temporary restraining order (TRO), and Judge Ronald Buckwalter of the U.S. District Court for the Eastern District of Pennsylvania, granted a limited TRO.

The grounds? The terms "indecent" and "patently offensive" were unconstitutionally vague and required interpretation. Following this action, a panel of three judges from the Third Circuit Court of Appeals was convened to examine this case. Attorney General Janet Reno agreed not to investigate or prosecute violators until the judges rendered their decision. The court said Reno could file charges for violations in the future, if the provisions were found to be constitutional.[14]

Some case highlights are as follows:

1. The government said the ACLU was overreacting to the CDA's language since speech would only be restricted in narrow circumstances. The government cited supporting factors. For example, a telecommunications facility/service would not be in violation of the CDA if it acted in good faith and took "reasonable, effective, and appropriate actions to restrict or prevent access" to inappropriate materials by a minor.[15]

2. The ACLU did not dispute that obscene, pornographic, and harassing material was inappropriate on the Internet. It argued the CDA reached too far and violated the First Amendment since it "effectively bans a substantial category of protected speech from most parts of the Internet."[16]

3. The government indicated that "shielding minors from access to indecent materials is the compelling interest supporting the CDA."[17] According to the court, "whatever the strength of the interest the government has demonstrated in preventing minors from accessing 'indecent' and 'patently offensive' material online, if the means it has chosen sweeps more broadly than necessary . . . , it has overstepped onto rights protected by the First Amendment."[18] The court then had to answer the question: Did the government overstep it bounds?

4. What may be objectionable to one person in one city may not be objectionable to another person in another city. *Angels in America,* for instance, a play about sexuality and AIDS, may be unacceptable in certain communities. But the play won two Tony awards and a Pulitzer Prize, and some adults could view it as appropriate material for older minors.[19]

If introduced online, however, this same material would fall under the CDA's purview. Other nonobscene material, ranging from certain *National Geographic* photographs to sculptures to textual descriptions, could follow suit and would be subject to the "CDA's criminal provisions."[20]

5. The ACLU had to demonstrate it was likely to prevail on the case's merits. If the case went to court, the ACLU would most likely win.

6. The ACLU had to convince the court that if the TRO were not granted, irreparable harm would occur. Citing past case law, the ACLU claimed halting protected speech, even for a brief time, constituted irreparable harm.

7. Past courts have ruled that criminalizing protected speech has a chilling effect on the speaker's freedom of expression, also resulting in irreparable harm.

8. The ACLU had to establish that a TRO was in the public's best interest. Citing past case law, it was alleged that an open marketplace of ideas and the preservation of protected speech are always in the public's interest.

### The Ruling

The court eventually ruled the CDA was overly intrusive, and the government's claim of a compelling interest in restricting these materials from children was outweighed by the right to free speech.[21]

Criminal prosecution resulting from a violation could also subject the content provider to public damnation and would create a financial hardship. In fact, content providers could back away from questionable material out of fear of prosecution if they make a mistake as to what constitutes indecent and/or patently offensive material.[22]

The terms *indecency* and *patently offensive* are also inherently vague. The government could not identify the specific community standards that would be used to judge questionable online material.[23] The court further

noted that Congress could have supported the development of blocking software, thus placing the decision-making process in the hands of parents and educators.

Congress could have also followed a well-traveled path by looking toward the print media for guidance. Nonobscene, but indecent and/or patently offensive books and magazines, are available everywhere.[24] The government, however, is not primarily responsible for controlling a minor's exposure to these materials. The decision making rests squarely on the shoulders of adults.

The court also indicated that the Internet "may fairly be regarded as a never ending worldwide conversation," and the government cannot use the CDA to "interrupt that conversation. As the most participatory form of mass speech yet developed," the Internet merits the highest level of protection.[25]

Furthermore, while the lack of government regulation "has unquestionably produced a kind of chaos, . . . The strength of the Internet is that chaos. Just as the strength of the Internet is chaos, so the strength of our liberty depends on the chaos and . . . the unfettered speech the First Amendment protects."[26] Thus, the court held the indecency and patently offensive provisions of the CDA unconstitutional.[27]

The court granted the TRO and instructed Attorney General Janet Reno, and other parties, that they were "enjoined from enforcing, prosecuting, investigating, or reviewing any matter" concerning these matters.[28]

The Attorney General subsequently appealed the case and in June 1997 the U.S. Supreme Court heard *Reno et al. v. American Civil Liberties Union et al* (117 S.Ct. 2329. 1997). In a 7–2 decision with Justice John Paul Stevens writing for the majority, the Supreme Court upheld the lower court, finding the indecency and patently offensive portions of the CDA violated free speech protected by the First Amendment.[29]

Justice Stevens wrote, "We are persuaded that the CDA lacks the precision that the First Amendment requires when a statute regulates the content of speech. In order to deny minors access to potentially harmful speech, the CDA effectively suppresses a large amount of speech that adults have a constitutional right to receive and address to one another."[30]

The Court found those portions of the CDA unconstitutionally vague, overbroad, and not sufficiently narrowly tailored based on the fact that, ". . . the scope of the CDA is not limited to commercial speech or commercial entities. Its open-ended prohibitions embrace all nonprofit entities and individuals posting indecent messages or displaying them on their own computers in the presence of minors. The general, undefined terms 'indecent' and 'patently offensive' cover large amounts of nonpornographic material with serious, educational or other value."[31]

## CHILD ONLINE PROTECTION ACT

Despite this strong endorsement of free speech, Congress passed the Child Online Protection Act (COPA), also known as "CDA-II," in October 1998. COPA was more narrowly drawn than the CDA. Nevertheless, free speech advocates still opposed it on First and Fifth Amendment grounds.

COPA made it a crime to permit anyone under the age of 17 to access materials, which were construed as harmful to minors, from the web. Any communication, image, recording, or information that was obscene, or met other criteria, would fall under this aegis. The latter included materials that:

- "the average person, applying contemporary community standards, would find pandering to . . . prurient interests";

- depict ". . . in a manner patently offensive with respect to minors, an actual or simulated sexual act . . . or a lewd exhibition of the genitals . . ."; and
- "taken as a whole, lacks serious literary, artistic, political, or scientific value for minors."[32]

Additionally, an individual who "devotes time, attention, or labor to such activities [e.g., making such materials available to minors], as a regular course of . . . trade or business, with the objective of earning a profit . . ." would be liable.[33]

An individual, however, who acted in good faith to restrict minors from accessing proscribed online material, would be exempt from prosecution. This would require the use of a credit card, adult access code or other reasonable identification scheme.[34]

In sum, COPA would apply in the following situations:

1. When material harmful to minors is knowingly communicated.
2. When the communication is made for commercial purposes.
3. When the material is communicated via the web.
4. When you are the person actually posting the illegal material.

It should also be noted that COPA would still permit adults to receive "illegal" materials. This would be possible through the use of one of the aforementioned identification systems.[35]

### Objection

Despite COPA's built-in exemptions, the ACLU, the Electronic Frontier Foundation, and other parties, filed suit before the U.S. District Court for the Eastern District of Pennsylvania in November 1998. They requested a TRO to halt enforcement of

THE LAW

COPA. Much like the CDA, U.S. District Court Judge Lowell A. Reed, Jr., had to answer pertinent questions. These included the following:

- Based on the presented information, was it likely that the ACLU and the other suing parties would win if the case went to trial?
- Would less harm result by granting the TRO or by denying it?[36]

After deliberating, Judge Reed granted a preliminary TRO against government enforcement of the COPA. The TRO was initially granted for 14 days and then extended for an additional 2 months until relevant testimony could be heard.[37]

## The Case

The ACLU argued COPA was unconstitutionally vague under the First and Fifth Amendments. It was also overbroad, created an economic and technological burden for online speakers, would chill speech, and violated speech protected for minors. Attorney General Reno and the Justice Department countered and argued that COPA was constitutional. It was narrowly tailored, only targeted materials harmful to minors, and would not prevent adults from gaining access to such materials through a technologically and economically feasible identification system.[38]

Eventually, Judge Reed ruled the government had failed to show that COPA required the least restrictive means available to prevent access to materials that were harmful to minors. Furthermore, "The plaintiffs' . . . fears of prosecution under COPA will result in the self-censorship of their online materials in an effort to avoid prosecution, and this court has concluded . . . such fears are reasonable given the breadth of the statute."[39] Reed concluded by

writing that no one, including the government, has an interest in enforcing an unconstitutional law and that the public is not served by the enforcement of such a law.[40] Based on this rationale, Judge Reed formally granted a TRO against enforcement of the COPA (February 1, 1999).

The Attorney General appealed the case and it was heard by the Third Circuit which affirmed Judge Reed's decision but for an entirely different legal reason.[41] The court of appeals ruled that COPA was unconstitutionally overbroad because it depended on the use of "contemporary community standards" (the views of the average person in the community) to determine if material was harmful to minors.

Since the online universe has no specific geographic boundaries, the court reasoned that any material could be classified as harmful to minors if the standard of the most "puritan communities in any state"[42] were used. The language of COPA would chill speech and would lead to the potential deletion of vast amounts of protected material.[43] Therefore, the court of appeals allowed Judge Reed's injunction to stand. The Attorney General appealed the Third Circuit's decision and the U.S. Supreme Court agreed to hear *Ashcroft v. ACLU*.[44]

In a narrow decision, the Supreme Court ruled that this use of contemporary community standards, in and of itself, did not violate the First Amendment.[45] The Justices also stated they would not examine other legal questions concerning COPA's constitutionality because the Third Circuit had not yet done so. Consequently, the Supreme Court sent the case back to the court of appeals to answer all relevant legal questions concerning COPA's constitutionality. In the meantime, Judge Reed's injunction against enforcing COPA remained intact.[46]

In March 2003, following the instructions of the Supreme Court, the Third Circuit reexamined COPA (*ACLU v. Ashcroft*. No.

THE LAW

99-1324. 2003). The court of appeals had to determine if the U.S. district court's preliminary injunction against COPA constituted an abuse of judicial discretion or an obvious legal error.[47]

The court of appeals did find that the government has a compelling interest in protecting minors from harmful materials. But COPA's provisions were not narrowly tailored to achieve this goal.[48] The court also

- indicated that COPA's requirement to prevent minors from accessing harmful materials, "automatically impacts nonobscene, sexually suggestive speech that is otherwise protected for adults";[49]
- examined the use of the term "minor." It cited that no distinction is made between material inappropriate for an infant, a five year old or someone about to turn 17;[50]
- examined COPA's definition of the term "commercial purposes." COPA "imposed content restrictions on a substantial number of 'commercial,' nonobscene speakers in violation of the First Amendment."[51] The court added that COPA's affirmative defenses for accessing material, such as requiring the use of a credit card or adult personal identification number, would likely deter many adults from accessing content harmful to minors thus chilling speech for adults;[52] and
- indicated that COPA failed to employ the least restrictive means to protect minors from harmful material (e.g., blocking and filtering software).[53] COPA's language also constituted overbreadth since it "places significant burdens on web publishers' communication of speech that is constitutionally protected as to adults and adults' ability to access such speech."[54]

In conclusion, the Third Circuit ruled that the district court did not abuse its judicial discretion or commit a clear legal error in granting the preliminary injunction against COPA, since the ACLU would likely succeed on the merits in establishing that COPA was unconstitutional.[55] Therefore, the court of appeals affirmed the issuance of a preliminary injunction.[56]

## Other Issues

We should note that the CDA was not the only online obscenity case. While the CDA ran into a roadblock, partly due to its indecency and patently offensive provisions, other obscenity regulations still apply. To-date, most of these online criminal obscenity cases have involved online interstate transportation (from one state to another state) of child pornography and have been decided by lower courts.

The most highly publicized case is *United States v. Robert and Carleen Thomas.* Husband and wife, Robert and Carleen Thomas, were convicted of federal obscenity charges (18 U.S.C. Section 1462/1465, 1996) in the U.S. District Court for the Western District of Tennessee. They used their bulletin board system to transport obscene computer files from California to other states. The recipients included an undercover agent in Tennessee.[57]

The Thomas' appealed, and the Court of Appeals for the Sixth Circuit agreed to review the case. After careful scrutiny of the lower court decision, the Sixth Circuit Judges ruled the convictions should stand.[58]

## FILTERING

A sidebar online obscenity issue also arose. Should mandatory software be installed on public library computers to block access to Internet sites with sexually explicit content? Various library boards installed or considered installing blocking software to prevent minors from accessing various categories of speech including, "hate speech, criminal

THE LAW

THE LAW

activity, sexually explicit speech, adult speech, violent speech, religious speech, and even sports and entertainment."[59]

The ACLU, the American Library Association, and others have opposed such software use. They believe that library computers provide links to diverse and global information, which is subsequently made available to all segments of the American public. This includes users who might not otherwise have access to the Internet.[60]

Recent statistics indicate that 60% of the nearly 9000 U.S. public libraries provide Internet access to their patrons. A Nielsen survey further reveals that many people specifically go to the library to use the Internet.[61] In many instances, users opt to go online because it affords them more privacy when accessing information on assisted suicide and other, sensitive and controversial topics.

Library blocking software opponents generally agree that children should not be able to gain access to sexually explicit information, but believe that parents, teachers, and librarians should guide their Internet use—not restrictive and oftentimes flawed filtering software.[62] Proponents include library boards, the American Family Association, parents, librarians, and politicians.[63] They indicate the Internet can provide information, images, and chat rooms that are inappropriate for minors—the online world can be a dangerous place for children. They believe that blocking software could help protect minors from these pitfalls.

### Cases and Congress

The first case to examine the constitutionality of using blocking software on public library computers was heard in 1998 by the U.S. District Court for the Eastern District of Virginia. In *Mainstream Loudoun et al. v. Board of Trustees of the Loudoun County Library,* Judge Leonie M. Brinkema had to decide if the library board policy requiring mandatory blocking of child pornography, obscene material, and material harmful to juveniles, violated the First Amendment.[64]

Citing *ACLU v. Reno,* the precedent-setting online obscenity case, Judge Brinkema ruled that this policy was unconstitutional for these reasons:

- The policy was not narrowly tailored.
- There were less restrictive options to prevent minors from accessing online sexually explicit material.
- The policy restricted adults from obtaining materials that were deemed inappropriate for children.[65]

The Loudoun County Library Board considered appealing the decision, but the ACLU convinced them that their policy was unconstitutional. Instead, the board developed a new policy which allows both adults and minors to use either a filtered or an unfiltered computer.[66] While the ACLU has also convinced other library boards not to implement blocking software, library communities in some states continue the practice of library computer Internet filtering.

Congress has also played a role in this controversy through filtering and blocking software bills. Finally, in December 2000, Congress passed the Children's Internet Protection Act or CIPA. CIPA required libraries and schools that received federal funding for telecommunications expenses, to install filtering software on computers used by adults or children. Congressional intent was to block access to online obscenity and child pornography for adults and for children under the age of 17. Children would also be denied access to material deemed harmful to minors but protected for adults. Authorized library/school representatives could, however, unblock a given computer if it was being used by an adult for research or other legitimate purposes.[67]

After its passage, the American Library Association (ALA), the Multnomah County Public Library, and other groups challenged CIPA's constitutionality. These parties filed suit in U.S. District Court for the Eastern District of Pennsylvania in 2002 citing several legal flaws in CIPA's language.[68] The sequence of events is as follows:

1. The district court had to determine if current filtering technology would block only the prescribed, and not other, online material.[69]
2. After examining different filtering systems, the court ruled the current generation of programs could not meet this criterion.[70]
3. The court indicated the government did have a compelling interest in protecting children under the age of 17 and others who use public libraries, but the court had to decide if the use of filtering software was the least restrictive method of achieving CIPA's goals.
4. The court determined there were less restrictive methods of denying access to these materials. Thus, the court held the CIPA was unconstitutional.[71]
5. It should be noted the court did not speak to the issue concerning schools and school libraries accepting federal funding for computers.

The government appealed the lower court decision and due to the language of CIPA, the appeal went directly to the Supreme Court for review.

In June 2003 the Supreme Court handed down a 6–3 decision reversing the lower court on the grounds CIPA's filtering requirement did not violate the First Amendment. Writing for the Court, Chief Justice Rehnquist said filtering pornographic and other web sites harmful to minors, did not halt freedom of expression because, "A public library does not acquire Internet terminals in order to create a public forum for web publishers to express themselves, any more than it collects books in order to provide a public forum for the authors of the books to speak."[72]

Rehnquist also said public libraries must render decisions concerning which traditional publications are "suitable and worthwhile material" for their collections and must be permitted to make similar decisions concerning the appropriateness of material from the Internet.[73]

In response to the lower court's claim filtering would result in overblocking, the Supreme Court said "such concerns are dispelled by the ease with which patrons may have the filtering software disabled. When a patron encounters a blocked site, he need only ask a librarian to unblock it or (at least in the case of adults) disable the filter."[74]

The Court likewise concluded the constitutional rights of public library patrons were not violated by the filtering requirement but instead, the requirement only served to ensure public funds were spent for their authorized purpose.[75]

It should be noted, however, while Justices Breyer and Kennedy were among the six who voted to reverse the lower court, each wrote a separate opinion explaining why they did so. This is significant because both Justices voted, in part, on the assumption public librarians would unblock filtered sites for adults without delay.[76]

Justices Stevens, Souter, and Ginsburg cast dissenting votes for similar reasons. Justice Stevens, for example, said CIPA was unconstitutional because, "[a] federal statute penalizing a library for failing to install filtering software on every one of its Internet-accessible computers would unquestionably violate the First Amendment."[77]

In sum, the decision in this case marks the first time the Supreme Court has upheld the constitutionality of legislation designed to restrict Internet content. Unlike the CDA

THE LAW

and COPA, the Supreme Court held CIPA did not violate the First Amendment. The ACLU and other civil liberty groups disagree with the Court's decision and hope CIPA's constitutionality will be revisited if unblocking filtered sites for adults doesn't occur. Such a problem could make Justices Breyer and Kennedy reconsider their judicial stance on the constitutionality of CIPA.

In addition to federal legislation, more than half of the States have enacted or considered state Internet censorship legislation. To-date, federal courts have struck down various Internet censorship laws, but the future of such legislation is unknown.

## INTERNATIONAL RESPONSE TO ONLINE OBSCENITY

### CompuServe

Besides the United States, other countries have developed their own rules governing sexually explicit online material. For an online company, this situation has become a legal nightmare. If an operation is international in scope, how do you comply with each country's laws?

For example, in December 1995, the German government informed CompuServe that a number of its sexually oriented newsgroups contained materials that were unsuitable for children and violated German law. A German prosecutor insisted that CompuServe block this access in Germany.

CompuServe informed the German government it was not responsible for, and had no authority over, a newsgroup's content. Nevertheless, CompuServe's employees working in Germany were threatened with arrest if blocking was not initiated.[78]

CompuServe's CEO had to decide if he should comply with German law or keep the newsgroups and risk his employees' potential arrest. In the end, lacking the

technological means to isolate German users, CompuServe was forced to block access to these newsgroups for all its subscribers. Basically, German obscenity standards were globally imposed.[79]

To some individuals, the German government's decision was an echo of its past, a time when publishers were ordered to halt distribution of certain information based on its content. Some saw the German prosecutor's action as a "digital book-burning" with global implications.[80] "What could have been the most democratic medium in the world all of a sudden becomes the lowest common denominator of what is acceptable to all governments."[81]

The Germans denied they forced CompuServe to block the newsgroups and claimed they simply informed the company that local German law would be enforced. They also said they were not trying to make their standard a global legal standard.

CompuServe was criticized for its actions. Some individuals believed the company should have anticipated this problem. It had more than a million non-U.S. users and a history of similar problems with Germany prior to the 1995 incident.[82]

Given this past history, critics claimed that CompuServe and other U.S. online services should have been prepared for this latest legal salvo. CompuServe's lack of insight was costly, and its technicians were under pressure to quickly find a solution to this problem.

***The Case.*** While CompuServe blocked these newsrooms until suitable parental control software could be deployed, the Chief German Prosecutor indicted Felix Somm, the Managing Director of CompuServe, Germany, in 1997. The charge?—allowing the trafficking of violence, child pornography, and bestiality via various CompuServe newsgroups during 1995 and 1996.[83] These indictments mark the first time that Western authorities have filed charges against a commercial online service based on the content of material the service did not create.[84]

In May 1998, Somm was found guilty on 13 counts of assisting in the dissemination of pornographic writings and for negligent dissemination of publications morally harmful to youths *(People of Munich v. Felix Somm)*.[85] The court ruled that Somm had violated German law that criminalizes making ". . . 'simple' pornography (softcore nudity) available to children and 'hardcore' (child pornography and bestiality) available to anyone, . . . and with allowing the dissemination of Nazi, neo-Nazi materials, and violent computer games . . . in Germany."[86]

Judge Wilhelm Hubbert found Somm guilty on all 13 counts of complicity, handing down a two year suspended jail sentence with probation. CompuServe appealed the conviction, and in November 1999, a three-day hearing was convened in Munich.[87]

At this time, Chief Judge Laszlo Ember reversed the lower court overturning Somm's conviction. He ruled that, at the time of the violation, CompuServe did not have the technology to block the publication of the cited material. The judge further held that Somm had done all he could and he was merely a slave to his parent company. Finally, Judge Ember cited Germany's newly enacted Multimedia Law as an additional legal rationale for reversing Somm's conviction since it would only hold a service provider legally liable for material if they have the technical capability to block it.[88]

### Other Countries

Online censorship of obscenity, pornography, and other controversial material has not been restricted to any one country or part of the world, however. Government control over online content is a global problem.

For instance, France was the site of another Internet censorship case that occurred in 2000. Members of the French Union of Jewish Students filed suit in French court against Yahoo for allowing Nazi memorabilia to be sold over Yahoo's Internet auction site. Under French law, displaying such memorabilia is illegal, and the French court (Tribune de Grande de Paris. *LICRA v. Yahoo!, Inc.* No. RG 00/05308. November 2000), subsequently ordered Yahoo to block access to the site in France.

Yahoo executive Timothy Koogle was also personally accused of justifying and displaying the memorabilia of those committing crimes against humanity. While the court found Koogle not guilty, had he been convicted, the punishment would have been 5 years in prison and a U.S.$50,000 fine. Prior to the conclusion of the trial, however, the prosecutor requested that Koogle not be punished even if he was found guilty. Nonetheless, the French had censored the Internet in their country and had put an American executive on trial because of the content of his company's web site.[89]

Various other governments are attempting to control user access to sexual, political, and religious online information.

In Saudi Arabia, despite the existence of private ISP's, all Internet traffic must go through government servers that filter sites deemed "harmful to Islamic values."[90] Iran has mandated the use of filters to block access to web sites with sexual, religious, and political content. In one case, medical students cannot gain access to sites that contain information concerning the human anatomy.[91] Burma law requires computer owners to declare that they have a computer. Failure to do so can result in a 15-year maximum prison sentence.[92] Syria bans Internet access to all "individuals."[93]

Internet access in China has also been tightly controlled. Users are monitored, and according to Chinese law, are expected to

THE LAW

register with government authorities. Violations are treated seriously with possible imprisonment. For instance, in 1999, a computer technician received a two-year jail sentence for providing a dissident web site with the e-mail addresses of 30,000 Chinese online subscribers. Chinese officials have also shut down 300 cybercafes to keep users from accessing controversial information.[94]

In Turkey, an Internet user received a 10-month suspended sentence for using an online forum for criticizing the police for treating a group of blind protesters roughly. Russian authorities, for their part, are developing a plan that would monitor all information sent via the Internet within Russian borders.[95]

In sum, proposed and adopted Internet regulation is an international situation—from the United States to Russia. It is also ironic. For the first time in human history, millions of individuals can readily exchange ideas and information by tapping into the Internet and other information sources. But now, government regulation may stifle this free and open forum.

## THE V-CHIP AND THE FIRST AMENDMENT

Another technology with content censorship implications is the V-chip. As is described earlier in this chapter, the government is concerned about the availability of material harmful to minors. However, the Internet is not the only source of such material. Television and cable programming offer children and adults alike a wealth of sexually oriented and violent content. Based on this fact, the Clinton Administration was instrumental in influencing the broadcast and cable industries to implement technology that would enable a parent to electronically block out specific programs, channels, or time slots. This technology, dubbed the "v-chip" (violence chip), was extremely controversial because of its potential for abuse.

### Background

There have been public outcries about violence in the media for years. For example, it is estimated that American children witness thousands of televised murders and acts of violence by the time they're 14 years old.[96] Studies have also pointed to a significant correlation between viewing violent programming and committing violent acts.[97] In one case, "children exposed to violent video programming at a young age have a higher tendency for violent and aggressive behavior later in life than children not so exposed . . ."[98]

Congress addressed this issue as a part of the 1996 Telecom Act. The television and cable industries were expected to design technology that would give adults control over a child's television viewing options.

V-chip proponents applauded the new technology and claimed it empowered the viewer, particularly the parents of young children. They also claimed that it did not violate the First Amendment because it was analogous to the use of a remote control or to parental decision making.[99]

But opponents were troubled by potential First Amendment abridgements. Programming would have to be rated for the v-chip technology to work. Ratings categories would also have to be created, and programs would have to be appropriately slotted.

The television and cable industries had to develop a "voluntary" ratings system that would identify violent, sexually oriented, and indecent programming inappropriate for children. Programming would not be rated however, based on political or religious content.[100] If they failed, the FCC would develop its own guidelines, based on the recommendations from an advisory committee composed of parents, industry representatives, and others. The committee would be politically balanced and would represent diverse viewpoints.[101]

FCC intervention was not necessary, however, since the television and cable industries created guidelines prior to the deadline. In March 1998 the Commission formally adopted the "TV Parental Guidelines" and a multicategory content ratings system was developed. The system included the following categories: TV-Y (all children), TV-7 (older children), TV-G (general audience), TV-PG (parental guidance suggested), TV-14 (parents strongly cautioned), and TV-MA (mature audience).[102] In addition to developing a multicategory content ratings system, the industries also agreed to provide on-screen ratings icons and establish an oversight monitoring board.

Broadcasters were supposedly caught off guard about the v-chip. Industry representatives claimed that the amount of violent and sexually oriented programming aired by the commercial networks had actually declined. They supported this contention with a UCLA study that was conducted on the industry's behalf.[103] The industry had also earmarked monies to develop acceptable v-chip alternatives without requiring the use of a program ratings system.[104]

In operation, a program would be rated, and this information would be placed in the television signal's VBI.[105] When the program is received by a viewer's television set, the chip would determine if that particular program content had been blocked.[106]

Finally, the actual cost of the plan's implementation was disputed. According to Congressman Edward Markey (D-Massachusetts), who was instrumental in the 1996 Telecom Act's final design, a television set could be equipped with a v-chip device for $1. He also indicated that like the closed-caption chip, v-chip technology would eventually get cheaper.[107]

Markey's contention were contested by the Electronic Industries Association (EIA), a trade organization for television set manufacturers. The EIA indicated that "the chip could add as much as $72 to some low-end television sets ($20 to $50 for additional memory capability and another $10 or more for channel-blocking capabilities)."[108]

### Monitoring

Because the v-chip plan called for a ratings system, programming had to be monitored and assigned to certain categories.[109] Opponents indicated that this system violated the First Amendment. There was also a practical concern.[110] Unlike the film industry, which only has to rate some 700 to 800 films a year, thousands of hours of television programming would have to be evaluated.[111] Some questioned the networks' capabilities to evaluate so much programming. Congressman Markey indicated the networks had the capability and also suggested the television and cable industries should examine the Motion Picture Association of America's (MPAA) film ratings system.[112]

### The V-Chip and "Other" Programming.

The television and cable industries became alarmed at the "snowball" effect associated with the v-chip ratings plan. Initially, legislation called for regulating "violent" content only. However, the Act's actual language expanded this scope and also encompassed the regulation of "violent, sexual, and indecent material."[113]

This type of regulatory "creep" concerned First Amendment advocates. But during a meeting between the television industry and President Clinton, news and sports programming were exempt from the ratings process.[114] For journalists, this would protect their rights to make story content decisions. It also made legal sense. Any move to expand v-chip jurisdiction to include

THE LAW

news program content would never withstand constitutional scrutiny.[115]

Congressman Markey agreed with this analysis and indicated that controlling newsroom decisions was not the Act's intent.[116] He also promised to protect First Amendment rights, but insisted that a viewer's rights, parents and children in particular, must be given equal protection.[117]

More ominously, though, at least in the eyes of v-chip opponents, was his opinion that news programming was too violent:

I don't think anyone will deny that the rule of thumb in too many newsrooms in the United States is "if it bleeds, it leads." Too many newsrooms use that as their guide because they think it might draw more viewers, in the same way that other drivers slow down when they see a car crash on the highway.[118]

Based on this viewpoint, could regulation be extended to news in the future? This statement also has the potential to "chill speech." A broadcaster may think twice before airing an important but violent story.

In 1998, the FCC formally adopted the ratings system developed by the television and cable industries. The TV-G (General Audience) category, for instance, indicated that, "Most parents would find this program suitable for all ages . . . most parents may let younger children watch this program unattended . . ."[119] the rating would appear on the television screen and would trigger the v-chip.

Some proponents hailed this development as an empowering tool that would place television control back in the hands of consumers and parents. There were, however, two categories of opponents. The first group thought the ratings were too diluted—they did not provide adults with enough detailed information.

The second group viewed this entire scenario as a First Amendment violation. As one writer put it, "the thought of putting a device in every television set in the country with an umbilical that can lead back to the government, scares the hell out of us."[120]

Finally, other issues are still unresolved. Will future legislation be more restrictive? Although news is currently exempt, will this hold true down the road? As was the case with the 1996 ratings, which didn't satisfy numerous critics, would the ratings system be a moving target that could change every few years?

## CONCLUSION

As is evident, the issue of online obscenity and the First Amendment is multifaceted. On one hand, there is the compelling need to protect children. On the other, how do you do so while safeguarding other users' First Amendment rights? In the case of libraries, the answer may lie in the development of enhanced filtering software and the implementation of a dual standard system—one that would keep the software "on" for children and, potentially, turned "off" for adults.

The next two chapters explore other First Amendment issues. They range from cyberstalking to hate speech.

In one sense, the Internet can be viewed as a marvelous window to view and explore our universe. But for some individuals, this window can be shattered through its inappropriate use.

THE LAW

## REFERENCES/NOTES

1. Brock Meeks, "The Obscenity of Decency," 1995, 1, downloaded.

2. Ibid.

3. Telecommunications Act of 1996. P.L. 104–104, 66.

4. Electronic Frontier Foundation, *Constitutional Problems with the Communications Decency Amendment: A Legislative Analysis,* June 16, 1995, downloaded from http://www.eff.org/.

5. Ibid.

6. *Miller v. California.* (413 U.S. 15. 1973).

7. *Sable Communications v. Federal Communications Commission.* (492 U.S. 115. 1989).

8. Electronic Frontier Foundation, *Constitutional Problems,* 4, downloaded from http://www.eff.org/.

9. Ibid.

10. *American Civil Liberties Union v. Attorney General Janet Reno,* U.S. District Court Eastern District of Pennsylvania, June 11, 1996. No. 96–963. No. 96–1458. 14.

11. Ibid., 16.

12. Ibid., 3.

13. Ibid., 3.

14. Ibid., 2

15. Ibid., 4.

16. Ibid., 33.

17. Ibid., 31.

18. Ibid., 33.

19. Ibid., 32. *Note:* 25% of all new HIV cases in the United States occur in the 13- to 20-year-old age group.

20. Ibid., 32. *Note:* This section was concerned with the community standards/vagueness issue.

21. Ibid., 35.

22. Ibid., 36. *Note:* The speaker was also charged with the responsibility for screening potentially inappropriate online material. Therefore, content providers would have been required to decide what would constitute inappropriate materials for minors.

23. Ibid., 37.

24. Ibid., 37.

25. Ibid., 65.

26. Ibid., 66.

27. Ibid., 65.

28. Ibid., 65.

29. *Reno et al. v. American Civil Liberties Union et al.* 117 S.Ct. 2329. 1997, downloaded from www.aclu.org.

30. Ibid.

31. Ibid.

32. *American Civil Liberties Union et al. v Janet Reno et al.* No. 98–5591. February 1, 1999. U.S. District Court Eastern District of Pennsylvania, downloaded from www.aclu.org.

33. Ibid.

34. Ibid.

35. "ACLU Files Suit Challenging the Child Online Protection Act," *Tech Law Journal,* October 23, 1998, downloaded from www.techlawjournal.com.

36. *American Civil Liberties Union v. Janet Reno.* No. 98–5591. U.S. District Court Eastern District of Pennsylvania, November 20, 1998, downloaded from www.aclu.org.

37. Ibid.

38. *ACLU v. Reno.* No. 98–5591. February 1, 1999, downloaded from www.aclu.org.

39. Ibid.

40. Ibid.

41. *Ashcroft v. American Civil Liberties Union.* No. 00–1293 at 8. 2002, downloaded from LEXIS.

42. Ibid.

43. Ibid., 9.

44. Following the 2000 presidential elections, John Ashcroft replaced Janet Reno as Attorney General.

45. Ibid., 22.

46. Ibid.

47. Ibid., 20, 21.

48. Ibid., 22.

49. Ibid., 23.

50. Ibid., 26.

51. Ibid., 32.

52. Ibid., 35.

53. Ibid., 44.

54. Ibid., 47.

55. Ibid., 21.

56. Ibid., 54.

57. *United States v. Robert Thomas and Carleen Thomas* 1996 WL. 30477 at 1. Sixth Circuit Tennessee. January 29, 1996.

THE LAW

58. Ibid.

59. ACLU. Cyber-Liberties. "Censorship in a Box: Why Blocking Software Is Wrong for Public Libraries," 1998, downloaded from www.aclu.org.

60. Ibid.

61. Ibid.

62. Ibid.

63. Ibid.

64. *Mainstream Loudoun v. Board of Trustees of the Loudoun County Library*. No. 97-2049-A. 1998, *Tech Law Journal*, downloaded from www.techlawjournal.com.

65. Ibid.

66. ACLU. Cyber-Liberties. "Censorship in a Box: Why Blocking Software Is Wrong for Public Libraries," 1998, downloaded from www.aclu.org.

67. Children's Internet Protection Act. Pub. L. 106–554.

68. *American Library Association et al. v. United States*. No. 01–1303/No. 01–1322. 2002, downloaded from LEXIS.

69. Ibid., 5.

70. Ibid., 6.

71. Ibid., 9.

72. *United States v. American Library Association et al.* 2003 U.S. LEXIS 4799 at 11. No. 02–361. 2003, downloaded from LEXIS.

73. Ibid.

74. Ibid., 12.

75. Ibid., 14.

76. *United States v. American Library Association et al.* case Syllabus at 2/3, downloaded from LEXIS.

77. Ibid., 15.

78. Christopher Stern, "CompuServe Shuts Down Sex After German Protest," *Broadcasting & Cable* (January 8, 1996), 69.

79. Steven Vonder Haar, "Censorship Wave Spreading Globally," *Inter@ctive Week* 3 (February 12, 1996), 6.

80. Ibid.

81. Stern, "CompuServe Shuts Down Sex After German Protest," 69.

82. Ibid.

83. In the Name of the People Judgment of the Local Court Munich in the Criminal Case v. Somm, Felix Bruno. Local Court [Amtsgericht] Munich. File No. 8340 Ds 465 Js 173158/95.1998, downloaded from www.cyber-rights.org.

84. Stuart Biegel. "Indictment of CompuServe Official in Germany Brings Volatile Issues of Cyber-Jurisdiction Into Focus." UCLA Online Institute For Cyberspace Law and Policy. 1997, downloaded from www.gse.ucla.edu.

85. *People v. Somm.*

86. Andrew Gelman. "Sting Time for CompuServe in Germany: Online Service Head Indicted for Pornography." 1997, downloaded from www.excitesearch.netscape.com.

87. "Update: CompuServe Ex-Official's Porn-Case Conviction Reversed," Associated Press. November 17, 1999. Cyber-Rights & Cyber-Liberties (UK), downloaded from www.Cyber-Rights.org.

88. Edmund L. Andews. "German Court Overturns Pornography Ruling Against CompuServe," *New York Times* on the web, November 18, 1999, downloaded from www.NYTimes.com.

89. BBC. "Yahoo Boss Cleared Over Nazi Site." February 11, 2003. 1–2, downloaded. *Note:* The U.S. judicial system ruled that the French court did not have jurisdiction in this case.

90. Radio Free Europe/Radio Liberty. "20 Enemies of the Internet." August, 1999, 3, downloaded.

91. Letters of Protest and Press Release. "20 Enemies of the Internet," downloaded from www.rsf.fr/uk/.

92. Radio Free Europe/Radio Liberty. "20 Enemies of the Internet." August 1999, 2, downloaded.

93. Ibid., 3.

94. Letters of Protest and Press release. "20 Enemies of the Internet," downloaded from http://www.rsf.fr/uk/.

95. Special Issues and Campaigns. Freedom of Expression on the Internet, Human Rights Watch, 1999, downloaded from www.hrw.org.

96. U.S. Congress, Telecommunications Act of 1996, 104th Cong., 2nd session, section 551 (January 3, 1996).

97. Barri Lazar, "The V-Chip: Censorship or Technological Miracle," unpublished paper, April 23, 1996, 14.

98. Telecom Act, section 551.

99. "Why the Markey Chip Won't Hurt You," *Broadcasting & Cable* (August 14, 1995), 11.

100. Telecom Act, section 551

101. Ibid.

102. FCC, "Commission Finds Industry Video Programming Rating System Acceptable: Adopts Technical Requirements to Enable Blocking of Video Programming [the v-chip]." Report No. GN 98-3. March 12, 1998.

103. Christopher Stern, "Face-off on the V-Chip," *Broadcasting & Cable* (October 2, 1995), 19.

104. Ibid.

105. See footnote 1, Chapter 17, for VBI definition.

106. The set would be equipped with this function.

107. Art Brodsky, "The V-Chip Moves Toward Reality," *Broadcast Engineering* 37 (September 1995), 81.

108. Ibid.

109. "Why the Markey Chip Won't Hurt You," 11.

110. Ibid.

111. Sarah Haag, "The V-Chip: Censorship or Technological Miracle?," unpublished paper, April 23, 1996, 8.

112. Kent R. Middleton and Bill F. Chamberlin, *The Law of Public Communication* (White Plains, NY: Longman Publishing, 1994), 354. *Note:* The MPAA "rates films as, for example, 'G' for general audience and 'R' for restricted to those under 17 unless accompanied by a parent/guardian."

113. Telecom Act, section 551.

114. Christopher Stern, "Valenti Pledges Government-Free Ratings," *Broadcasting & Cable* (April 8, 1996), 14. *Note:* The meeting took place shortly after the Act's passage.

115. "Why the Markey Chip Won't Hurt You," 12.

116. Ibid., 15.

117. Ibid.

118. Ibid., 12.

119. NAB web site, downloaded 12/96.

120. "Why the Markey Chip Won't Hurt You," 15.

## SUGGESTED READINGS

### Case Law

*American Civil Liberties Union v. Ashcroft.* No. 99-1324. Third Circuit Court of Appeals. March 6, 2003. 1–53, downloaded from LEXIS.

*American Civil Liberties Union v. Attorney General Janet Reno.* No. 96-963. No. 96-1458. U.S. District Court for the Eastern District of Pennsylvania. June 11, 1996. 1–76, downloaded from LEXIS.

*Ashcroft v. American Civil Liberties Union.* No. 00-1293. U.S. Supreme Court. May 13, 2002. 1–22, downloaded from LEXIS.

*Miller v. California.* 413 U.S. 15. 1973. 1–25, downloaded from www.bc.edu.

Tribune de Grande de Paris. *LICRA v. Yahoo!, Inc.* No. RG. May 22, 2000. 1–2, downloaded from www.juriscom.net/.

Tribune de Grande de Paris. *LICRA v. Yahoo!, Inc.* No. RG 00/05308. November 2000. 1–21, downloaded from www.juriscom.net.

*United States v. American Library Association.* No. 02–361. June 23, 2003. 1–17, downloaded from LEXIS.

*United States v. Robert Thomas and Carleen Thomas.* 1996 WL. 30477. Sixth Circuit Tennessee. January 29, 1996, 1–68.

### Legislation, Journals, Magazines, and Online Documents

ACLU. "*ACLU v. Reno*, Round 2: Broad Coalition Files Challenge to New Federal Net Censorship Law." October 22, 1998. 1–2, downloaded from www.aclu.org.

THE LAW

ACLU. "*ACLU v. Reno*, Round 2: Internet Censorship Battle Moves to the Appeals Court." April 2, 1999. 1, downloaded from www.aclu.org.

ACLU. "*ACLU v. Reno*, Round 2: Rejecting Cyber-Censorship, Court Defends Online 'Marketplace of Ideas'." February 1, 1999, 1–2, downloaded from www.aclu.org.

ACLU. Civil Liberties. "Online Censorship in the States." 1998, 1–4, downloaded from www.aclu.org.

ACLU. "Federal Court Rejects Government Censorship in Libraries, Citing Free Speech Rights of Patrons." May 31, 2002, 1–2, downloaded from www.aclu.org.

"ACLU Files Suit Challenging the Online Protection Act." *Tech Law Journal*. October 23, 1998, 1, downloaded from www.techlawjournal.com.

ACLU. "Violence Chip: Why Does the ACLU Oppose the V-Chip Legislation Currently Pending in Congress." ACLU Freedom Network. 1996, 1–7, downloaded from www.aclu.org.

Biegel, Stuart. UCLA. Online Institute for Cyberspace Law and Policy. "Indictment of CompuServe Official in Germany Brings Volatile Issues of Cyber-Jurisdiction into Focus." April 27, 1997, 1–3, downloaded from www.gse.ucla.edu.

Bitol. Solange. "The V-Chip: Point-and-Click Parenting." March 16, 1998, 1–2, downloaded from www.freedomforum.org.

Boles, Garry. "Online Obscenity: Decency by Law." *Inter@ctive Week* 2 (April 10, 1995), 15–16, downloaded from http://gort.ucsd.edu.

Cantom, William. FCC. "RE: CS Docket No. 97-55." September 10,1997, 1–6, downloaded from www.freenix.fr/neitizen.

Children's Internet Protection Act. Pub. L. 106-554. 1–22, downloaded from www.ifea.net.

Crocker, Steven. "The Political and Social Implications of the Net." July 5, 1994, 1–14, downloaded from http://textfiles.group.

Electronic Frontier Foundation. *Constitutional Problems with the Communications Decency Amendment: A Legislative Analysis.* June 16, 1995, 1–10, downloaded from www.eff.org.

Elmer-Dewitt, Philip. "On a Screen Near You: Cyberporn." July 3, 1995, 1–3, downloaded from www.academiclibrary.com.

Federal Communications Commission. "The V-Chip: Putting Restrictions on What Your Children Watch, Even When You're Not There." 1–3, downloaded from www.fcc.gov.

Freedom Forum Online. "Justice Department Appeals Ruling That Blocked Enforcement of COPA." April 5, 1999, 1–2, downloaded from www.freedomforum.org.

"Loudoun Library Board Decides Not to Appeal Filtering Decision." *Tech Law Journal*. April 21, 1999, 1–3, downloaded from www.techlawjournal.com.

Radio Free Europe/Radio Free Liberty. "20 Enemies of the Internet." August, 1999, 1–5, downloaded from www.rferl.org.

Reuters. "Judge: Yahoo Not Bound by French Nazi Ban." November 7, 2001, 1–2, downloaded from http://news.cnet.com/.

Steiner-Threlkeld, Tom, and Brock Meeks. "Free Speech Defenders Rally in Cyberspace." *Inter@ctive Week* 2 (March 13, 1995), 36–37, 59–60, downloaded from http://gort.ucsd.edu.

## GLOSSARY

*Children's Internet Protection Act (CIPA):* This regulation required federally funded public libraries to install filtering software on computers to block Internet material that was pornographic or harmful to minors.

*Children's Online Protections Act (COPA):* This regulation prohibited Internet content that was harmful to minors.

*Chilling Effect:* Any regulation or threat of regulation that makes the speaker think twice about providing information that might be considered inappropriate by some. In such a situation, the speech is halted before it ever has a chance to be heard.

*Communications Decency Act (CDA):* This regulation was a part of the Telecom Act of 1996.

THE LAW

It controls the use of obscene as well as indecent material online.

*Indecent Material:* Nonobscene language that is sexually explicit or includes profanity.

*Obscene Speech:* Language that appeals to prurient interest, is patently offensive, and lacks serious literary, artistic, political, or scientific value.

*Spectrum Scarcity:* Theory based on the belief that since the electromagnetic spectrum is finite, only a limited number of individuals could receive a license.

# 20 Other First Amendment Issues: Libel, Hate Speech, Cyberstalking, and Copyright

## INTRODUCTION

The online universe is an important part of our information and communications infrastructure. Yet with the growing list of applications and information pools, there are potential legal hot spots. Therefore, besides examining online obscenity as described in Chapter 19, it is also important to explore other First Amendment issues including online libel, hate speech, cyberstalking and copyright.

## ONLINE LIBEL

Internet service providers (ISPs) offer e-mail, Bulletin Boards, electronic news reports, and other services to customers. But what happens when a third party uses an ISP's service to make libelous remarks? Is the ISP legally responsible for defamatory remarks it did not create? Over the past decade, numerous courts have examined this complex legal question and a body of law concerning online libel exists. This chapter will provide a historical roadmap of such case law.

Determining the culpability of an ISP for third party libelous remarks requires two elements:

- defining online libel and
- determining if an online service should be treated as a distributor or a publisher.

But before we begin, three terms, as used in this chapter, must be defined.

- **Libel:** Any expression that damages your reputation within the community, causes others to disassociate themselves from you, or attacks your character or professional ability.[1]
- **Publisher** and **Distributor:** The CompuServe and Prodigy cases, covered in the next section, reflect how online libel has been handled. A critical point was determining if a service was either a publisher (originator) or an information distributor. A publisher of a libelous remark is legally responsible for the act. A distributor is not legally responsible if it was made by a third party without the distributor's knowledge or consent.[2]

### Cases

***CompuServe.*** In *Cubby Inc. v. CompuServe Inc.*, the U.S. District Court for the Southern District of New York (19 Med. L. Rptr. 1525. 1991), had to decide if CompuServe

THE LAW

had acted as a publisher or distributor of libelous material. Specifically, was CompuServe legally responsible for remarks published by a third party in *Rumorville USA*, a component of the Journalist Forum? *Rumorville* was a daily, electronic newspaper included on the forum.

CompuServe independently contracted with a third party to " 'manage, review, create, delete, edit, and otherwise control the contents' of the Journalism Forum 'in accordance with editorial and technical standards and conventions of style established by CompuServe.' "[3] Furthermore,

- CompuServe had no contractual or direct tie with the forum or *Rumorville*.
- The third party accepted full responsibility for *Rumorville's* content.
- The setup made it technically unfeasible for CompuServe to monitor *Rumorville* prior to its uploading.
- CompuServe did not receive money for *Rumorville* or pay a fee for its inclusion on the service. The only fee was the standard subscriber charge.[4]

In presenting its case, CompuServe did not question whether the statements were libelous. Rather, it argued that like a news vendor, bookstore, and library, it was a distributor and not a publisher. Based on current law, CompuServe could not be legally responsible for the libel committed by its independent contractor unless it played an active part in the libel.[5]

Cubby, in turn, alleged that CompuServe was a publisher and was responsible for publishing the libel. Based on the language of *Cianci v. New Times Publishing Co.* (639 F.2d. 54), anyone who repeats or "republishes" the original libel is considered a publisher who is equally guilty and can also be sued.[6]

The court disagreed and found that CompuServe should be treated as a distributor. It compared CompuServe's services to an electronic, for profit library that provided its subscribers with access to numerous publications. The court further held that it was impossible for CompuServe to be aware of the content of all its information services.

For instance, although CompuServe can decline to carry a publication, once it does decide to carry one, it has little or no editorial control over its content. This is especially true when the publication is part of a forum that is managed by an unrelated company.

CompuServe has no more editorial control over a publication than does a public library, bookstore, or newsstand, and it would be no more feasible for CompuServe to examine every publication it carries for potentially defamatory statements than it would be for any other distributor to do so.[7]

The court cited *Smith v. California* (361 U.S. 147) and other cases as precedents. In *Smith*, the U.S. Supreme Court struck down a law that would hold a bookstore owner legally responsible for carrying an obscene book even though the owner was unaware of the book's content. "Every bookseller would be placed under an obligation to make himself aware of the contents of every book in his shop. . . . And the bookseller's burden would become the public's burden, for by restricting him the public's access to reading matter would be restricted."[8] Consequently, the court found that CompuServe had acted as a distributor and not a publisher. As such, it was not guilty of libeling Cubby.

***Prodigy.*** In *Stratton Oakmont et al. v. Prodigy Inc.* (63 USWL. 2765. 1995), the New York Supreme Court examined whether Prodigy was the publisher or distributor of the online libel of Stratton Oakmont, Inc. Specifically, did Prodigy act as the publisher of the online libel that appeared on *Money Talk*, a financial investment and information resource?

The court was also asked to determine if *Money Talk's* board leader (BL) had acted in an official capacity as Prodigy's representative. BLs sign an agreement with Prodigy that spells out their responsibilities. These include an obligation to work with members using the service and to follow Prodigy's procedures.

Stratton Oakmont also claimed that Prodigy promoted itself as a service that exercised some degree of editorial control. For example:

Users are requested to refrain from posting notes that are "insulting" and are advised that notes that harass other members or are deemed to be in bad taste or grossly repugnant to community standards, or are deemed harmful to maintaining a harmonious online community, will be removed when brought to Prodigy's attention; the Guidelines all expressly state that although "Prodigy is committed to an open debate and discussion on the bulletin boards, . . . this doesn't mean that anything goes."[9]

Prodigy also employed a software system that prescreened bulletin board content for offensive material. BLs could also use an emergency deletion function.[10] Consequently, because Prodigy exercised this degree of editorial control, Stratton Oakmont asked the court to declare the company a publisher and not just a distributor. As a publisher, Prodigy would be legally responsible for repeating or republishing the libel originating on *Money Talk*. In its defense, Prodigy claimed the court ruling in *Cubby v. CompuServe* set a precedent. It should be treated as a distributor.

In reaching its decision, the court distinguished the CompuServe case from the Prodigy case on three points:

1. CompuServe had no opportunity to monitor their boards' content prior to distribution.

2. CompuServe had not promoted itself as a service that exercised Prodigy's type of editorial control.
3. Prodigy used an automatic screening program and designed BL guidelines.[11]

Based on these factors, the court ruled that Prodigy was a publisher of the libel and was legally responsible. According to the court, Prodigy "has virtually created an editorial staff of Board Leaders who have the ability to continually monitor incoming transmissions and in fact do spend time censoring notes."[12]

The court further held that Prodigy's own policies, technology, and personnel decisions were responsible for the service's classification as a publisher. Its choice to gain the benefits of editorial control also "opened it up to a greater liability than CompuServe and other computer networks that make no such choice."[13]

The court then turned its attention toward deciding if *Money Talk's* BL had acted as Prodigy's agent, for the purpose of the libel. The court said, "Where one party retains a sufficient degree of direction and control over another, a principal-agent relationship exists."[14] Because Prodigy required BLs to follow their guidelines and performed administrative duties, the court ruled the BL was Prodigy's legal agent.

### The 1996 Telecom Act

The Prodigy decision had a chilling effect in the online community, even though Prodigy was differentiated from CompuServe by its operational guidelines. The 1996 Telecom Act, however, included a "Good Samaritan" blocking and screening clause concerning offensive material: "No provider or user of an interactive computer service shall be treated as the publisher or speaker of any information provided by another information content provider."[15] An infor-

THE LAW

mation content provider is an information producer.

Similarly, a provider (e.g., Prodigy) or a user would not be legally responsible for "any action voluntarily taken in good faith to restrict access to or availability of material that the provider or user considers to be obscene, lewd, lascivious, filthy, excessively violent, harassing, or otherwise objectionable, whether or not such material is constitutionally protected . . . ."[16] This is exercising editorial control.

Libelous remarks, another key point, are said to fall within the "otherwise objectionable material" clause of this section of the act. If this holds up, legal scholars, media organizations, and public interest groups have indicated the Prodigy decision would be overturned.[17]

In sum, the "Good Samaritan" defense may have prevented Prodigy from being punished for trying to police its service that is acting as a publisher. To encourage this type of behavior, Congress also refused to hold Prodigy and other such companies to a higher standard than companies that do not police their services.[18]

However, despite this defense, an online service that was aware of a libelous re-mark and intentionally and/or recklessly "repeated" or "republished" it, would not be protected. An online service could also be held legally responsible for any libelous statement that it originated.

**Cases.** Since the passage of the 1996 Telecom Act, online libel cases have been heard by federal and state courts. Three important federal cases are *Kenneth M. Zeran v. America Online* (958 F. Supp. 1124. 1997. 129 F. 2d 327. 1997), *Sidney Blumenthal et al. v. Matt Drudge and America Online* (992 F. Supp. 44. 1998), and *Christianne Carafano v. Metrosplash.com* (No. CV 01-0018 DT. 2002. 2002 U.S. Dist. LEXIS 10614). *Zeran v. AOL* was heard by the U.S. District Court for

the Eastern District of Virginia and by the Fourth Circuit Court of Appeals. The case was appealed to the U.S. Supreme Court but was not accepted for review.[19] *Blumenthal v. Drudge and AOL* was heard by the U.S. District Court for the District of Columbia and was appealed to the District of Columbia Circuit Court.[20] *Carafano v. Metrosplash.com* was heard by the U.S. District Court for the Central District of California and is awaiting appeal before the Ninth Circuit Court of Appeals.[21]

***Zeran v. AOL.*** In *Zeran v. AOL* the district court had to decide if America Online (AOL), a service provider, could be held legally responsible for libelous remarks made by a third party against Kenneth Zeran. An AOL subscriber, posing as Kenneth Zeran, advertised the sale of tee shirts and other items that contained slogans praising the murder of the 168 Oklahoma City bombing victims. This offensive online message, which included Zeran's home telephone number, was posted without his knowledge or consent. Zeran became aware of the message when he began receiving abusive and threatening telephone calls from outraged AOL users.[22]

Upon discovering the message, Zeran contacted AOL demanding its prompt removal and a retraction identifying him as the victim of the online hoax. AOL deleted the message but refused to post a retraction—this did not fall within company policy. Unfortunately for Zeran, additional notices of a similar nature continued to appear on AOL for several more days and the threatening telephone calls continued.[23]

Zeran responded by suing AOL for negligence. He claimed that AOL failed to adequately respond to the online hoax after it was contacted. He also contended that AOL was legally responsible for the libelous message posted by an unknown third party on its service.[24]

Zeran based his legal claim on Virginia state law that permits legal action against the distributor of libelous material if the information provider knew or should have known of its existence. AOL, on the other hand, argued to the court that the 1996 Telecom Act preempted Virginia state libel law and as such, no service provider could be held legally responsible for the libelous remarks of a third party.[25]

After considering each party's complaint, the U.S. district court ruled in favor of AOL on the grounds that treating the service provider as the publisher of the online libel violated Section 230 of the Telecom Act. Zeran responded by appealing the case to the Fourth Circuit Court of Appeals.[26]

On appeal, Chief Judge Wilkinson affirmed the decision of the lower court, holding that, "by its plain language, Section 230 creates a federal immunity to any cause of action that would make service providers liable for information originating with a third party user of the service."[27] Thus, the Fourth Circuit Court of Appeals upheld the language of the 1996 Telecom Act, and the U.S. Supreme Court refused to accept the case for review.

### Blumenthal v. Matt Drudge and AOL.

In 1998, another important online libel case challenging the 1996 Telecom Act was heard by a federal court. The case involved top Clinton aide Sidney Blumenthal, political reporter Matt Drudge, and AOL, the service provider for Drudge's electronic reports.[28] This case is of particular importance since it marks the first time the courts have examined Section 230 in a case where the online service provider paid the content provider for its information.[29]

Blumenthal filed suit against Matt Drudge and AOL for allegedly libelous remarks made in the Drudge Report. AOL subsequently removed all the remarks from past editions of the Report, which were stored in its electronic archives, and Drudge retracted the story.[30] Nevertheless, Blumenthal proceeded with the lawsuit.

Blumenthal contended that AOL was also responsible since:

- The company was more than a mere service provider. Drudge was paid a flat monthly royalty payment of $3000 by AOL for his online reports.[31]
- The remarks did not come from an unknown person sent over the Internet anonymously.[32]
- "AOL reserves the right to remove, or direct [Drudge] to remove, any content which, as reasonably determined by AOL . . . violates AOL's . . . terms of service."[33]

AOL responded by filing a court motion to have itself dismissed from the suit. This would leave Drudge solely responsible for the allegedly libelous remarks. Citing Section 230 of the 1996 Telecom Act, AOL argued that as a service provider it was immune from third party remarks.[34]

U.S. District Court Judge Paul Friedman granted AOL's request for dismissal from the suit, but included an opinion that was highly critical of AOL. Friedman clearly agreed, however, that the language of Section 230 protected AOL from liability. He wrote that Congress had provided, ". . . immunity even where the interactive service provider has an active, even aggressive role in making available content prepared by others."[35] Sidney Blumenthal appealed Judge Friedman's decision but agreed to settle the case in May 2001 due to staggering legal bills.[36]

### Carafano v. Metrosplash.com.

Following the court's decision in Zeran and Blumenthal, legal scholars and the online community assumed that the final chapter had been written on Section 230 and Internet company immunity. However, a 2002 case decided by the U.S. District Court for the

THE LAW

Central District of California may have rekindled the old debate.

In *Carafano v. Metrosplash.com*, Christianne Carafano, an actress with the stage name "Chase Masterson" (*Star Trek: Deep Space Nine*), filed suit against Metrosplash.com and its dating site, Matchmaker.com. The grounds for the suit included invasion of privacy, misappropriation of the right to publicity and negligence.[37]

An unknown third party posed as Carafano and posted her name, photograph, contact information, and false and defamatory sexual information about her. Matchmaker.com had designed an extensive questionnaire that all members, including the third party posing as Carafano, were required to answer. This individual's information was subsequently posted without Carafano's knowledge. This triggered obscene messages, and Carafano responded by filling the lawsuit.

The lawsuit specified that based on the questionnaire's answers and profile, which were then posted, Matchmaker.com was an information content provider partially responsible for the libel. In response, Matchmaker.com claimed immunity under Section 230.

After review, the court dismissed all the claims.[38] However, Judge Dickran Tevizian delivered a potentially devastating blow to Internet companies when he ruled that Matchmaker.com was not protected under Section 230 because it had acted as an information provider and was partially responsible for the third party's libelous actions. As of this writing, the case is on appeal before the Ninth Circuit (March 2003). The court's decision could have key First Amendment implications for Internet companies under certain circumstances.

**State Cases.** Three other online libel cases, *Kempf v. Time, Inc.* (No. BC 184799. California. 1998), *Ben Ezra, Weinstein & Co.*

*v. America Online* (No. 97–485. N.M. 1999), and *Alexander Lunney v. Prodigy Services Co. et al.* (250AD 2d 230. N.Y. 1994. 99 N.Y. Int. 0165. 1999) were heard in 1998 and 1999 by state courts.

In *Kempf v. Time*, a California court dismissed a libel suit against the service provider, Time Inc., on the grounds that it was not involved in creating or developing the libelous content in question and was, in fact, protected according to the language of Section 230 of the 1996 Telecom Act.[39]

In 1999, a New Mexico court deciding *Ben Ezra, Weinstein & Co. v. AOL,* ruled that as a service provider, AOL was not legally responsible for false and libelous online remarks concerning the company's financial situation. The court, citing Section 230 of the 1996 Telecom Act, ruled in AOL's favor stating it had Internet Service Provider immunity.[40]

Later in 1999, the Court of Appeals of New York heard *Lunney v. Prodigy*. Judge Rosenblat, writing for a unanimous court, upheld a lower New York state court's dismissal of an online libel suit against Prodigy on the grounds that it was not the "publisher" of vulgar and threatening e-mail messages sent to a third party by an imposter using Alexander Lunney's name. The New York Court of Appeals based this decision citing Section 230 of the Communications Decency Act.[41] Lunney appealed the decision to the U.S. Supreme Court but the Justices refused to hear the case, letting the Court of Appeals decision stand. (2000 U.S. LEXIS 3037.) Other state cases have also been heard upholding Section 230.[42]

## Summary

As of this writing, since the passage of the 1996 Telecom Act, courts have upheld the immunity protection afforded to Internet Service Providers based on the language of Section 230. However, the end result of an ongoing challenge before the Ninth Circuit is yet unknown. The current blanket protection for Internet Service Providers may have created a mixed bag of benefits, depending on which side of the fence you are on. Thus far, a company has been protected from carrying libelous statements made by a third party. On the flip side, some individuals believe the clause has removed all responsibilities from a company, even if its policies permit the removal of content, as was the case with AOL. The decision of the Ninth Circuit may answer these questions.

## HATE ON THE INTERNET

### Overview

As was the case with online libel, hate on the Internet also presents a First Amendment dilemma. Hate crime is a serious problem in America. In 2001 alone, over 12,000 individuals were victims of a crime because of their race, ethnicity, national origin, religion, sexual orientation or gender.[43] Even more disturbing, however, is the ease with which the Internet has made online terror just a mouse click away.

In response to the proliferation of such crimes, 42 states and the District of Columbia have enacted hate crime statutes that protect victims to varying degrees. However, no single, comprehensive federal law specifically aimed at protecting victims against all categories of hate crimes exists. This makes dispensing uniform justice for hate crimes impossible.[44] Instead, prosecutors have been forced to rely on specific sections of a host of federal laws when hate crimes have been committed.[45]

For instance, under 28 USC 994, The Hate Crimes Sentencing Enhancement Act, anyone convicted of a crime against an individual because of his or her actual or perceived race, color, religion, national origin,

ethnicity, gender, sexual orientation or disability, will receive an increased sentence.[46]

18 USC 245 is also frequently used to prosecute hate crimes. Section 245 "allows for the prosecution of whoever, by force or threat of force, willfully injures, intimidates, or interferes with, or attempts to injure, intimidate or interfere with any person or class of persons because of his or her race, color, religion or national origin."[47] However, it should be noted that hate crimes based on sexual orientation, gender, ethnicity or disability are not punishable under this law.

This "gap" in federal protection has prompted congressional action in the form of hate crime prevention bills over the past several years. One such legislative example, HR 80, entitled The Hate Crime Prevention Act of 2003, was being debated by the 108th Congress.[48] If signed into law, the Act would close this gap and provide comprehensive federal protection against hate crime.

This Act is based on the legal premise that hate crimes and their threat, violate interstate commerce in the following ways:

- An individual who is the target of a hate crime may be forced to move across state lines to escape becoming the victim of violence.
- The perpetrator of a hate crime may across state lines to commit hate-motivated violence against the victim.
- The perpetrator may use articles, including computers that have traveled in interstate commerce, to commit a hate crime.

In regard to the hate crime and communications technology, this book's focus, a person who receives bias-motivated threatening e-mails may feel forced to move across the country, may leave their job or refuse to buy a home. Moreover, the instrument used to perpetrate the threat, the computer (as well as the communication by computer) has traveled in interstate commerce. Specifically, HR 80 criminalizes the offense if, "the defendant or victim travels in interstate or foreign commerce, uses a facility or instrumentality of interstate or foreign commerce, or engages in any activities affecting interstate or foreign commerce."[49]

Finally, if HR 80, or a similar future Act, is signed into law, full federal protection against bias-motivated hate crimes will exist. Victims will no longer be forced to depend on inadequate federal and/or state statutes to receive justice.

### E-Mail Based Cases

In the absence of such federal protection, only a few Internet hate crimes have been prosecuted. Some examples are given in the following subsections:

### U.S. v. Richard Machado (U.S. Dist. Crt. Central Dist CA. No. 96-00142.AHA).

Richard Machado, a naturalized citizen from El Salvador, was a 19-year-old student at the University of California until he was dismissed for failing grades. While at the university, Machado blamed Asian students for a number of problems and, he even adopted "Asian-hater" as his e-mail name. Once expelled, Machado used a campus computer lab and e-mailed racially motivated death threats to 62 Asian-American students. Machado also concluded the e-mail message by saying if each individual did not immediately leave campus permanently, he would make it his life career to find, hunt down, and kill them.[50]

Machado was subsequently charged under the language of 18 USC 245 for violating the civil rights of his victims. Machado's e-mail death threats violated a federally protected activity, namely, the right to attend school without the fear of death. Prosecutors alleged that Machado victim-ized these students based on their race, ethnic, and national origin.

During the pretrial stage of his prosecution, Machado admitted that his e-mail messages constituted threats and he would have been afraid if he had received such a message. But during the U.S. district court trial, he changed his stance and said that the e-mail messages were simply a joke not meant to be taken seriously. The prosecution called nine of the victims to testify, however, and they indicated they feared for their lives.

After listening to the testimony, the jury was at a stalemate and could not reach a verdict. This forced the Judge to declare a mistrial.

The U.S. Attorneys' Office decided to retry Machado. The end result was a guilty verdict and 1 year imprisonment followed by a 1-year period of supervised release. This verdict made Richard Machado the first person convicted under federal law for making bias-motivated hate threats via the Internet.[51]

### U.S. v. Kingman Quon.

A second violation of 18 USC 245 occurred when Kingman Quon, a young Chinese-American man from California sent bias-motivated death threats via e-mail to Hispanic-Americans at Cal State Los Angeles, Xerox Corporation, the IRS, and other institutions. Quon's e-mail message used demeaning racial slurs and said he hated the Hispanic race and wanted them to die. He concluded his message by saying that he was going to come and kill them.

Quon was subsequently identified, arrested, and charged with violating his victim's federally protected rights to enjoy employment and attend school regardless of their race or ethnicity. Quon's victims said they believed the threats were legitimate, and they feared for their lives. Quon admitted the e-mails were intended to intimidate his victims, who were targeted because of their national origin.

Quon plead guilty to the charges and received a 2-year prison sentence. Furthermore, upon release from prison, Quon was prohibited from using a computer for a period of 1 year. Kingman Quon's sentence marked the second time that an individual had been incarcerated under federal law for using the Internet to make bias-motivated hate threats.[52]

**Ryan Wilson and ALPHA HQ.** In this case, Secretary of Housing and Urban Development (HUD), Andrew Cuomo, used the language of the Fair Housing Act (42 USC 3617) to protect Bonnie Jouhari, one of his employees. Jouhai assisted individuals embroiled in housing discrimination disputes. She was also white, had a bi-racial daughter and was a highly visible, antihate activist in Philadelphia.

A local Philadelphia man, Ryan Wilson, became aware of her activities. Using his web site, ALPHA HQ, he posted or placed for review:

- death threats against her;
- Jouhari's photo with accompanying racial slurs;
- an animated picture of Jouhari's office exploding; and
- a message that read, "Traitors like this should beware, for in our day, they will be hung from the neck from the nearest lamppost."[53]

Soon after the ALPHA HQ posting, Jouhari was subject to threatening telephone calls, slashed car tires, Ku Klux Klan flyers on her car windshield, and dead flowers thrown in her yard. Jouhari and her daughter feared for their lives.

In response, Jouhari filed charges with federal, state, and local law enforcement agencies. But she was not able to obtain legal protection, and thus, she fled the State and moved half way across the country for safety's sake.[54]

Meanwhile, the Pennsylvania Attorney General, Michael Fisher, filed a complaint in state court for Jouhari on the grounds that the ALPHA HQ death threats violated a state civil statute against ethnic-oriented intimidation. A permanent injunction, which barred the posting of threats against Bonnie Jouhari or any other human rights employees working in the state of Pennsylvania, was subsequently granted.[55] Following the state's action, Ryan Wilson removed the web site and no further action was taken.

The Justice Department also investigated Jouhari's complaint. But it made little progress and HUD was told they could pursue the incident. At this point, Andrew Cuomo charged Ryan Wilson as the head of ALPHA HQ, with violating HUD's Fair Housing Act, creating the first federal civil rights case targeting a web site. Cuomo charged that the bias-motivated death threats on the web site violated the Fair Housing Act by:

1. Preventing Jouhari from being able to perform her job, which was to enforce that very Act.
2. Making her feel unsafe in her own home.

A HUD Administrative Law Judge convened a hearing, and Ryan Wilson choose not to attend. The Judge held that his absence was an admission of truth concerning the charges. After reviewing the facts, the Judge also ruled that Ryan Wilson had discriminated against Jouhari under Section 818 of the Fair Housing Act. Jouhari and her daughter were subsequently awarded $1.1 million in damages. Wilson was also required to pay over $55,000 in fines to HUD.[56] Nevertheless, Bonnie Jouhari and her daughter were still forced to flee their home, leaving their family and friends because of an Internet hate crime and its underlying philosophy.

### The Nuremberg Files

In 1995, a computer programmer, Neal Horsley, created an anti-abortion web site. The home page featured a line of red blood dripping down the page and a Nazi Nuremberg Trial scene with an abortion doctor sitting in the witness chair.[57]

Horsley called his site the "Nuremberg Files" in light of the 1940's Nuremberg trials in Germany. According to Horsley, many Nazi murderers escaped conviction, at the time, because of the lack of evidence. In contrast, his site would provide the necessary evidence to convict abortionists of mass murder if and when abortion became illegal.

While pursuing this goal, Horsley collected extensive dossiers on over 200 doctors who performed abortions throughout the United States. The data were then made available to anyone who accessed his web site. The information included names, "wanted posters" style photos, home addresses, social security numbers, license plate numbers, names of spouses and children, divorce records, fingerprints, and surveillance photos and videos. He obtained the information from various U.S. anti-abortion informants. Several anti-abortion activists who sent the "wanted posters" to Horsley for his web site also had print copies made and distributed.

The site also listed page after page of the names of doctors nationwide, who perform abortions. The names of doctors not yet killed or wounded appeared in black print. The names of wounded doctors appeared in gray print. Doctors, who had been murdered, had a line drawn through their name shortly after they had been killed.

The doctors whose names appeared on the web site lived in constant fear for their lives. They believed that the web site, in effect, encouraged people to kill or harm them. Most had FBI and/or local law enforcement protection 24 hours a day and wore bulletproof vests. At least five of the doctors on the list were either killed or wounded.[58]

Finally, some of the doctors and Planned Parenthood filed a lawsuit against the anti-abortion activists who made the "wanted posters" and against the Nuremberg Files web site, which also posted the threatening "wanted posters." Neal Horsley was not named in the lawsuit, however. The case was *Planned Parenthood of the Columbia/ Willamette Inc. et al. v. American Coalition of Life Activists et al.*[59]

The U.S. District Court heard the case for the District of Oregon and pitted doctors and abortion clinics against an anti-abortion group with a past record of threatening, and encouraging others to commit, actual violence against abortion providers. This case was a civil suit with 8 jurors who had to determine if the printed and on on-line "wanted posters" constituted protected speech (speech permitted under the First Amendment) or a "true threat" to the targeted doctors.[60]

The plaintiffs claimed the "wanted posters" violated the Freedom of Access to Clinics Entrances Act (FACE) and the Racketeering Influenced and Corrupt Organizations Act (RICO). FACE prohibits, "whoever by force or threat of force or by physical obstruction, intentionally injures, intimidates, or interferes with, or attempts to interfere with . . . anyone who provides or obtains reproductive services."[61] RICO prohibits an "organized conspiracy to commit or attempt to commit extortion or coercion resulting from fear due to threats to do bodily harm . . ."[62]

The plaintiffs asked that the defendants' actions be punished in two ways. First, they requested a permanent injunction against further distribution of the printed and the online "wanted posters." Second, they asked the jury to award them compensatory damages (out-of-pocket expenses) and punitive damages (meant to punish and make an example out of their actions).

The defendants claimed the "wanted posters" in both print and online form constituted free speech and were protected under the language of the First Amendment.

The jury found in favor of the plaintiffs on the grounds that the "wanted posters" did constitute a true threat to abortion providers and therefore, did not constitute protected speech. Based on this conclusion, a permanent injunction against the print and the online "wanted posters" was granted. The injunction did not ban the Nuremberg Files web site completely. Only the "wanted posters" were banned. The jury also awarded the plaintiffs $109 million in damages.[63]

The defendants appealed the case to the Ninth Circuit where a panel of three judges overturned the district court on First Amendment grounds. The Judges said that the anti-abortionists activist could not be held legally liable because the "wanted posters" did not directly authorize or threaten violence. Planned Parenthood requested an *en banc* hearing (a hearing by a quorum of all the Judges in the Circuit instead of the normal panel of only 3 Judges) by 11 Judges of the Ninth Circuit Court of Appeals, which was granted. In a split 6–5 decision, the *en banc* court ruled the "wanted posters" were not protected speech under the First Amendment because they did constitute illegal threats against abortion providers. However, they ordered the U.S. district court Judge to reduce the amount of money awarded for punitive damages.[64]

Finally, in December of 2002, the anti-abortion activists asked the Supreme Court to hear the case. The Supreme Court has first asked the Solicitor General Theodore Olson to file a brief expressing the Bush Administration's views on the issue. No further action has occurred as of this writing.[65]

## CYBERSTALKING

In addition to hate on the Internet, other issues continue to crop up. Online stalking, called "cyberstalking," for example, has become more prevalent. As implied, you could be the target of harassing e-mail and an individual who follows you around in cyberspace—impersonating you in a chat room to incite other members to react to typically volatile information. As the target, you could receive hostile e-mail responses. This is only the tip of the proverbial iceberg.

You should also think about how you use e-mail. It has, to a great extent, become the electronic/written equivalent of picking up a telephone. But now, you can reach tens, hundreds, or even thousands of sites and people at the push of a button. Cyberstalkers have tapped this capability to harass and threaten others.

Another factor is e-mail's anonymity. As stated in a government report, "whereas a potential stalker may be unwilling or unable to confront a victim in person or on the telephone, he or she may have little hesitation sending harassing or threatening electronic communications to a victim."[66] Geographical and time elements are also removed, and you can sign-up for free e-mail without any verification on the company's part.

Another factor is this crime's fairly recent appearance. In some cases, a local law enforcement agency may be ill equipped (training-wise) to handle the situation. In response, cyberstalking support groups have appeared on the Internet.

But legally, what can you do? As of 2002, 41 states had cyberstalking laws that specif-

THE LAW

ically criminalized stalking via the Internet, e-mail or other electronic means.[67]

However,

- victims of cyberstalking in states without specific laws remain unprotected;
- even if the victim lives in a state with a cyberstalking law, legal protection is not guaranteed; and
- while some federal protection may exist under Title 18 Section 875 of the U.S. Code, the majority of cyberstalking incidents cannot be prosecuted based on the law's language.

For the latter, the language of Section 875 may appear to provide federal legal protection against cyberstalking: "whoever transmits in interstate commerce any communication containing any threat to kidnap any person or any threat to injure the person or another, shall be fined under this Title or imprisoned not more than 5 years or both."[68] But in reality, unless a specific, direct threat is made, the law does not apply. If the cyberstalker harasses or terrorizes the victim but does not make a specific, direct threat (e.g., I'm going to kill you), Section 875 cannot be used to protect the victim. Since most cyberstalking incidents fall into the harassment category, a comprehensive federal cyberstalking law is necessary.

## COPYRIGHT

As was the case with online stalking, copyright in regarding the Internet also presents a First Amendment dilemma. As described in Chapter 4, a copyright protects certain intellectual property classes. This section of the book expands on this preliminary information.

### Introduction

Under the terms of the Copyright and Patent Clause of the Constitution (Article I, Section 8), Congress has the authority to regulate "Intellectual Property," thus providing copyright protection for the authors of various creative works. The first federal copyright law in the U.S. was enacted in 1790 to protect authors of books, maps, and charts. Since then, Title 17 of the United States Code has been revised numerous times to reflect the evolving issues surrounding copyright law.[69] For instance, legislative and judicial action have been taken to protect materials that are available via the Internet.

### Legislative Action

Internet copyright legislation that has increased the protection of an author's work include the No Electronic Theft Act, The Digital Millennium Copyright Act (DMCA), and the Sonny Bono Copyright Term Extension Act (CTEA).

*No Electronic Theft Act.* The No Electronic Theft Act, signed into law by President Clinton in December 1997, closed a loophole in the illegal distribution of software. Criminal infringement would occur when a party infringed on a copyright in a willful manner for commercial advantage or private financial gain.[70] But this was not a prerequisite for infringement. Infringement would also occur if, during any 180 day period, one or more copies of a copyrighted work having a retail value of more than $1000 was illegally reproduced or distributed via electronic means, regardless if the party committing the action gained from the infringement.[71]

*The Digital Millennium Copyright Act.* The Digital Millennium Copyright Act

(DMCA) was created in the late 1990s. Two important sections for our discussion are the sections:

- World Intellectual Property Organization (WIPO) Copyright Treaties Implementation Act and
- Online Copyright Infringement Liability Limitation Act[72]

The WIPO section of the Digital Millennium Copyright Act guarantees adequate legal protection and legal penalties against circumvention of effective technological measures employed by authors to control the use of their work. Manufacturing and trafficking in such technology is prohibited as well.[73]

The WIPO can trace its roots to international treaties ratified in 1998 by the United States and 120 other United Nations' countries. The treaties were written to protect copyright works fixed in traditional and nontraditional mediums. What was the goal of the treaties—to further develop rights previously established by the terms of the Berne Convention and a World Trade Organization agreement.[74]

The new Act would more securely protect a copyright owner's right to reproduce, communicate to the public, and adapt their work. Copyright owners were also granted the right to prohibit the commercial rental of their computer programs and musical recordings and to make their work available online, among other stipulations.[75]

The Online Copyright Infringement Liability Limitation section of the Act was designed to encourage service providers and copyright owners to work together to discover and deal with online copyright infringement. The Act also gives service providers a clearer picture of their legal responsibilities.

Service providers are also protected from copyright infringement liability if they simply transmit information over the Internet. However, they are required to remove material from a web site if copyright infringement is apparent.[76]

This section of the Act also contains strong language to protect copyright owners against potential abuses that could arise from the latest technological advancements. For example, it criminalizes any interference with antipiracy measures built into commercial software.[77]

***The Sonny Bono Copyright Term Extension Act.*** The Sonny Bono Copyright Term Extension Act (CTEA) concerned the duration of copyright protection for the author(s) of a creative work and was signed into law by president Clinton in October 1998. The Act extended a copyright's duration for works created on or after January 1978. Under the CTEA, the protection was extended

- from the life of the author plus 50 years to the life of the author plus 70 years;
- from 75 years to 95 years from the date of publication, or 120 years from the date of creation (whichever expired first), for works made for hire (corporate works); and
- from 75 years to 95 years for works published before January 1, 1978.[78]

While this legislation brought the United States in line with European Union (EU) provisions, it was highly controversial. Opponents claimed Congress did not have the authority to grant such extensions. The CTEA would also keep copyrighted works from passing into the public domain where anyone could use them, thus violating the people's First Amendment right to freedom of expression. Based on these legal grounds,

THE LAW

the Sonny Bono Copyright Term Extension Act was challenged in Federal Court in *Eldred v. Ashcroft.*[79]

After winding its way through the lower federal court system, *Eldred v. Ashcroft* was heard by the Supreme Court in January 2003. In a 7–2 decision, the Court upheld the constitutionality of the Sonny Bono Copyright Term Extension Act (CTEA). Writing for the majority, Justice Ruth Bader Ginsburg said Congress had the authority to extend copyright duration and this did not violate First Amendment free speech rights. Justice Ginsburg wrote that, "The CTEA is a rationale enactment; we are not at liberty to second-guess congressional determinations and policy judgments of this order, however debatable or arguably unwise they may be . . . Accordingly, we cannot conclude that the CTEA . . . is an impermissible exercise of Congress' power under the Copyright clause."[80]

To sum up, the No Electronic Theft Act, Digital Millennium Copyright Act, and the Sonny Bono Copyright Term Extension Act were written, in part, in response to the new technologies and their applications. It is also an evolving field, as electronic services, including those on the Internet as described in a later chapter, continue to evolve.

### Judicial Action

Case law has also been decided based on copyright infringement issues. In *Playboy v. Frena* and *Playboy v. Hardenburgh,* for instance, bulletin board services (BBSs) illegally posted copyrighted photographs from *Playboy* magazine.

A BBS is an electronic communications system that generally predates web sites. They are used for distributing information and can serve as communications conduits for people who dial into their web sites via modems.

In each instance, the service provider had not obtained copyright clearance to post the photos. *Playboy* sued and the courts ruled that *Playboy* had the exclusive rights to display the works. Therefore, copyright infringement had occurred.[81]

In another case, *Sega v. MAPHIA,* a BBS owner knowingly and intentionally offered subscribers access to Sega video games and the hardware to copy the games. Based on these facts, the court found in favor of Sega, holding that copyright infringement had occurred.[82]

In a different vein, a student who operated his own BBS made software, valued in the millions of dollars, available to his subscribers. He was sued in federal court by the government. But the court ruled he could not be held legally liable for copyright infringement since he had not charged his subscribers for the software. The aforementioned No Electronic Theft Act, which was passed at a later date, has since eliminated this loophole.[83]

In *Tasini et al. v. New York Times et al.*, the court examined the posting of copyrighted articles of freelance writers on electronic databases and/or CD-ROMs without an author's permission and/or additional payments. The case was heard at the U.S. District Court and the Court of Appeals level with two different outcomes.

The U.S. District Court ruled in favor of the *New York Times* and the other publishers. The Court indicated the additional publication of the works did not constitute copyright infringement. However, on appeal, the circuit court overturned this decision holding that posting the works via the Internet and/or on CD-ROMs was an additional use. Without the author's consent, it constituted a copyright infringement.[84]

The *New York Times* appealed the decision to the Supreme Court (*New York Times v. Tasini.* No. 00201. 2001). In a 7–2 decision, the Court affirmed the Second Circuit, ruling that "Both the print publishers and the electronic publishers . . . have infringed the copyrights of freelance authors."[85]

## Entertainment Industry and Copyright

*Introduction.* The technology that has revolutionized a consumer's ability to illegally copy protected music has resulted in litigation. Three representative cases include *Recording Industry Association of America (RIAA) v. Diamond Multimedia Systems* (180 F.3d 1072, 1999), *UMG Recordings et al. v. MP3.com, Inc.*, (92 F. Supp. 2d 349, 2000), and *A&M Recordings, Inc., et al. v. Napster, Inc.*, (284 F. 3d 1091, 2002).

*RIAA v. Diamond.* The RIAA, representing the major record labels, control 90% of the U.S. distribution market. It filed suit in federal court to halt the manufacture and distribution of Diamond Multimedia's portable music player, the "Rio." As described in a previous chapter, this type of system is used to playback MP3 files.[86]

The RIAA filed suit on the grounds that Rio violates the language of the Audio Home Recording Act of 1992 (17 USC sub-ch. B, Section 1002 [a][1] and [a][2]) "because it does not employ a Serial Management System (SCMS)" that sends, receives, and acts upon information about the generation and copyright status of the files, it plays."[87] For the latter, Rio does not incorporate a copy protection system.

Diamond Multimedia argued the Audio Home Recording Act did not apply to Rio for two reasons. First, the Act regulates "digital audio recording devices," and Rio was not such a device. Second, while Rio did employ the use of a hard drive for file storage, hard drives were exempt from the Act.[88]

The Ninth Circuit Court of Appeals, affirming the lower court, held that "Rio is not a digital recording device subject to the restrictions of the Audio Home Recording Act of 1992."[89]

*UMG Recordings v. MP3.com.* UMG v. MP3.com was a pioneer case that focused on the distribution of copyrighted music via the Internet and fair use. This new and rampantly developing legal issue questions whether or not the distribution of copyrighted music via the Internet violates fair use. Fair use is the "limited copying of copyrighted work, usually done for 'productive' purposes such as news reporting, criticism, and comment."[90]

MP3.com purchased "tens of thousands" of UMG, Sony, Warner Brothers, and the other plaintiffs' CDs and subsequently copied them to its server without prior authorization. (These companies held the CD's copyright.)[91] But to gain access to this information, subscribers had to prove they already owned a CD version by either inserting their CD in their computer's CD-ROM drive or by purchasing a copy of the CD from a cooperating online music retailer.[92] At this point, the subscriber could download the file(s) from MP3.com and "store, customize, and listen to the recordings contained on their CD's from anyplace where they have an Internet connection."[93] Thus, MP3.com extended a subscriber's use of his or her recordings (contained on the CDs they owned).

MP3.com indicated its system fell under the legal umbrella of fair use. UMG and the other plaintiffs strongly disputed this claim and stated that MP3.com was simply "re-playing for the subscriber converted versions of the recordings it copied, without authorization, from plaintiffs' copyrighted CDs."[94]

The court, in examining the merits of the case, considered the four factors of fair use:

1. The purpose/character of the use (commercial versus educational/nonprofit);
2. The nature of copyrighted work;
3. The amount/substantiality of portion used in relation to copyright work as a whole; and
4. The effect of the use on the potential market for or value of the copyrighted work.[95]

The court subsequently determined that:

- On point 1, there was no dispute that the purpose of the use was commercial.
- On point 2, the works were "creative recordings," not "factual description" works that might enjoy fair use protection, thus negating the fair use defense.
- On point 3, there was no dispute the entire copyrighted works were used. Therefore, fair use did not apply.
- On point 4, MP3.com's actions violated the plaintiffs' right to license their copyrighted musical recordings to others for reproduction.[96]

Based upon these reasons, the court ruled the "defendant's fair use defense is indefensible and must be denied as a matter of law."[97] Finally, in another legal proceeding (2000 U.S. Dist. LEXIS 17907. November 2000), Judge Jed S. Rakoff ordered MP3.com to pay UMG and the other plaintiffs, $53,400,000 in damages plus court costs and attorneys' fees.[98]

***A&M Records v. Napster* (284 F. 3d 1091. 2002, 239 F. 3d 1004. 2001, 2000 U.S. App. LEXIS 18688. 2000, 2001 U.S. Dist. LEXIS 2186. 2001, 2000 U.S. Dist. LEXIS 20668. 2001, 114 F. Supp. 2d 896. 2000, 2000 U.S. Dist. LEXIS 6243. 2000).** The most high profile court case about accessing copyrighted music from the Internet is *A&M v. Napster*. College student Shawn Fanning's music files service, "Napster," became a household word.

Napster's creators claimed their goal was to provide the online music community with an innovative way to gain access to a vast number of music files. In brief, the Napster system enabled a user to search for, request, and download MP3 files. Users could also play a downloaded song and participate in a Napster-sponsored chat room.[99]

The music industry was stunned by the ease and efficiency with which Napster enabled users to download copyrighted music without actually purchasing CDs. Napster's disregard for copyright law, and its potential economic impact, sent shock waves through the music industry.

In response, 17 record companies went to federal court to obtain an injunction against Napster on the grounds that the service violated copyright law. This lawsuit was the beginning of years of extremely complex court decisions sending the case back and forth between the U.S. District Court for the Northern District of California and the Court of Appeals for the Ninth Circuit.

The record companies claimed the following:

1. Napster was guilty of direct copyright infringement, which occurs when a person directly copies, performs or violates the copyright owner's exclusive rights without obtaining permission to do so.
2. Napster was guilty of contributory infringement. This occurs when the copyright violator has knowledge that they have caused a third party, or has contributed to a third party's, infringement.
3. Napster was guilty of vicarious infringement. This occurs when the right and ability to supervise a third party copyright violator's actions exists and when the supervising party benefits from the infringement.[100]

Napster countered these accusations by claiming that:

- It was protected under the terms of fair use (Section 107 Copyright Act).
- It was protected under the terms of the "Safe Harbor" clause, which protects Internet Service Providers (Section 512 DMCA).
- It was protected under the language of the Audio Home Recording Act (17 USC. Subchapter B. Section 1002).
- It was protected against prior restraint under the First Amendment.[101] Prior restraint in this case would mean denying users access to all songs including the songs of an unsigned artist. Likewise, access to Napster's New Artist Program, message boards and chat rooms would also be denied, all before a final decision concerning Napster's legality occurred.[102]

After numerous complex rulings and modifications by the District Court and the Court of Appeals, the follow conclusions were reached:

1. Napster was liable of direct, contributory, and vicarious copyright infringement.
2. Napster's service was not protected by the fair use provision.
3. Napster's protection according to the Safe Harbor clause must be further considered.
4. Napster does not fall under the protection of the Audio Home Recording Act because music on hard drives does not constitute digital music recordings.
5. Napster and it's users were not subjected to a denial of their First Amendment rights.[103]

The Ninth Circuit subsequently upheld an injunction against Napster's continued violation of the copyrights held by the record companies. The Ninth Circuit, however, modified the injunction on the grounds that it was constitutionally overbroad.[104]

The modified injunction, "obligates Napster to remove any user files from the system's music index if Napster has reasonable knowledge that the file contains plaintiffs' copyrighted works. Plaintiffs, in turn, must give Napster notice of specific infringing files. For each work sought to be protected plaintiffs must provide the name of the performing artist, the title of the work, a certification of ownership, and the name(s) of one or more files that have been available on the Napster file index containing the protected copyrighted work."[105]

In the wake of this court decision, Napster, while in the process of trying to design a legal service in alliance with German music company Bertelsmann, went bankrupt. Bertelsmann later tried to buy Napster but a U.S. Bankruptcy Court denied their bid.[106]

But in March 2003, a Santa Clara-based CD-burning software company, Roxio, purchased Napster's web site and Intellectual Property.[107] Roxio plans to relaunch Napster as a legitimate service offering users the ability to pay a per-song individual fee as well as a monthly subscription-based downloading service. Roxio negotiated for the right to license the music of the top five record companies, which would subsequently be available through its new service.[108]

In another twist, Apple computer introduced its iTunes Music Store in 2003. Designed much like the Roxio service, Mac users could download songs for $.99 each. You could browse a large selection and subsequently download your choices. However, there was a copyright protection scheme. As described by Apple, "The iTunes Music Store is fast and convenient for you, and fair to the artists and record companies. In a nutshell, you can play your music on up to three computers, enjoy unlimited synching with your iPods [Apple player], burn unlimited CDs individual songs, and burn unchanged playlists up to 10 times each."[109] When first launched, the service became an immediate hit.

***Linking, Deep Linking, and Framing.*** As indicated in the Internet chapter, Chapter 17, you can gain access to information through links. But is linking, especially deep linking, legal?

Deep linking is the practice of linking from one web site to another web site bypassing the linked site's home page. Thus the user may not view this page, which may include legal and privacy information and, in some instances, advertisements.

Early deep linking cases have been settled out of court without definitively spelling out if this practice violates copyright law. (See: *Ticketmaster v. Tickets.com* and *Ticketmaster v. Microsoft*).[110] The same scenario played out for other cases when the act of providing a link to another site has generally not been considered a violation. (See *Shetland Times, Ltd. v. Dr. Jonathan Wills and Zetnews, Ltd.*, 1996. *Ticketmaster v. Microsoft*, No. 97-3055 DDP. D. Cal. April 1997. *Ticketmaster v. Tickets.com*, 2000 U.S. Dist. LEXIS 4553. D.C. Cal. 2000.)

But in 2003, the courts handed down two important decisions further clarifying these practices. (*Ticketmaster v. Tickets.com*. U.S. Dist Crt. Ctrl Dist. CA. 2003 U.S. Dist. LEXIS 6483. March 6, 2003; and *Leslie Kelly v. Ariba Soft Corp.* No. 00-55521. DC No. CV-99-00560. GLT. July 9, 2003.) In *Ticketmaster v. Tickets.com*, the court ruled that linking does not violate copyright law.[111] In

*Kelly v. Ariba Soft*,[112] the court held that deep linking in the form of "thumbnail" links (small versions of larger images) constituted fair use of a copyrighted work.[113] Future case law will serve to develop a body of law more capable of answering such complex copyright questions.

Another copyright issue closely associated with linking is framing. Through framing,

- you may gain access to a different web site; but
- you may not realize you are retrieving information from this new site (e.g., site B), since it may appear to be a component of the original site (e.g., site A).[114]

Through framing, the other site's appearance, (site B) has been visually altered (i.e., it may not be viewed as a stand-alone page). This alteration has raised a legal question. Does such an action create a derivative work, thus violating the copyright law?[115]

## CONCLUSION

In conclusion, the online universe has provided users with a host of new communication tools. But the technological advancements have come with a price. The same tools that can be used to send an e-mail to a friend or allow you to search the Internet for vast amounts of information, can also result in death threats, cyberstalking, and copyright infringement. As described in the chapter, this fact has resulted in a large body of legislative and judicial action.

## REFERENCES/NOTES

1. Kent Middleton and Bill Chamberlin. *The Law of Public Communication*. (White Plains, NY: Longman Publishing, 1994), 70.

2. *Cubby Inc. v. CompuServe Inc.*, U.S. District Court for the Southern District of New York (19 Med. L. Rprt. 1527. 1991).

3. Ibid., 1526.

4. Ibid., 1526.

5. Ibid., 1528.

6. Ibid., 1527.

7. Ibid., 1528.

8. Ibid., 1528.

9. *Stratton Oakmont et al. v. Prodigy Inc., Supreme Court of New York* (63 USWL. 2767. 1995).

10. Ibid., 2767.

11. Ibid., 2767.

12. Ibid., 2770.

13. Ibid., 2770.

14. Ibid., 2772.

15. Telecommunications Act of 1996, 71.

16. Ibid.

17. Kenneth Salomon and Todd Gray, "Online Content Liability," *American Council of Education,* February 14, 1996, 6.

18. Ibid.

19. U.S. Supreme Court. Cert. Pet. 97–1488. 1997. "Zeran v. America Online," *Tech Law Journal,* 1998, downloaded from www.techlawjournal.com.

20. "Judge Friedman Grants Summary Judgment to AOL in Defamation Case," *Tech Law Journal,* April 24, 1998, downloaded from www.techlawjournal.com.

21. *Christianne Carafano v. Metrosplash.com.* 2002. U.S. Dist. LEXIS 10614, downloaded from LEXIS.

22. *Kenneth M. Zeran v. AOL.* 958 F. Supp. 1124. 1997, downloaded from www.Loundy.com.

23. Ibid.

24. Ibid.

25. Ibid.

26. *Kenneth M. Zeran v. AOL.* 129 F 2d 327. 1997, downloaded from www.Law.Emory.Edu.

27. Ibid.

28. *Sidney Blumenthal et al. v. Matt Drudge and AOL.* 992 F. Supp. 44. *Tech Law Journal,* 1998, downloaded from www.techlawjournal.com.

29. "Judge Friedman Grants Summary Judgment to AOL in Defamation Case," *Tech Law Journal,* April 24, 1998, downloaded from www.techlawjournal.com.

30. Ibid.

31. Ibid.

32. *Sidney Blumenthal v. Drudge and AOL,* downloaded from www.techlawjournal.com.

33. Ibid.

34. Ibid.

35. Ibid.

36. Howard Kurtz. "Clinton Aide Settles Libel Suit Against Matt Drudge—At a Cost." Washingtonpost.com. May 2, 2001. C01, downloaded from www.washingtonPost.com.

37. *Carafano v. Metrosplash.com.* 2002 U.S. Dist. LEXIS 10614, downloaded from LEXIS.

38. The reason did not pertain to our present discussion.

39. The Perkins Coie Internet Digest. "Defamation," 1999, downloaded from www.perkinscoie.com.

40. Ibid.

41. *Alexander Lunney v. Prodigy.* 99 N.Y. Int. 0165 at 3. 1999, downloaded from LEXIS.

42. See: *Stoner v. Ebay.* No. 305666. Sup. Crt. CA. 2000, *Schneider v. Amazon.com* No. 46791-31. 31 P. 3d 37. Wash. Ct. App. 2001, *Doe v. AOL* No. 97-2587 So. 3d 385. 4th D. Ct. App. Fla. 2001.

43. FBI, Department of Justice. "Hate Crimes Fact Sheet." November 25, 2002. 1, downloaded from www.FBI.gov.

44. Christopher Wolf. "Racists, Bigots and the Law of the Internet." July 2002, 3, downloaded from www.gigalaw.com.

45. 18 U.S.C. 245, 42 U.S.C.1983, 18 U.S.C.241, 18 U.S.C.242, 18 U.S.C.247, 42 U.S.C.3617, 28 U.S.C.994.

46. Sentence would be increased by one-third; the "normal" punishment for assault is 3 years imprisonment. Hate-motivated assault would result in a 4-year sentence.

47. 18 U.S.C.245.

48. H.R. 80. Hate Crimes Prevention Act of 2003. January 7, 2003, downloaded from http://thomas.loc.gov/.

49. Ibid., sect. 4[2][B].

50. Department of Justice. United States Attorneys Office. Central District of California. Thomas Mrozek. May 4, 1998. 1, downloaded from www.FBI.gov.

51. Machado appealed a complex procedural matter not related to the facts of the case. The

THE LAW

Ninth Circuit (195 F. 3d 454. 1999) affirmed the lower court, however.

52. Department of Justice. U.S. Attorneys Office. Thomas Mrozek. June 28, 1999, 1, downloaded from www.FBI.gov.

53. *Secretary v. Wilson*. United States of America Department of Housing and Urban Development. *Office of Administrative Law Judges v. Ryan Wilson*. HUDALJ 03-98-0692-8. July 19, 2000, 4, downloaded from www.HUD.gov/.

54. Ibid.

55. Pennsylvania Office of the Attorney General. Michael Fisher. "Fisher Sues Hate Group for Terroristic Threats on the Web." October 20, 1998, 1–2, downloaded from www.attorneygeneral.gov/.

56. *Secretary v. Wilson*. HUD, downloaded from www.HUD.gov/.

57. The Nuremberg Files, downloaded from www.abortioncams.com/atrocity/.

58. *Planned Parenthood et al. v. American Coalition of Life Activists et al.* 945 F. Supp. 1355. U.S. Dist. LEXIS 16387 at 1/2. 1996, downloaded from LEXIS.

59. *Planned Parenthood et al. v. American Coalition of Life Activists et al.* 945 F. Supp. 1355. 1996, 23 F. Supp. 2d 1182. 1998, 1999 U.S. Dist. LEXIS 4332, 244 F. 3d 1007. 2001, 268 F. 3d 908. 2001, 290 F. 3d 1058. 2002, 2002ES. App. LEXIS 13829. 2002, 123 S. Ct. 715. 2002, downloaded from LEXIS.

60. *Planned Parenthood v. American Coalition of Life Activists*. 945 F. Supp. 1355, 1996, downloaded from LEXIS.

61. 18 U.S.C. 248.

62. 18 U.S.C. 1962.

63. *Planned Parenthood v. American Coalition of Life Activist*. 290 F. 3d 1058. 2002. U.S. App. LEXIS 9314, downloaded from LEXIS.

64. "Appeals Court." *Alameda Times-Star* (Alameda, CA). May 17, 2002, 1/2, downloaded from LEXIS.

65. *Planned Parenthood v. American Coalition of Life Activists*. 123 S. Ct. 715. 2002, downloaded from LEXIS.

66. U.S. Department of Justice, *1999 Report on CyberStalking: A New Challenge for Law Enforcement and Industry,* downloaded from www.usdoj.gov.

67. Harry A. Valetk. "A Guide to the Maze of Cyberstalking Laws." July 2003, 2, downloaded from www.gigalaw.com.

68. 18 U.S.C. 875.

69. Kent Middleton, Robert Trager, and Bill Chamberlin. *The Law of Public Communication*. (Boston, MA: Allyn & Bacon Publishing Co., 2002), 216.

70. No Electronic Theft Act. Pub. Law 105–147. 111 Stat. 2678. 1997, downloaded from www.GSE.UCLA.Edu.

71. Ibid.

72. H.R. 2281. Bill Summary and Status from the 105th Congress. October 28, 1998, downloaded from http://thomas.loc.gov/.

73. Digital Millennium Copyright Act. Pub. Law 105-304. October 1998, downloaded from http://thomas.loc.gov.

74. Kent Middleton et al., *The Law of Public Communication*. (Boston, MA: Longman Publishing, 2000), 224.

75. Ibid., 225.

76. The Digital Millennium Copyright Act. UCLA, Online Institute for Cyberspace Law and Policy, downloaded from www.gse. UCLA.edu.

77. Ibid.

78. Sonny Bono Copyright Term Extension Act. Pub. L. 105–298, 11 Stat. 2827, Section 101–106, downloaded from www.techlawjournal.com.

79. *Eric Eldred v. John Ashcroft*. 123 S. Ct. 769. 2003, downloaded from LEXIS.

80. Ibid., 783.

81. *Playboy Magazine v. George Frena*. 839 F. Supp. 1552. 1993. *Playboy Magazine v. Russ Hardenburgh*. 982 F. Supp. 503. 1997.

82. *Sega v. MAPHIA*. 857 F. Supp. 679. 1994.

83. *United States v. David LaMacchia*. 871 F. Supp. 535. 1994.

84. *Jonathan Tasini et al. v. New York Times et al.* 972 F. Supp. 804. 1997. *Jonathan Tasini et al. v. New York Times et al.* 192 F. 3d 356. 1999, downloaded from LEXIS.

85. *New York Times Co. Inc., et al. v. Tasini et al.* No. 00201 at 3. June 2001, downloaded from LEXIS.

86. *Recording Industry Association of America (RIAA) v. Diamond Multimedia Systems.* 180

F. 3d 1072 at 2. 1999, downloaded from LEXIS.

87. Ibid., 3.

88. Ibid., 4.

89. Ibid., 8.

90. Kent Middleton, William Lee, and Bill Chamberlin. *The Law of Public Communication*. (Boston, MA: Allyn & Bacon Publishing Co., 2003), 235.

91. *UMG Recordings et al. v. MP3.com, Inc.* 92 F. Supp. 2d 349 at 2. 2000, downloaded from LEXIS.

92. Ibid.

93. Ibid.

94. Ibid., 2/3.

95. Ibid., 3/4.

96. Ibid., 4–7.

97. Ibid., 9.

98. Ibid., 1.

99. *A&M Records v. Napster.* 2001 U.S. Dist. LEXIS 2186 at 3. 2001, downloaded from LEXIS.

100. *A&M v. Napster.* 114 F. Supp. 2d 896, 2000, downloaded from LEXIS; and Middleton, Trager, and Chamberlin. *The Law of Public Communication.* (Boston, MA: Allyn & Bacon Publishing Co., 2002), 232.

101. *A&M v. Napster.* 114 F. Supp. 2d 896. 2000, downloaded from LEXIS.

102. Ibid., 36/37.

103. *A&M v. Napster.* 114 F. Supp. 2d 896. 2000, downloaded from LEXIS.

104. *A&M v. Napster.* 284 F. 3d 1091. 2002, downloaded from LEXIS.

105. Ibid., 10.

106. "Last Remains Auctioned Off." Online Reporter. Gale Group, Inc. *Business and Industry.* 2002, 1, downloaded from LEXIS.

107. Ibid.

108. Reuters. "Napster Set to Return." February 25, 2003, 1, downloaded from www.CNN.com.

109. Downloaded from www.apple.com/music/store/shop.html.

110. Margaret Smith Kubiszyn. Perkins Coie Online Document. 1, downloaded from www.perkinscoie.org/.

111. *Ticketmaster v. Tickets.com.* U.S. Dist. Crt. Ctrl. Dist. CA. 2003 U.S. Dist. LEXIS 6483 at 7. March 6, 2003, downloaded from LEXIS.

112. *Note:* Ariba Soft changed its name during the court proceedings and is now Ditto.com.

113. "Appeals Court Lifts 'Black Cloud' on Deep Linking." *Washington Internet Daily.* Vol. 4, No. 131. July 9, 2003, 1, downloaded from LEXIS.

114. Perkins Coie. "Framing." 3, downloaded from www.perkinscoie.org/.

115. Ibid.

## SUGGESTED READINGS

**Case Law**

*A&M v. Napster.* 284 F. 3d 1091. 2002. 1–8. 239 F. 3d 1004. 2001. 1. 2000 U.S. App. LEXIS 18688. 2000.1. 2001 U.S. Dist. LEXIS 2186. 2001.1–6. 2000 U.S. Dist. LEXIS 20668. 2001.1–7. 114 F. Supp. 2d 896. 2000.1–45. 2000 U.S. Dist. LEXIS 6243. 2000.1–15, downloaded from LEXIS.

*Carafano v. Metrosplash.com.* No. CV 01-0018 DT. 2002, 1–27, downloaded from LEXIS.

*Cubby v. CompuServe,* U.S. District Court for the Southern District of New York 19 Med. L. Rprt. 1527. 1525–1533 1991, 1–12, downloaded from LEXIS.

*Eldred v. Ashcroft.* 123 S. Ct. 769. 2003, 1–78, downloaded from LEXIS.

*Lunney v. Prodigy et al.* 250 AD 2d 230. NY 1994. 99 NY Int. 0165. 1999. 2000 U.S. LEXIS 3037. 1–12, downloaded from LEXIS.

*Miller v. California* 413 U.S. 15. 1973, 1–40, downloaded from LEXIS.

*New York Times et al. v. Tasini et al.* No. 00201. 2001, 1–43, downloaded from LEXIS.

THE LAW

*Planned Parenthood et al. v. American Coalition of Life Activists et al.* 945 F. Supp. 1355. 1996, 1–57. 23 F. Supp. 2d 1182.1998, 1–24. 1999 U.S. Dist. LEXIS 4332.1–45. 244 F. 3d 1007. 2001, 1–17. 268 F. 3d 908. 2001, 1–3. 290 F. 3d 1058.2002, 1–90. 2002 E.S. App. LEXIS 13829.2002, 1–4. 123 S. Ct. 715. 2002, 1, downloaded from LEXIS.

*RIAA v. Diamond Multimedia.* 180 F. 3d 1072. 1999, 1–15, downloaded from LEXIS.

*Stratton Oakmont et al. v. Prodigy Inc.*, Supreme Court of New York 63 USWL. 2765–2773. 1995, 1–4, downloaded from LEXIS.

*UMG et al. v. MP3.com* 92 F. Supp. 2d 349. 2000, 1–7. 200 U.S. Dist. LEXIS 17907. 2000, 1–2, downloaded from LEXIS.

U.S. Department of HUD. *Office ALJ. v. Ryan Wilson.* HUDALJ 03-98-0692-8. July 19, 2000, 1–35, downloaded from www.HUD.gov.

*U.S. v. Machado.* U.S. Dist. Crt. Cntrl. Dist. CA. No. 96-00142.AHA.1-5, downloaded from LEXIS.

*U.S. v. Quon.* 1999. (Guilty Plea: No trial).

## Legislation

Computer Security Act of 1987. P.L. 100–235. 101 Stat. 1724. 1724–1730.

Fair Housing Act. 42 U.S.C. 3617. 3601–3619, downloaded from http://caselaw.lp.findlaw.com.

Freedom of Access to Clinics Entrances Act. 18 U.S.C.A 248. 1–2, downloaded from http://new.crosswalk.com.

Hate Crimes Prevention Act of 2003. HR 80. 108th Congress. 2003, 1–5, downloaded from http://thomas.loc.gov/.

Racketeering Influenced Corrupt Organizations Act. 18 U.S.C. 1962, 1–123, downloaded from http://caselaw.lp.findlaw.com.

Sonny Bono Copyright Term Extension Act. Pub. L. 105–298. 11 stat. 2827. sect. 101–106, downloaded from www.techlawjournal.com.

Telecommunications Act of 1996. P.L. 104–104. 1–76.

## Journals, Magazines, and Online Documents

Amalfe, Christine A., and Kerrie R. Heslin. "Court Starts to Rule on Online Harassment." *The National Law Journal.* January 24, 2000, downloaded from LEXIS.

Brown, Peter. "Is Hollywood Holding Its Breath on Copyright Protection." *Digital Television* 2 (May 1999), 44.

CNN.com. "Napster Set to Return." February 25, 2003, 1, downloaded from www.CNN.com.

Crocker, Steven. "The Political and Social Implications of the Net." July 5, 1994, 1–14.

DiLeleo, Edward. "Functional Equivalency and Its Applications for Freedom of Speech on Computer Bulletin Boards." *Columbia Journal of Law and Social Problems* (winter 1993), 199–247.

Faucher, John. "Let the Chips Fall Where They May: Choice of Law in Computer Bulletin Board Defamation Cases." *Davis Law Review* (summer 1993), 1045–1078.

Godwin, Mike. "Internet Libel: Is the Provider Responsible?" *Internet World* (November/December 1993), 1–3.

Godwin, Mike. "Libel, Public Figures, and the Net." *Internet World.* (June 1994), 1–4.

Jones, Christopher. "Digital Media Is Spinning Toward a Distribution Revolution," *NewMedia* 9 (June 6, 1999), 26–34.

Labriola, Don. "Getting Through the Media Maze." *Presentations* 8 (September 1995), 23–29.

Lewis, Anthony. "Staving Off the Silencers." *New York Times Magazine* December 1, 1991, 72–75.

Roberts, Jon L. "A Poor Man's Guide to Copyrights, Patents and Trade Secrets." *Advanced Imaging* 9 (September 1994), 13–14.

Valetk, Harry A. "A Guide to the Maze of Cyberstalking Laws." July 2002, 1–2, downloaded from www.gigalaw.com/.

Wolf, Christopher. "Racists, Bigots and the Law of the Internet." July 2002, 1–3, downloaded from www.gigalaw.com/.

## GLOSSARY

*Administrative Law Judge (ALJ):* A judge that presides over a federal administrative agency hearing (e.g., FCC, FTC, and HUD).

*Chilling Effect:* Any regulation or threat of regulation that makes the speaker think twice about providing information that might be considered inappropriate by some. In such a situation, the speech is halted before it ever has a chance to be heard.

*Cyberstalking:* Stalking a person(s) via the Internet, e-mail or by other electronic means.

*Hate Crime:* Any crime perpetrated against a person(s) because of his or her actual or perceived race, color, religion, national origin, ethnicity, gender, sexual orientation or disability.

*Indecent Material:* Nonobscene language that is sexually explicit or includes profanity.

*Interstate Commerce:* Trafficking, trading, or transportation of persons or property between or among States (e.g., transporting goods between New York and Ohio).

*Libel:* Any expression that damages your reputation within the community, causes others to disassociate themselves from you, or attacks your character or professional ability.

*True Threat:* A serious, direct threat of violence.

# 21 New Technologies: Wiretapping, Privacy, and Related First Amendment Issues

In addition to the online legal issues examined in previous chapters, the use of communications technologies to perform surveillance functions is of equal importance. While criminal and terrorist activities have been monitored for years, the privacy implications remain a major source of controversy. Of particular concern is the collateral damage caused when individuals *not* under investigation are caught in the surveillance net and don't realize it. In sum, this chapter will examine the various issues surrounding the legal "tug of war" between the need for security versus the need for privacy.

## WIRETAPPING AND ENCRYPTION

### Introduction

In the past, the structure of the telephone system made it easier to wiretap someone's telephone line. But the introduction of digital communication, fiber-optic systems, and *encryption techniques*, where information is rendered unintelligible without the proper decoding mechanisms, have made electronic eavesdropping more complicated. For most people, the added security is welcome news. But for law enforcement agencies, their job has become more complicated. Consequently, different legislation had been floated to ensure an agency's continued access to this information, potentially through an electronic back door, which would allow an agency to gain access to the targeted information.[1]

Elements of the business community and other organizations have been opposed to this type of legislation on privacy and economic grounds. Although a court order must be obtained to initiate wiretapping, as has been the practice, legislation may make the communications network more susceptible to illegal wiretapping. The system would also be more open to potential government and nongovernment abuses. Ultimately, instead of using technology to build a more secure communications system, which could carry voice as well as computer transactions, it could potentially be compromised.

A network could also be more expensive to build, and a manufacturer would have to contend with more red tape. Other concerns center on the stifling of technological developments and international implications.[2] As other nations developed more secure networks, U.S. systems could be burdened by this organizational hierarchy and a communications structure that could be vulnerable to government and potentially illegal eavesdropping.

The government, for its part, has defended its stance in the name of law enforcement and national security. The prosecution and conviction of a criminal/terrorist could hinge on, for instance, recorded telephone calls. The inability to tap into this and other forms of information would hinder such an operation.

Finally, friction over encryption entered the international market. Tight export standards made it difficult for U.S. companies to compete in this arena. Legislative initiatives sought a balance between individual/business rights and law enforcement/national security concerns.[3]

### Wiretapping

In 1994, the Communications Assistance for Law Enforcement Act (CALEA), was signed into law. The CALEA was partly born from a government concern—as stated, sophisticated communications technologies would significantly hinder the government's ability to conduct legal surveillance activities. But the ACLU, The Electronic Privacy Information Center, and other public interest groups, questioned the CALEA's potential Fourth Amendment privacy implications.[4]

In brief, telecommunications carriers must guarantee they can isolate and intercept all wire and electronic communication, once proper legal authority has been obtained.[5] The CALEA also requires carriers to deliver intercepted communications and call identifying information to the government. The FCC also has the authority to create rules establishing technical standards in accordance with the Act's assistance capabilities requirements.[6]

It should also be noted that the Department of Justice and FBI petitioned the FCC to expand their capabilities to gather surveillance information. This request was strongly opposed by public interest groups.[7] Despite this opposition, however, the FCC adopted rules granting the FBI additional powers. In response, the U.S. Telecom Association and others, argued the FCC had exceeded its statutory authority by expanding the type of call identifying information that carriers had to make acces-

sible to law enforcement agencies (e.g., call forwarding).[8]

The Court of Appeals for the District of Columbia Circuit agreed, in part, with this legal challenge in *United States et al. v. Federal Communications Commission et al.* (227 F. 3d 450.2000). Based on the ruling, these sections of the FCC's CALEA order were vacated or struck down. The requirement to provide law enforcement with the location of wireless telephones was, however, upheld. Therefore, the FCC had to redesign its CALEA Order to reflect this ruling.[9]

### Encryption

The Clinton Administration took a hard line approach to the domestic and international use of encryption technology. There were concerns that

- law enforcement and national security agencies would be significantly hindered in their criminal and terrorist investigations and
- foreign enemies could use the technology and its applications against the United States.

This led to the adoption of restrictive encryption policy.

But business, industry, and academia opposed this stance, and legislative efforts designed to relax encryption restrictions were drafted. Eventually, these efforts helped bring about a policy shift, and a liberalized viewpoint was adopted.

President Clinton issued an Executive Order transferring the responsibility for regulating encryption technology from the Department of State to the Department of Commerce. The Order also amended the existing Export Administration Regulations (EAR). The goals of the new policy were twofold: to support law enforcement/national security and to promote user privacy

THE LAW

and electronic commerce.[10] The policy was designed to accomplish these tasks in various ways.

1. All mandatory technical reviews of encryption products would be concluded in advance of their sale.[11]
2. The postexport reporting system would be significantly simplified.[12]
3. A process would be created for the government to review strong encryption products[13] that could then be exported to nonterrorist nations.[14]
4. A new category of products, "retail encryption commodities and software," was created. Products falling under this heading, which were available over-the-counter, by telephone sales, and through other mass sales means, could be exported and re-exported to any nonterrorist end-user.[15]
5. An "open source" approach to software development was adopted, leading to unrestricted encryption source code that could be exported or re-exported without review, as long as licensing fees or royalties were not involved. (*Note:* Other restrictions may also apply.)[16]
6. Telecommunications and ISPs were permitted to use any encryption product, "to provide encryption services, including public key infrastructure services for the general public."[17]

This evolutionary process continued under the Bush Administration. In 2002, U.S. policy was brought in line with the European Union, Japan, and other trade and security partners. The modification permitted the export and re-export of enhanced mass-market encryption products, following a 30-day review process. Once the review had been completed, no postexport reporting was required.[18]

Finally, in January 2003, the Department of Commerce announced an additional modification. This change allowed for the export of general-purpose microprocessors "used worldwide in technology and commercial applications such as personal computers and cell phones."[19] No license was required to export such products unless they were going to

- terrorist countries,
- military end-users, and/or
- end-users in countries that pose national security concerns.[20]

Cases concerning the constitutionality of the U.S. encryption policy have been heard. These are included in the following subsections.

***Karn v. Department of State.*** The District Court examined if the State Department had jurisdiction over the export of a diskette containing cryptographic software. The court held that under the terms of the Arms Export Control Act, Karn was not entitled to a judicial review of the legality of State Department jurisdiction over such software. The court further held that even if the cryptographic software was considered "speech," controlling its export would not violate the First Amendment because the regulation was content neutral, narrowly tailored, within the government's legitimate interest to control the export of defense articles, and was based on a rational premise.[21]

In response, Karn appealed the case to the Washington, DC Circuit. In the interim, the president issued the aforementioned Executive Order. Based on this change, the court of appeals remanded, or sent the case back, to the district court for judicial consideration in light of the new encryption rules.[22]

***Bernstein v. United States.*** The case was heard on appeal from the U.S. District Court for the Northern District of California to the Ninth Circuit. The case

THE LAW

examined the constitutionality of the Department of Commerce policy, which made it illegal to export encryption programs of a certain sophistication, without first securing a license from the department. The policy also prohibits the posting of such programs on the Internet since proscribed groups could subsequently download them.

The Ninth Circuit, in a 2–1 decision, upheld the lower court decision that said government prohibition of the export of strong encryption violated the First Amendment.[23] Bernstein could legally post his encryption program's source code on the Internet without obtaining permission. Senior Ninth Circuit Judge Betty Fletcher concluded that code is free speech and should be afforded the same full First Amendment protections guaranteed for traditional language.[24] However, in response to grave concerns by the government, the Justice Department requested that the case be heard *en banc* by a panel of 11 Court of Appeals judges from the Ninth Circuit.[25]

It was remanded, instead, to the U.S. District Court for the Northern District of California in 2000. To-date, Bernstein, who filed additional judicial complaints to the district court in 2002, still awaits a response to his legal challenge.[26]

In sum, while legislative initiatives continue to establish a legal framework for surveillance and encryption law, case law will be heard in an effort to interpret the intricate details of such legislation in new and legally unique circumstances. As in other situations where technological advances come into contact with the law, numerous legal questions remain.[27]

## SURVEILLANCE BEYOND THE YEAR 2000

As had been the case with CALEA in the 1990s, the new millennium ushered in even more troubling technological advancements with privacy implications. Two controversial surveillance applications are the FBI's Carnivore (DCS1000) system and Congressional Legislation in the form of the USA PATRIOT Act.

## Carnivore/DCS1000

*Introduction.*   While most people use the Internet for legal purposes, it also functions as a powerful communications tool for those engaged in organized crime, drug trafficking, espionage, and terrorism. This was the impetus for the creation of the FBI's *Carnivore*, also called *DCS1000*, program.[28]

Carnivore is an e-mail/electronic communications surveillance program that requires Internet Service Providers (ISPs) to attach a box to their networks so the FBI can monitor the traffic that passes through their facility.[29] Much like wiretapping a telephone to gain access to conversations for law enforcement applications, Carnivore follows suit with e-mail and can analyze millions of messages per second.

*Critics.*   According to the FBI, Carnivore only captures the information the agency is legally authorized to collect. This contention has been challenged by civil liberty groups, members of Congress, and others since Carnivore has access to all the traffic, including the electronic communications between individuals who are not under investigation. They indicated that Carnivore's technical capabilities could be abused, and as a consequence, they question if innocent electronic communications could also be archived for review at some later date.[30]

Critics were also troubled by the Carnivore program's breadth. While the FBI may only be interested in one person's electronic activities, the system intercepts all e-mail/electronic communications. The ACLU, for

THE LAW

one, compared this action to sending FBI agents "into the post office to rip open each and every mailbag in search for one person's letters."[31] The ACLU also indicated that "Carnivore is like the telephone company being forced to give the FBI access to all the calls on its network when it only has permission to seek the calls for one of its subscribers."[32] It is also unnecessary since ISPs currently provide law enforcement agencies with information if a court order is issued.[33]

***Supporters.*** The FBI subsequently defended the use of Carnivore/DCS1000. According to the agency, procedural safeguards to protect privacy rights were implemented. These included the establishment of a judicial review process, and the system was designed to "surgically" intercept and only collect legally authorized e-mail and other electronic communications.[34] For example,

- A high-level Justice Department official must initially approve Carnivore surveillance.
- Carnivore can only be used for specifically identified felony crimes and if the FBI can show probable cause for doing so.[35]
- The FBI Carnivore request must meet certain criteria, including the identification of the offense(s) being committed and a description of the type of communication to be intercepted.[36]
- The FBI must demonstrate that a standard, less intrusive means of gathering the information, would not work or would be too dangerous.
- If judicial approval for Carnivore surveillance is granted, the Order is limited to 30 days.
- Progress reports must be submitted to the judge.[37]

***Conclusion.*** In light of our sophisticated communications systems, law enforcement agencies must have access to sophisticated surveillance tools. Yet, when Carnivore's existence became public knowledge, there was an immediate backlash, including the positions articulated by the civil liberties groups.

In response, the Justice Department agreed to allow the Illinois Institute of Technology to conduct a technical review of the program's surveillance capabilities. A goal was to determine if Carnivore could operate without violating privacy rights.[38]

But the study's results were inconclusive, and as of this writing, distrust over Carnivore's use continues. This is particularly true in light of the privacy concerns raised by other planned and existing government initiatives.[39]

As stated elsewhere in the book, it is also a balancing act. *How do you provide a law enforcement agency with the means to gain access to information while ensuring that an individual's civil liberties and privacy are protected?* It is a difficult question to answer, but one that is crucial in any democratic society.

## THE USA PATRIOT ACT

Following the horrific terrorists attacks in the United States on September 11, 2001, Americans as well as most of the rest of the world, were stunned by the vulnerability of the United States within its own borders. The ensuing grief and fear of future terrorist activity lead governmental officials to propose, and the majority of the American public to tolerate, the potential reduction of certain civil liberties.

For example, Congress drafted legislation that significantly expanded the information tracking and gathering powers of various government bodies. The outcome was the enactment of the "Uniting and Strengthening America by Providing Appropriate Tools Required to Intercept and Obstruct Terrorist"(USA PATRIOT) Act of 2001 (The USA PATRIOT Act. Pub. L. No. 107–56).

Despite the major modifications of existing law, the legislation was speedily passed with very little discussion. In fact, President George W. Bush signed the Bill into law on October 26, 2001, just 7 weeks after the national tragedy.

In the past, privacy law concerning the tracking and gathering of communications was designed to strike a balance between protecting an individual's rights while allowing government authorities to identify and intercept criminal communications. Based on the degree of "intrusiveness" caused by these activities, varying degrees of privacy protection were provided. These guarantees were spelled out in Title III of the Omnibus Crime Control and Safe Streets Act of 1968 and in Chapter 121 and 206 of Title 18 of the U.S. Code.[40]

Title III provided the highest level of privacy protection, making it illegal for government authorities to eavesdrop on face-to-face, telephone, computer or other forms of electronic communications in nearly all instances. A narrow exception to this restriction enabled law enforcement to request a court order from a "high-level" Justice Department official to secretly track and gather such communications in "serious" criminal cases. If granted, these court orders were narrow in scope and required notification of the surveillance once the order expired.[41] Chapter 121 provided less privacy protection than Title III, followed in turn, by Chapter 206.

The USA PATRIOT Act, which is 342 pages long, alters the degree of privacy protection afforded to communication by government authorities.[42]

Specifically, the act expands tracking and gathering capabilities involving:

- *Search warrants:* that authorize the search and seizure of property consisting of evidence of the commission of a crime, contraband, or items used in the commission of a crime.
- *Wiretaps:* that constitute electronic eavesdropping by government authorities based on a court order.
- *Subpoenas:* that compel a party to appear, give testimony, and produce required documents.
- *Pen register and trap and trace orders:* that authorize the collection of incoming and outgoing telephone numbers from a specific telephone.[43]

Furthermore, the Foreign Intelligence Surveillance Court, a secret court empowered to grant U.S. intelligence agencies with the authority to conduct surveillance of foreign nations and U.S. citizens under special circumstances, has similar expanded capabilities.[44]

## Tracking and Gathering of Domestic Communication by Law Enforcement

In operation, the USA PATRIOT Act allows the government to:

1. Collect information about the web browsing, e-mail, and person-to-person activities of Americans, including individuals who are not under criminal investigation. A judge located anywhere in the United States must simply be "notified" and agree that such surveillance may result in obtaining information relevant to an ongoing criminal investigation.[45]
2. Obtain stored voice-mail with only a search warrant.
3. Obtain pen register and trap and trace orders for electronic communications, including e-mail.[46]

4. Employ *sneak and peak warrants* that enable law enforcement agencies to surreptitiously enter homes and other private areas to search and inspect items or evidence. In some instances, physical property and electronic communications can be seized. Notice of such a search could be delayed for an undetermined period of time.[47]
5. Conduct surveillance of "computer trespassers" through ISPs, universities, or network administrators without a judicial order.[48] An example of a "computer trespasser" could be someone downloading copyrighted MP3 music files.

### Foreign Intelligence Investigations and the USA PATRIOT Act

Under the terms of the USA PATRIOT Act, certain prior restriction on gathering foreign intelligence information within U.S. borders have been relaxed. According to the language of the Act, the government has the authority to:

1. Share information gathered in the process of conducting a domestic criminal investigation (e.g., Grand Jury testimony and wiretaps) with intelligence bodies.[49]
2. Conduct *roving surveillance* of a suspect's wire or electronic communications, concerning the actions under investigation, regardless of the suspect's location. For example, a roving wiretap would allow the interception of communications as the individual moves from phone booth to phone booth.[50]
3. Obtain a surveillance or a search order even if gathering foreign intelligence is only a "significant" reason for such a request rather that "the" reason.[51]

4. Obtain pen register and trap and trace orders for e-mail and telephone communications.[52]
5. Obtain access to "any tangible item" (e.g., a letter or a diskette) without a court order.[53]
6. Obtain nationwide search warrants for domestic or international terrorist investigations making it easier to "venue shop" for a court most likely to grant the request.[54]

In sum, no one can deny that Americans and America was changed forever on September 11, 2001 and expanded governmental powers have been one result. But this scenario also raises critical questions—how do you strike a constitutional balance between the need to protect the public and the need to preserve civil liberties? This answer is particularly important in light of potential legislation, including USA PATRIOT II that could expand the USA PATRIOT Act's provisions. The answer may also enhance, or could potentially curtail, our civil liberties.[55]

### CONCLUSION

Conducting wire and electronic monitoring to thwart criminal and terrorist activity is a legitimate use of surveillance and related communications technologies. However, careful vigilance is essential to ensure the individual privacy rights of those not under investigation. Continued legislation and judicial oversight may be necessary if privacy safeguards are to be maintained.

THE LAW

## REFERENCES/NOTES

1. ABC News, "FBI Pushing for Enhanced Wiretap Powers," Transcript from *Nightline*, show #2870 (May 22, 1992), 3.

2. Sam Whitmore, "Drop a Dime and Stop Some Spooky Legislation," *PC Week* 9 (May 18, 1992), 100.

3. U.S. Senator Patrick Leahy, "Leahy Introduces Encryption Communications Privacy Act," News Release. Senator Patrick Leahy. March 5, 1996, p. 1. Downloaded from http://leahy.senate.gov/text/press/199603/19960305.html.

4. Surreply Comments of the ACLU, the EPIC, the EFF, and Computer Professionals for Social Responsibility, Before the Federal Communications Commission, CC Docket No. 97–213, downloaded from www.aclu/Congress.

5. CALEA. 47 USC 1001–1010. *Tech Law Journal*, downloaded from www.techlawjournal.com.

6. FCC. FCC Proposes Rules to Meet Technical Requirements of CALEA, CC Docket No. 97–213. October 22, 1998, downloaded from www.FCC.gov.

7. Surreply Comments off ACLU, EPIC, EFF, and Computer Professionals.

8. *U.S. Telecom Association et al. v. FCC et al.* 2000 U.S. App. LEXIS 19967 at 2, downloaded from LEXIS.

9. Ibid.

10. "BXA Encryption Export Regulations." January 12, 2000, 1, downloaded from www.eff.org.

11. Ibid.

12. Ibid.

13. *Note:* U.S. identified terrorist nations have included Libya and North Korea among others.

14. BXA. "Encryption Export Regulations." January 12, 2000, 1, downloaded from www.eff.org.

15. Ibid., 3.

16. Ibid., 2.

17. Ibid., 4.

18. U.S. Department of Commerce. "U.S. Encryption Export Control Policy Frequently Asked Questions." June 6, 2002, 2, downloaded from www.bxa.doc.gov.

19. U.S. Department of Commerce. "New Regulation Streamlines Export Controls." *Commerce News.* January 14, 2003, 1, downloaded from www.bxa.doc.gov.

20. Ibid.

21. *Karn v. Department of State.* 925 F. Supp. 1. 1996.

22. *Karn v. Department of State.* No. 96–5121. January 21, 1997, 1, downloaded from LEXIS.

23. *Bernstein v. Department of State.* U.S. Dist. Crt. No. D. CA. No. C–95–0582MHP, downloaded from www.eff.org/.

24. *Bernstein v. United States.* 97–16686. CV–9700582.1999.

25. Electronic Privacy Information Center. "EPIC Alert." Vol. 6.16. October 12, 1999, 3, downloaded from www.epic.org/.

26. Electronic Frontier Foundation. Media Advisory. "Professor Pushes for Revised Encryption Regulations." January 7, 2002, downloaded from www.eff.org/.

27. In another case, the court examined the constitutionality of State Department regulation controlling the use and transfer of sophisticated encryption software. The court ruled that according to the language of the Export Administration Regulations, the State Department's policy did not abridge the First Amendment since it only controlled the distribution of the encryption software itself and did not impact on ideas concerning encryption. See 38. *Junger v. Christopher/Junger v. Daley.* 8 F. Supp. 2d 708. 1998.

28. *Note:* The FBI, because of the name's negative connotation, changed Carnivore's name.

29. Internet Service Providers (ISPs) provide users with Internet access.

30. ACLU. "Letter to Reps. Canady and Watts on the FBI's E-Mail Surveillance System, 'Carnivore'." July 11, 2000, 2, downloaded from www.aclu.org.

31. ACLU. "Urge Congress to Stop the FBI's Use of Privacy-Invading Software." January 5, 2003, 1, downloaded from ACLU.

32. Ibid.

33. Ibid.

34. "FBI Carnivore Diagnostic Tool." Statement of Dr. Donald M. Kerr, FBI. Before the House of Representatives Committee on the Judiciary. July 24, 2000, 1, downloaded from www.fbi.gov/.

35. Ibid.

36. Ibid.

37. Ibid.

38. Erich Luening. "Don't Be Fooled: DCS1000 Still a 'Carnivore' at Heart." ZDNET. February 8, 2001, 1, downloaded from http://zdnet.com.

39. Please see Chapter 1 for more details.

40. Charles Doyle. "The USA PATRIOT Act: A Sketch." Congressional Research Service. Order Code R21203. April 18, 2002, 2, downloaded from www.house.gov/.

41. Ibid.

42. Ibid.

43. Electronic Frontier Foundation (EFF). "Analysis of the Provisions of the USA PATRIOT Act That Related to Online Activity." October 31, 2002, 5, downloaded from www.eff.org.

44. Ibid.

45. Ibid., 2.

46. Congressional Research Service. 2, downloaded from www.house.gov/.

47. Center for Democracy and Technology (CDT). "Anti-terrorism Legislation Gutting Privacy Standard Becomes Law." Policy Post Vol.7, No.11. October 26, 2002, 1, downloaded from www.cdt.org.

48. Ibid.

49. Center for Democracy and Technology. "Summary and Analysis of Key Sections of USA PATRIOT Act of 2001." 2. October 31, 2001, downloaded from www.cdt.org.

50. Ibid., 3.

51. Congressional Research Service, 3.

52. Ibid.

53. Ibid.

54. Center for Democracy and Technology. "Summary and Analysis of Key Sections of USA PATRIOT Act of 2001." 8, downloaded from www.cdt.org.

55. David Cole. "What PATRIOT II Proposes to Do." Georgetown University Law Center. (February 10, 2003), downloaded from LEXIS.

## SUGGESTED READINGS

ACLU. "Urge Congress to Stop FBI's Use of Privacy-Invading Software." January 5, 2003, 1, downloaded from www.aclu.org.

Center for Democracy and Technology. "Anti-terrorism Legislation Gutting Privacy Standard Becomes Law." October 26, 2002. vol. 7 No. 11, pp. 1–4, downloaded from www.cdt.org.

Department of Commerce. "BXA Encryption Export Regulation." January 12, 2000, 1–5, downloaded from www.eff.org.

Department of Commerce. "U.S. Encryption Export Control Policy: FAQ." June 6, 2002, 1–9, downloaded from www.BXA.doc.gov.

Department of Commerce. "New Regulation Streamlines Export Controls." January 14, 2003, 1, downloaded from www.BXA.doc.gov.

Doyle, Charles. "The USA PATRIOT Act: A Sketch." Congressional Research Service. April 18, 2002, 1–5, downloaded from www.house.gov.

Electronic Frontier Foundation. "Analysis of the Provisions of The USA PATRIOT Act that Relate to Online Activity." October 31, 2002, 1–15, downloaded from www.eff.org.

Electronic Privacy Information Center. "Appeals Court to Review Bernstein Crypto Decision." October 12, 1999. Vol. 6.16, pp. 1–4, downloaded from www.epic.org.

Electronic Privacy Information Center. "The USA PATRIOT Act." October 31, 2001, 1–10, downloaded from www.epic.org.

Kerr, Donald. "FBI Carnivore Diagnostic Tool." Statement of Dr. Donald M. Kerr before the House of Representatives Committee on the Judiciary. January 24, 2000, 1–2, downloaded from www.cdt.org.

THE LAW

USA PATRIOT Act. Pub. L. No. 107–56. 107th Congress.2001, 1–91, downloaded from http://216.110.42.179/docs/usa.act.final.102401.html.

*United States Telecom Association et al. v. FCC et al.* 2000 U.S. App. LEXIS 19967. 1–23, downloaded from LEXIS.

## GLOSSARY

*Carnivore/DCS1000:* A powerful online e-mail/electronic surveillance program requiring ISPs to attach a box to their networks so the FBI can monitor vast volumes of traffic passing through the ISP's facility.

*Foreign Intelligence Surveillance Court:* Secret court composed of seven judges housed in the Department of Justice. The court meets in secret and authorizes secret surveillance requests. The court does not publish its orders or opinions or provide any public record of its decisions. The subjects of surveillance approved by the court are unaware of the investigation.

*Pen Register and Trap and Trace Order:* Authorizes the collection of incoming and outgoing telephone numbers from a specific telephone.

*Roving Wiretap:* Allows interception of communications, as the individual under investigation moves from place to place such as from phone booth to phone booth.

*Search Warrant:* Authorizes the search and seizure of property consisting of evidence of the commission of a crime, contraband, or items used in the commission of a crime.

*Sneak and Peak Warrant:* Enables law enforcement agencies to surreptitiously enter homes and other private areas to search and inspect items or evidence. In some instance, physical property and electronic communications can be seized. Notice of the search could be delayed for an undetermined period of time.

*Strong Encryption:* Encryption method using a very large number for its cryptographic key making it much more difficult to illegally break the encryption code.

*Subpeona:* Compels a party to appear, give testimony, and produce required documents.

*Wiretap:* Electronic eavesdropping by government authorities based upon a court order.

# Afterword

This afterword is a general summary and explores some potential technological trends.

On the political front, the Internet could nurture an electronic democracy while providing access to vast information resources. Optical media will serve as information and electronic publishing sources in their own right.

PC developments will blossom, and new systems may sport GUIs and other interfaces that will make the human–computer link even more transparent. Speech recognition and wireless systems may also play prominent roles in this field, and when appropriate, could even be extended to everyday appliances.[1]

The convergence between technologies will continue, and the resulting synergy will be fertile ground for new products and applications. One such area is the convergence between computers and audio-video equipment, as is the case with the HDTV/digital television market.

Entire communications systems may be similarly influenced. As introduced in an earlier chapter, we are witnessing the development of integrated information and entertainment highways. This may be matched by a growing number of intra- and interindustry mergers.

Digital technology will continue to play a key role in our future information and communications systems. But as described at different points in this book, the analog world is still very much alive.

Personal media options will also grow. By using desktop publishing and video systems, we become producers and not just consumers. This facet of the communication revolution cannot be overstated. We now possess a remarkable set of personal communications tools. You can let your imagination soar as you create a 3-D graphic, and you can gain access to tailored information and entertainment pools.

By extension, our communication options will deepen, especially in the area of wireless systems. We may also hold a videoconference from either the office or home. For more people the home may, in fact, become their offices.

Information will remain an important resource. As our information society matures, the production of, and the demand for, information will increase. We'll also develop new ways to manipulate and use this information. The tools include existing and newer configurations (e.g., virtual reality) that may use faster communications and information highways (e.g., Internet2).

But despite this growth of the information sector, it's also important to remember that heavy industries are still vital—someone has to make the steel and other products we use.

The integration of and the correct balance between an information and industrial base could help propel a country toward the future. If the balance is lost, a country's economy could suffer.

## OTHER ISSUES

Some people believe the new technologies may be too much of a good thing. Will electronic clutter become too pervasive? How do you strike a balance between the need for quiet, reflective periods and your communications and information requirements?

With the possible proliferation of virtual reality systems and the Internet's growth, will people become more divorced from reality? What of potential electronic addictions and obsessions? Will more people spend more time online than in the real world? Are we catching a glimpse of Asimov's future world in our present society?

While this may appear to be a remote possibility, think about the potential. The Internet can give you access to an almost unlimited and ever-changing data pool. Some people have likened this potential addiction to smoking. But instead of one more cigarette, there may be one more web site to visit. And there may always be one more web site.

Although current and projected applications are exciting, they must also be examined with a dose of reality. One of the problems with predictions is that they often remain predictions even after a number of years. In one classic example, artists and authors in the 1950s and 1960s painted a bright picture for the space program. Moon bases would be sprawled over the lunar surface, and huge space stations, much like the one portrayed in *2001: A Space Odyssey*, would orbit the Earth. But many of these predictions have not materialized. Part of the problem was the lack of money. Another may have been the technical and/or political inability to transform these visions into a reality.

The same sequence of events could affect the new communications technologies. Although PCs can serve as educational tools, will schools be equipped with enough systems to make a difference? Even though we will have greater technical support to maintain the free flow of information, will government regulations intercede?

The First Amendment and its attendant rights will also be challenged and further defined for the electronic media. In one case, it has been suggested that CDA-type acts could be forestalled by taking reasonable means to protect children from potential, electronic online abuses. This could include, when properly implemented, filters that would prevent children from gaining access to certain web sites.

But as has been covered, constitutional questions are raised when mandatory blocking is adopted. Consequently, how do you reconcile the two viewpoints—the vital importance of protecting children while preserving the free access to information? Are more narrowly tailored regulations, in tandem with appropriate safeguards, the answer? In one earlier library dispute, computers were made available with and without filtering software. Another factor is parents and educators assuming greater responsibility in helping their children to navigate the Internet. In the end, this may be the most important and potent force to help ensure a child's safety.

In essence, technological changes will continue to clash with the legal system. In some cases, technological developments will outstrip the law's ability to react to these changes.

Some people also believe the legal system may overreact. As covered in Chapter 6, do you introduce tighter controls on imaging, which some have categorized as vague and a First Amendment violation? Or do you work with the media in regard to national security issues?

As an employer, you may have to make your own hard decisions. Do you adopt a published e-mail policy? Or do you use the legal system, as of this writing, to monitor your employees at your discretion?

Other ethical and legal questions will loom ever larger. Pertinent topics include intellectual property, digital image manipulation, and privacy.

Finally, it is important to remember that certain new technologies and their applications are like the proverbial double-edged sword. As has been explored at different points in this book, one side may be used to fight terrorism and to provide us with more detailed information about rapidly changing world events. The other side, however, may provide a government with the means to curtail its citizens' freedom.

As stated in the Preface, *the same tools used to protect our freedom have the potential to curtail our freedom.* As communicators, it's important to be aware of these issues as we make determinations as to a technology's appropriate or inappropriate use and application.

**Figure A.1**
*The disk mounted on the Voyager spacecraft that may eventually carry information about Earth to a distant star; it's potentially one of our first calling cards to another world. (Courtesy of NSSDC.)*

## REFERENCES/NOTES

1. For wireless technology, this could include linking a television set in the network. For a completely wireless home, your future coffeepot may similarly be tied-in, so the computer controlling your home's functions could turn it on before you get home from work.

# Index